THE CITY & GUILDS TEXTBOOK

LEVEL 3 DIPLOMA IN
PLUMBING STUDIES 6035
UNITS 201, 301, 303, 304 AND 306

THE CITY & GUILDS TEXTBOOK

LEVEL 3 DIPLOMA IN
PLUMBING STUDIES 6035
UNITS 201, 301, 303, 304 AND 306

MICHAEL B MASKREY

ANDREW HAY-ELLIS

TREVOR PICKARD

City&
Guilds

CONTENTS

ACKNOWLEDGEMENTS

City & Guilds would like to sincerely thank the following.

For invaluable plumbing expertise

John Hind

For taking photos

Jules Selmes and Adam Giles (Photographer and Assistant)

For supplying pictures for the front cover

Jules Selmes Photography

For help with photoshoots

Jocelynne Rowan, Baxi Training Centre, Dartford

Picture credits

Every effort has been made to acknowledge all copyright holders as below and the publishers will, if notified, correct any errors in future editions.

Alkaline Water p257; **Anglican Water** p214; **Arrow Valves** pp 207, 212, 217; **AquaTech Pressmain** pp 153, 167; **Ashley Bathrooms of Southport** p245; **Baxi** p415; **BD Sensors** p167; **Beggs and Partners** p449; **Bella Bathrooms** p240, p245; **Benchmark** p353; **Best 4 Pure Water** p183; **Boilershop.co.uk** p418; **Brimar Plastics Ltd** p154; **BSI** p7 (Permission to reproduce extracts from British Standards is granted by BSI Standards Limited (BSI). No other use of this material is permitted. British Standards can be obtained in PDF or hard copy formats from the BSI online shop: www.bsigroup.com/Shop or by contacting BSI Customer Services for hard copies only: Tel: +44 (0)20 8996 9001, Email: cservices@bsigroup.com); **City & Guilds/Jules Selmes/Baxi Training Centre** pp 1, 114, 346, 388, 389, 391-392, 393; **City & Guilds/Jules Selmes/Hackney College** p349; **Cotherm** p316; **Dales Water** p169; **Danfoss** p160; **Dewey Waters** p155; **Dundee Wharf** p322; **Ecologhouse.org** p447; **Ecoplay Nederland B.V.** p128; **ECP Group** p164; **Farm Products 4 U** p206; **Fernox** pp 470, 471; **Geotek Heating Ltd** p453; **Grundfos** p444; **Gurney & White/GW Supplies** p399; **Hansgrohe/Freewater UK Ltd** p129; **Heat and Plumb** pp 240, 242; **Horne** p306; **iStock** © JoeGough p211 © VitalyEdush, p222, © nsj-images p329, © redmal p409; **Jaystar Gas Services** p472; **Kensa Engineering Ltd** p81; **Keraflow** p161; **KSB** p191; **Masterwork Plumbing and Heating Inc** p351;

Mira Showers p302; **Monument** p253; **Myson** p434; **Not Just Taps** p436; **Ozone Solutions** p186; **PlumbNation** pp 240, 375, 399; **Plumbworld** pp 245, 366; **Plumbstore** p247; **Pricerunner** p245; **Purchase.ie** p248; **Radiators 4 You** pp 431-432; **RP Media** p424; **Silverline UK** p187; **Skirting Heating** p439; **Screwfix** pp 11, 15, 16, 17, 22, 26, 242; **Shutterstock** © JoLin p11, © Cynthia Farmer p45, © Palmaria p52, © Paul Orr p74, © Photoseeker p76, © Lari Saukkonen p82, © Benjamin Haas/Frank Fennema p95, © Tomislav Pinter p96, © CreativeNature.nl p97, © Peter Kunasz p108, © Sklep Spozywczy p148, © Roman Tsubin p143, © Singkham p351, © Leungchopan p411, © Rigamondis p448; **Skirting Heating** p435; **STC Plumbing** p395; **Sun Flow** p420; **SYR** p216; **Talis** p158; **Toolstation** p244, p400; **TP Pumps** p170; **Travis Perkins** p218; **UNECE** p18; **Wardle Drilling** p171; **Water Heaters Direct** p297; **Whisper Pumps** p192; **Wickes** p365; **Worcester Bosch Thermotechnology Ltd** p76; **World of Gardena** p226; **WRAS** p2.

Illustrations by Willam Padden.

It is usual, at this point in any book, to thank those people who have helped in some way in achieving the final product. The chapters of the book that I am responsible for have taken much research and many hours of writing. This has meant that I have not seen my two sons as much as I should have over the last year or so but they have never complained and their support has been unwavering throughout.

And so, my dedication has been the easiest part of the book to write. I would simply like to thank my two sons Scott and Joseph who have encouraged and, all too often, cajoled me into completing my units during this most difficult of years.

Thank you, Scott; thank you, Joseph! This one is simply for you.

Mike Maskrey

ABOUT THE AUTHORS

MICHAEL B MASKREY

My father was quite simply the best plumber I have ever seen. It was his enthusiasm for the trade that he loved that rubbed off on me at a very early age. I was working with him at weekends and school holidays from the age of 10!

In 1977, aged 16, fresh from school and armed with the little knowledge I had gained from my father, I started as an apprentice at a local plumbing firm in my home city of Nottingham where I gained a superb background of plumbing both industrial and domestic. In 1982, I joined my father's small plumbing firm where I stayed until his retirement in 1999 at the ripe old age of 73. He sadly passed away in 2006.

In 1988, I started teaching part-time at the Basford Hall College (now New College Nottingham), the same college where I had done my own training, initially teaching Heating and Ventilation and, soon after, Plumbing at both craft and advanced craft and later NVQ Level 2 and Level 3. My teaching career continued at Stockport College for 13 years. I now teach Building Services Engineering at HNC Level 4 and HND Level 5 (and occasionally plumbing!) at Doncaster College.

To the readers of this book, I say simply, to be an outstanding plumber requires three D's – Desire, Discipline and Dedication. Be the best plumber that you can possibly be and always strive to achieve excellence.

ANDREW HAY-ELLIS

When I first left school I joined a firm of chartered accountants (an excuse to play cricket) and after three years I moved to a small mechanical and electrical company primarily to look after the company's day-to-day accounts. As with many small companies, staff members often end up working wherever there is a need and so I soon found myself out on site with the engineers. Looking back to my childhood I always had an enquiring mind and electrical installation work provided the mental and physical challenges that I needed and was no longer finding in accountancy. Some thirty plus years later, having worked as an electrician, a supervisor and a project engineer on both commercial and industrial installations, I can honestly say this was the best decision I could have made.

An insatiable thirst for knowledge and new challenges led me back to college to improve my qualifications and this led to my next career change, moving from electrical contracting into teaching. I have now been teaching for over seventeen years, working both in private training and in further education. While my current roles as Director of Education for Trade Skills 4U and within City & Guilds are a more technical role, I still very much enjoy the thrill of teaching.

TREVOR PICKARD

I am an electrical engineering consultant and my interest in all things electrical started when I was quite young. I always had a battery powered model under construction or an electrical motor or some piece of electrical equipment in various stages of being taken apart to see how they operated. Looking back, some of my activities with mains electricity would certainly be considered as unacceptable today!

Upon leaving school in 1966 I commenced work with an electricity distribution company, Midlands Electricity Board (MEB) and after serving a student apprenticeship I held a series of engineering positions. I have never tired of my involvement with electrical engineering and was very fortunate to have had a varied and interesting career in the engineering department of Midlands Electricity and embraced its various changes through privatisation and subsequent acquisition. I held posts in Design, Safety, Production Engineering, as Production Manager of a large urban-based operational division, as General Manager of the Repair and Restoration department, and as General Manager of the Primary Network department (33kV–132kV).

My interest in electrical engineering has extended beyond the '9–5 job' and I have had the opportunity to become involved in the writing of standards in the domestic, European and international arena with BSI, CENELEC and IEC and have for many years lectured for the Institution of Engineering and Technology.

FOREWORD

I have been in the heating and plumbing industry all my working life, and was lucky enough to have an inspirational trainer when I first started out. That training set me on a career that has taken me through all sectors of the plumbing and heating industry.

Now, as a trainer myself, nothing gives me more pleasure than to witness what I call the 'light bulb' moment; that moment when, after drawing on many years of experience 'on the tools', after explaining how something works and the theory behind it, everything suddenly falls into place in a learner's mind. At that moment, you have learned something that will stay with you forever.

With this book, City & Guilds is providing an excellent foundation for a world-class skilled workforce. This will be an indispensable, comprehensive and relevant reference book for those entering today's plumbing trade and will help to inspire the next generation of heating engineers.

Steve Owen, National Training Manager, Baxi UK

Steve Owen was appointed National Training Manager for Baxi in 2006. He is responsible for all aspects of domestic technical training in the business, including internal and external customer training and partnerships with training centres and colleges. Steve first joined the company as a service engineer in 1990, moving into the training department in 2002. Prior to this, Steve completed an apprenticeship with British Gas before becoming self-employed.

HOW TO USE THIS TEXTBOOK

Welcome to your City & Guilds Level 3 Diploma in Plumbing Studies textbook. It is designed to guide you through your Level 3 qualification and be a useful reference for you throughout your career.

Each chapter covers a unit from the 6035 Level 3 qualification. Each chapter covers everything you will need to understand in order to complete your written or online tests and prepare for your practical assessments.

Throughout this textbook you will see the following features:

KEY POINT

LPG is heavier than air. Propane has a specific gravity of 1.5 and butane has a specific gravity of 2.

KEY POINT These are particularly useful hints that may assist in you in revision for your tests or to help you remember something important.

Backflow protection

Prevention of contamination of water through backflow or back siphonage.

DEFINITIONS Words in bold in the text are explained in the margin to aid your understanding. They also appear in the glossary at the back of the book.

SUGGESTED ACTIVITY

What should you check before moving an object from one place to another by manual handling?

SUGGESTED ACTIVITY These hints suggest that you try an activity to help you practice and learn.

ASSESSMENT GUIDANCE

When carrying out the practical assessment, always keep the work area clean and tidy. Dangerous working will mean you could fail.

ASSESSMENT GUIDANCE Guidance notes on how to approach your assessment.

 SmartScreen Unit 303
Handout 1

SMARTSCREEN References to online SmartScreen resources to help you revise what you've learnt in the book.

At the end of each unit are some 'Test your knowledge' questions. These questions are designed to test your understanding of what you have learnt in that unit. This can help with identifying further training or revision needed. You will find the answers at the end of the book.

Also at the end of each unit is an Assessment checklist, covering what you should know and be able to do on completing that unit. This checklist lists all learning outcomes and assessment criteria for the unit that you have just read, showing you where in the unit each outcome is covered. This enables you to understand what is required for your assessment and to go back and revise any area that you need a greater understanding of.

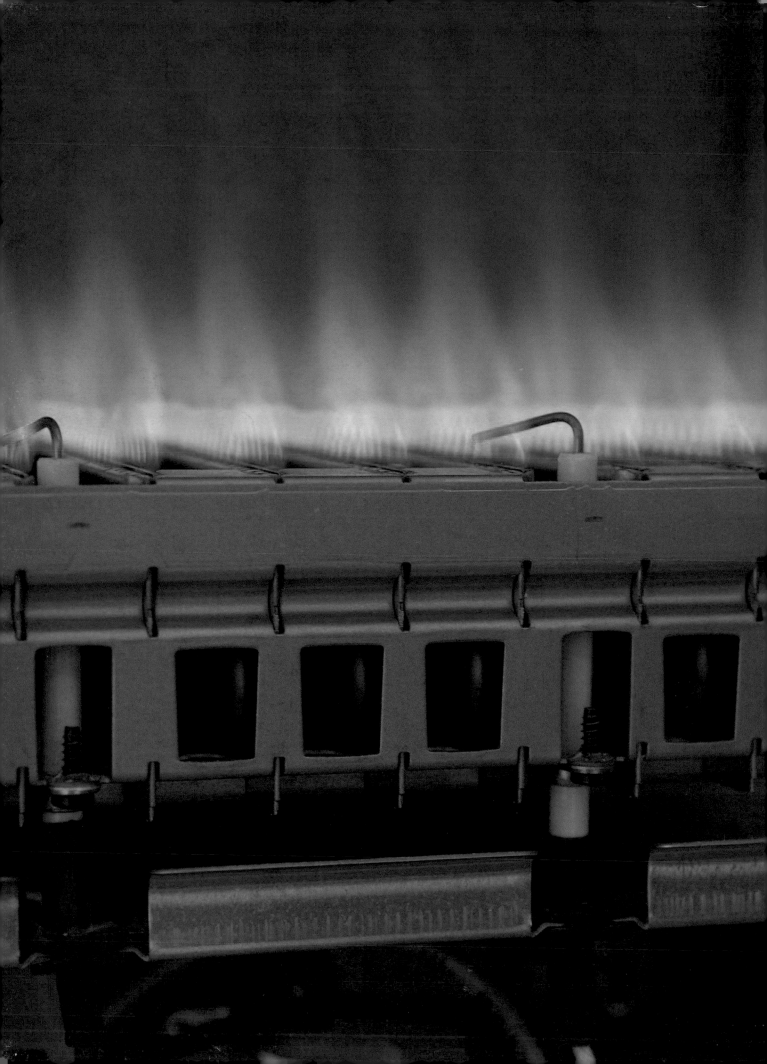

UNIT 201

Health and safety in building services engineering

Every year accidents at work result in the deaths of hundreds of people, with several hundred thousand more being injured in the workplace. In 2010/2011, Health and Safety Executive (HSE) statistics recorded 27 million working days being lost due to work-related illness and workplace injury.

Occupational health and safety affects all individuals in the workplace and all aspects of work in the complete range of working environments – hospitals, factories, schools, universities, commercial undertakings, manufacturing plants and offices. As well as the human cost in terms of pain, grief and suffering, accidents in the workplace also have a financial cost, such as lost income, insurance and production disturbance. The HSE puts this figure at £13.4 billion for the year 2010/2011. It is therefore important to identify, assess and control the activities that may cause harm in the workplace.

There are seven Learning Outcomes to this unit. The learner will:

1 Know health and safety legislation.
2 Know how to handle hazardous situations.
3 Know electrical safety requirements when working in the building services industry.
4 Know the safety requirements for working with gases and heat-producing equipment.
5 Know the safety requirements for using access equipment in the building services industry.
6 Know the safety requirements for working safely in excavations and confined spaces in the building services industry.
7 Be able to apply safe working practice.

This unit will be assessed by:

- workshop-based practical assessments
- online multiple-choice assessment.

Know health and safety legislation (LO1)

There are four assessment criteria for this Outcome:

1 State the aims of health and safety legislation.
2 Identify the responsibilities of individuals under health and safety legislation.
3 Identify statutory and non-statutory health and safety materials.
4 Identify the different roles of the Health and Safety Executive in enforcing health and safety legislation.

The basic concept of health and safety legislation is to provide the legal framework for the protection of people from illness and physical injury that may occur in the workplace.

There are two sub-divisions of the law that apply to health and safety; civil law and criminal law.

Civil law – deals with disputes between individuals, between organisations, or between individuals and organisations, in which compensation can be awarded to the victim. The civil court is concerned with **liability** and the extent of that liability rather than guilt or non-guilt.

Criminal law – is the body of rules that regulates social behaviour and prohibits threatening, harming or other actions that may endanger the health, safety and moral welfare of people. The rules are laid down by the government and are enacted by Acts of Parliament as **statutes**. The Health and Safety at Work etc Act 1974 (HSW Act) is an example of criminal law. It is enforced by the Health and Safety Executive (HSE) or local authority environmental health officers.

There are two sources of law that may be used in the above sub-divisions: common law and statute law.

Common law – is the body of law based on custom and decisions made by judges in courts. In health and safety, the legal definitions of terms such as 'negligence', 'duty of care', and 'so far as is reasonably practicable' are based on legal judgments and are part of common law.

Statute law – is the name given to law that has been laid down by Parliament as Acts of Parliament.

In terms of health and safety, criminal law is based only on statute law, but civil law may be based on either common law or statute law.

In summary, criminal law seeks to protect everyone in society and civil law seeks to recompense individuals to make amends for loss or harm they have suffered (ie provide compensation).

Liability

A debt or other legal obligation to compensate for harm.

Statute

A major written law passed by Parliament.

KEY POINT

'So far as is reasonably practicable' involves weighing a risk against the trouble, time and money needed to control it.

SmartScreen Unit 201

Presentation 1 and Worksheet 1

The main legal requirements for health and safety at work

The HSW Act is the basis of all British health and safety law. It provides a comprehensive and integrated piece of legislation that sets out the general duties that employers have towards employees, contractors and members of the public, and that employees have towards themselves and each other. These duties are qualified in the HSW Act by the principle of 'so far as is reasonably practicable'.

What the law expects is what good management and common sense would lead employers to do anyway: that is, to look at what the risks are and take sensible measures to tackle those risks. The person(s) who is responsible for the risk and best placed to control that risk is usually designated as the **duty holder**.

The HSW Act lays down the general legal framework for health and safety in the workplace with specific duties being contained in regulations, also called statutory instruments (SIs), which are also examples of laws approved by Parliament.

> **KEY POINT**
>
> The duty holder must be competent through formal training and experience and have sufficient knowledge to avoid danger. The appropriate level of competence differs for different areas of work.

Duty holder

The person in control of a danger.

Individuals' responsibilities under health and safety legislation

The HSW Act, which is an **enabling Act**, is based on the principle that those who create risks to employees or others in the course of carrying out work activities are responsible for controlling those risks. The HSW Act places specific responsibilities on:

- employers
- the self-employed
- employees
- designers
- manufacturers and suppliers
- importers.

This section will deal with the responsibilities of employers, the self-employed and employees.

Enabling Act

An enabling Act allows the Secretary of State to make further laws (regulations) without the need to pass another Act of Parliament.

The HSW Act makes provision for securing the health, safety and welfare of persons at work

Responsibilities of employers and the self-employed

Under the main provisions of the HSW Act, employers and the self-employed have legal responsibilities in respect of the health and safety of their employees and other people (eg visitors and contractors) who may be affected by their undertaking and exposed to risks as a result. The employers' general duties are contained in Section 2 of the Act. They are to ensure, 'so far as is reasonably practicable', the health, safety and welfare at work of all their employees, with particular regard to:

- the provision of safe plant and systems of work
- the safe use, handling, storage and transport of articles and substances
- the provision of any required information, instruction, training or supervision
- a safe place of work including safe access and exit
- a safe working environment with adequate welfare facilities.

These duties apply to virtually everything in the workplace, which therefore includes plumbing and electrical systems and installations, plant and equipment. An employer does not have to take measures to avoid or reduce a risk if that is technically impossible or if the time, trouble or cost of the measures would be grossly disproportionate to the risk.

Responsibilities of employees

Employees are required to take reasonable care for the health and safety of themselves and others. To achieve this aim, they have two main duties placed upon them:

- to take reasonable care for the health and safety of themselves and others who may be affected by their acts or omissions at work
- to cooperate with their employer and others to enable them to fulfil their legal obligations.

In addition there is a duty not to misuse or interfere with safety provisions.

Most of the duties in the HSW Act and the general duties included in the Management of Health and Safety at Work Regulations 1999 (the Management Regulations) are expressed as goals or targets that are to be met 'so far as is reasonably practicable' or through exercising 'adequate control' or taking 'appropriate' (or 'reasonable') steps. This involves making judgments as to whether existing control measures

ASSESSMENT GUIDANCE

The exam is multiple choice, but you will need to write for the practical assessment.

KEY POINT

Omitting/failing to do something required by law or where there is a duty to someone else can be just as bad in law as doing an illegal act or doing something negligently.

SUGGESTED ACTIVITY

Think of any jobs you have had in the past, such as part-time work in the holidays. What do you think were your responsibilities with regards to health and safety?

are sufficient and, if not, what else should be done to eliminate or reduce the risk. This risk assessment will be produced using approved codes of practice (ACoP) and published standards, as well as HSE or industry guidance on good practice where available.

Statutory and non-statutory health and safety materials

When the HSW Act came into force there were already some 30 statutes and 500 sets of regulations in place. The aim of the Health and Safety Commission (HSC) and the Health and Safety Executive (HSE) was to progressively replace the regulations with a system of regulation that expresses general duties, principles and goals, with any supporting detail set out in ACoPs and guidance.

Regulations

Statutory instruments (SIs) are laws approved by Parliament. The regulations governing health and safety are usually made under the HSW Act, following proposals from the HSC/HSE. This applies to regulations based on European Commission (EC) Directives as well as those produced in Great Britain.

The HSW Act and general duties in the Management Regulations set goals and leave employers the freedom to decide how to control the risks they identify. Guidance and ACoPs give advice.

However, some risks are so great or the proper control measures so costly that it would not be appropriate to leave employers to decide what to do about them. Regulations identify these risks and set out the specific action that must be taken. Often these requirements are absolute – they require something to be done without qualification. The employer has no choice but to undertake whatever action is required to prevent injury, regardless of cost or effort.

Some regulations apply across all workplaces. Such regulations include the Manual Handling Operations Regulations 1992, which apply wherever things are moved by hand or bodily force, and the Health and Safety (Display Screen Equipment) Regulations 1992, which apply wherever visual display units (VDUs) are used. Other regulations apply to hazards unique to specific industries, such as mining or the nuclear industry.

The following regulations apply across the full range of workplaces.

- **Control of Noise at Work Regulations 2005:** require employers to take action to protect employees from hearing damage.

> **ASSESSMENT GUIDANCE**
>
> Make sure you know the difference between statutory (legal) and non-statutory documents.

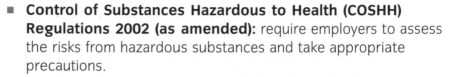

- **Control of Substances Hazardous to Health (COSHH) Regulations 2002 (as amended):** require employers to assess the risks from hazardous substances and take appropriate precautions.

- **Electricity at Work Regulations 1989:** require people in control of electrical systems to ensure they are safe to use and maintained in a safe condition.

- **Health and Safety (Display Screen Equipment) Regulations 1992:** set out requirements for work with VDUs.

- **Health and Safety (First-Aid) Regulations 1981:** require employers to provide adequate and appropriate equipment, facilities and personnel to ensure their employees receive immediate attention if they are injured or taken ill at work. These regulations apply to all workplaces, including those with fewer than five employees, and to the self-employed.

- **Health and Safety Information for Employees Regulations 1989:** require employers to display a poster telling employees what they need to know about health and safety.

- **Management of Health and Safety at Work Regulations 1999 (as amended):** require employers to carry out risk assessments, make arrangements to implement necessary measures, appoint competent people and arrange for appropriate information and training.

- **Manual Handling Operations Regulations 1992:** cover the moving of objects by hand or bodily force.

- **Personal Protective Equipment at Work Regulations 1992 (as amended):** require employers to provide appropriate protective clothing and equipment for their employees.

- **Provision and Use of Work Equipment Regulations 1998:** require that equipment provided for use at work, including machinery, is suitable and safe.

- **Reporting of Injuries, Diseases and Dangerous Occurrences Regulations 2013 (RIDDOR) (as amended):** require employers to notify the HSE of certain occupational injuries, diseases and dangerous events.

- **Workplace (Health, Safety and Welfare) Regulations 1992:** cover a wide range of basic health, safety and welfare issues such as ventilation, heating, lighting, workstations, seating and welfare facilities.

The following specific regulations cover particular areas, such as asbestos and lead:

- **Chemicals (Hazard Information and Packaging for Supply) Regulations 2009:** require suppliers to classify, label and package dangerous chemicals and provide safety data sheets for them.

ASSESSMENT GUIDANCE

As you can see, there are lots of different regulations that apply to the workplace. You should have a working knowledge of them.

SUGGESTED ACTIVITY

What personal protective equipment (PPE) could be worn to protect against damage to the eyes?

- **Construction (Design and Management) Regulations 2007:** cover safe systems of work on construction sites.

- **Control of Asbestos Regulations 2012:** affect anyone who owns, occupies, manages or otherwise has responsibilities for the maintenance and repair of buildings that may contain asbestos.

- **Control of Lead at Work Regulations 2002:** impose duties on employers to carry out risk assessments, prevent or control exposure to lead and monitor the exposure of employees.

- **Control of Major Accident Hazards Regulations 1999 (as amended):** require those who manufacture, store or transport dangerous chemicals or explosives in certain quantities to notify the relevant authority.

- **Dangerous Substances and Explosive Atmospheres Regulations 2002:** require employers and the self-employed to carry out a risk assessment of work activities involving dangerous substances.

- **Gas Safety (Installation and Use) Regulations 1998:** cover safe installation, maintenance and use of gas systems and appliances in domestic and commercial premises.

- **Work at Height Regulations 2005:** apply to all work at height where there is a risk of a fall liable to cause personal injury.

Approved codes of practice (ACoP)

ACoPs offer practical examples of good practice. They were made under Section 16 of the HSW Act and have a special status. They give advice on how to comply with the law by, for example, providing a guide to what is reasonably practicable. For example, if regulations use words such as 'suitable' and 'sufficient', an ACoP can illustrate what is required in particular circumstances. If an employer is prosecuted for a breach of health and safety law, and it is proved that they have not followed the provisions of the relevant ACoP, a court can find them at fault unless they can show that they have complied with the law in some other way.

Guidance and non-statutory regulations

The HSE publishes guidance on a range of subjects. Guidance can be specific to the health and safety problems of an industry or to a particular process used in a number of industries.

The main purposes of guidance are:

- to interpret and help people to understand what the law says
- to help people comply with the law
- to give technical advice.

Following guidance is not compulsory and employers are free to take other action, but if they do follow the guidance, they will normally be doing enough to comply with the law.

European law

In recent years much of Great Britain's health and safety law has originated in Europe. Proposals from the EC may be agreed by member states, which are then responsible for making them part of their domestic law.

Modern health and safety law in this country, including much of that from Europe, is based on the principle of risk assessment as required by the Management of Health and Safety at Work Regulations 1999.

Roles of the HSE in enforcing health and safety legislation

The Health and Safety Executive (HSE) and the Health and Safety Commission (HSC) were, until 2008, the two government agencies responsible for health and safety in Great Britain. The non-executive HSC was there to ensure that relevant legislation was appropriate and understood by conducting and sponsoring research, providing training, providing an information and advisory service, and submitting proposals for new or revised regulations (statutory instruments) and approved codes of practice (ACoPs).

The HSE was the operating arm of the HSC. It advised and assisted the HSC in its functions and had specific responsibility, shared with local authorities, for enforcing health and safety law. The HSC and HSE merged on 1 April 2008 and are now known simply as HSE.

Today, the HSE's aim is to prevent death, injury and ill health in Great Britain's workplaces and it has a number of ways of achieving this. Enforcing authorities may offer the duty holder information and advice, both face to face and in writing, or they may warn a duty holder that, in their opinion, the duty holder is failing to comply with the law. (For more on the term 'duty holder' see page 3.)

In carrying out the HSE's enforcement role, inspectors appointed under the HSW Act can:

- enter premises at any reasonable time, accompanied by a police officer if necessary
- examine, investigate and require the premises to be left undisturbed
- take samples, photographs and, if necessary, dismantle and remove equipment or substances
- review relevant documents or information such as risk assessments, accident books, or similar

SUGGESTED ACTIVITY

Find out who the duty holder is at your place of work. It might be more than one person.

SmartScreen Unit 201
Worksheet 1

- seize, destroy or make harmless any substance or article
- issue enforcement notices and start prosecutions.

An inspector may serve one of three types of notice:

- a prohibition notice tells the duty holder to stop an activity immediately
- an improvement notice sets out action needed to remedy a situation and gives the duty holder a date by which they must complete the action
- a Crown notice is issued under the same circumstances that would justify a prohibition or improvement notice, but is only served on duty holders in Crown organisations such as government departments, the Forestry Commission or the Prison Service.

Know how to handle hazardous situations (LO2)

There are fourteen assessment criteria for this Outcome:

1 Identify common hazardous situations found on site.
2 Describe safe systems at work.
3 Identify the categories of safety signs.
4 Identify symbols for hazardous substances.
5 List common hazardous substances used in the building services industry.
6 List precautions to be taken when working with hazardous substances.
7 Identify the types of asbestos that may be encountered in the workplace.
8 Identify the actions to be taken if the presence of asbestos is suspected.
9 Describe the implications of being exposed to asbestos.
10 State the application of different types of personal protective equipment.
11 Identify the procedures for manually handling heavy and bulky items.
12 Identify the actions that should be taken when an accident or emergency is discovered.
13 State procedures for handling injuries sustained on site.
14 State the procedures for recording accidents and near misses at work.

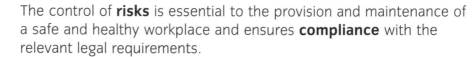

Risk

The chance (large or small) of harm actually being done when things go wrong (eg risk of electric shock from faulty equipment).

Compliance

The act of carrying out a command or requirement.

Hazard

Anything with the potential to cause harm (eg chemicals, working at height, a fault on electrical equipment).

KEY POINT

You must be able to identify hazards as part of a risk assessment. It should then be possible to eliminate, reduce, isolate or control the risk by the application of suitable control measures. The use of PPE should be a last resort.

 SmartScreen Unit 201

Presentation 2 and Worksheet 2

The control of **risks** is essential to the provision and maintenance of a safe and healthy workplace and ensures **compliance** with the relevant legal requirements.

Risk assessment starts with the need for **hazard** identification. Risk assessment is usually evaluated in terms of:

- the likelihood of something happening (ie whether an accident is going to occur)
- the severity of outcome (ie how serious the resulting injury would be).

To control these hazardous situations, duty holders need to:

- find out what the hazards are
- decide how to prevent harm to health
- provide control measures to reduce harm to health
- make sure the control measures are used
- keep all control measures in good working order
- provide information, instruction and training for employees and others
- provide monitoring and health surveillance in appropriate cases
- plan for emergencies.

Common hazardous situations on site

Housekeeping

This is one of the most important single items influencing safety within the workplace. Poor housekeeping not only causes an increase in the risk of fire, slips, trips and falls, but may also expose members of the public to risks created during building services engineers' work activities. The following are some examples of good housekeeping.

- Stairways, passages and gangways should be kept free from materials, electrical power leads and obstructions of every kind.
- Materials and equipment should be stored tidily so as not to cause obstruction and should be kept away from the edges of buildings, roofs, ladder access, stairways, floor openings and rising shafts.
- Tools should not be left where they may cause tripping or other hazards. Tools not in use should be placed in a tool belt or tool bag and should be collected and stored in an appropriate container at the end of each working day.

Tools should be kept neatly in a tool belt or bag

A badly managed site will inevitably lead to accidents

- Working areas should be kept clean and tidy. Scrap and rubbish must be removed regularly and placed in proper containers or disposal areas.

- Rooms and site accommodation should be kept clean. Soiled clothes, scraps of food, etc should not be allowed to accumulate, especially around hot pipes or electric heaters.

- The spillage of oil or other substances must be contained and cleaned up immediately.

- All flammable liquids, liquified petroleum gas (LPG) and gas cylinders must be stored properly.

ASSESSMENT GUIDANCE

When carrying out the practical assessment, always keep the work area clean and tidy. Dangerous working will mean you could fail.

Gas bottles should be stored properly

Slips, trips and falls

These are the most common hazards to people as they walk around the workplace. They make up over a third of all major workplace injuries. Over 10,000 workers suffered serious injury because of a slip or trip in 2011. Listed below are the main factors that can play a part in contributing to a slip- or trip-free environment.

Flooring

- The workplace floor must be suitable for the type of work activity that will be taking place on it.
- Floors must be cleaned correctly to ensure they do not become slippery, or be of a non-slip type that it keeps its slip-resistance properties.
- Flooring must be fitted correctly to ensure that there are no trip hazards and any non-slip coatings must be correctly applied.
- Floors must be maintained in good order to ensure that there are no trip hazards such as holes, uneven surfaces, curled-up carpet edges, or raised telephone or electrical sockets.
- Ramps, raised platforms and other changes of level should be avoided. If they cannot be avoided, they must be highlighted.

Stairs

Stairs should have:

- high-visibility, non-slip, square nosings on the step edges
- a suitable handrail
- steps of equal height
- steps of equal width.

Contamination

Most floors only become slippery once they become contaminated. If **contamination** can be prevented, the slip risk can be reduced or eliminated.

Contamination of a floor can be classed as anything that ends up on a floor, including rainwater, oil, grease, cardboard, product wrapping, dust, etc. It can be a by-product of a poorly controlled work process or be due to bad weather conditions.

Cleaning

Cleaning is important in every workplace; nowhere is exempt. It is not just a subject for the cleaning team. Everyone's aim should be to keep their workspace clear and deal with contamination such as spillages as soon as they occur.

Contamination

The introduction of a harmful substance to an area.

The process of cleaning can itself create slip and trip hazards, especially for those entering the area being cleaned, including those undertaking the cleaning. Smooth floors left damp by a mop are likely to be extremely slippery. Trailing wires from a vacuum cleaner or polishing machine can present an additional trip hazard.

People often slip on floors that have been left wet after cleaning. Access to smooth wet floors should be restricted by using barriers, locking doors or cleaning in sections. Signs and warning cones only warn of a hazard; they do not prevent people from entering the area. If the water on the floor is not visible, signs and cones are usually ignored.

Human factors

How people act and behave in their work environments can affect the risk of slips and trips. For example:

- Having a positive attitude toward health and safety, for example, dealing with a spillage instead of waiting for someone else to deal with it, can reduce the risk of slip and trip accidents.
- Wearing the correct footwear can also make a difference.
- Lack of concentration and distractions, such as being in a hurry, carrying large objects or using a mobile phone, all increase the risk of an accident.

Environmental factors

Environmental issues can affect slips and trips. The following points give an indication of these issues.

- Too much light on a shiny floor can cause glare and stop people from seeing hazards on floors and stairs.
- Too little light will prevent people from seeing hazards on floors and stairs.
- Unfamiliar and loud noises may be distracting.
- Rainwater on smooth surfaces inside or outside a building may create a slip hazard.

Footwear

Footwear can play an important part in preventing slips and trips. The following points highlight some relevant areas.

- Footwear can perform differently in different situations, for example, footwear that performs well in wet conditions might not be suitable where there are food spillages.
- A good tread pattern on footwear is essential on fluid-contaminated surfaces.

SUGGESTED ACTIVITY

It is very easy to trip or slip, which can result in an injury. What action should you take if you find that a plasterer has left material in a passageway?

- Sole tread patterns should not be allowed to become clogged with any waste or debris on the floor, as this makes them unsuitable for their purpose.

- Sole material type and hardness are key factors influencing safety.

- The choice of footwear should take into account factors such as comfort and durability, as well as obvious safety features such as protective toecaps and steel mid-soles.

Safe systems at work

A safe system of work is a work method that results from a systematic examination of the working process to identify the hazards and to specify work methods designed either to completely remove the hazards or control and minimise the relevant risks. Section 2 of the Health and Safety at Work etc Act 1974 (HSW Act) requires employers to provide safe plant and systems of work. Many of the regulations made under the HSW Act have more specific requirements for the provision of safe systems of work.

Competent person

Recognised term for someone with the necessary skills, knowledge and experience to manage health and safety in the workplace.

Safe systems of work should be developed by a **competent person**, that is, a person with sufficient training and experience or knowledge to assist with key aspects of safety management and compliance. Staff who are actively involved in the work process also have a valuable role to play in the development of the system. They can help to ensure that it is of practical benefit and that it will be applied diligently.

All safe systems of work need to be monitored regularly to ensure that they are fully observed and effective, and updated as necessary.

Safe systems of work are normally formal and documented but may be given as a verbal instruction. Examples of documented safe systems of work would be for asbestos removal, air-conditioning maintenance, working on live electrical equipment and portable appliance testing.

Method statements

ASSESSMENT GUIDANCE

Practise producing a method statement for a small electrical installation of your choice.

A method statement is a form of safe system of work and describes in a logical sequence exactly how a job is to be carried out, to be safe and without risk to health. It includes all the risks identified in the risk assessment and the measures to control those risks. The statement need be no longer than necessary to achieve these objectives effectively.

Permit-to-work procedures

A permit-to-work (PTW) procedure is a specialised written safe system of work that ensures that potentially dangerous work is done safely. Examples of such work include: work in confined spaces, **hot work**, work with asbestos-based materials, work on pipelines with hazardous contents, or work on high voltage electrical systems (above 1,000V) or complex lower voltage electrical systems.

A PTW procedure also serves as a means of communication between site/installation management, plant supervisors and operators and those who carry out the hazardous work. Essential features of PTW systems are:

- clear identification of who may authorise particular jobs (and any limits to their authority) and who is responsible for specifying the necessary precautions

- a PTW should only be issued by a technically competent person, who is familiar with the system and equipment, and is authorised in writing by the employer to issue such documents

- training and instruction in the issue, use and closure of permits

- monitoring and **auditing** to ensure that the system works as intended

- clear identification of the types of work considered hazardous

- clear and standardised identification of tasks, risk assessments, permitted task duration and any additional activity or control measure that occurs at the same time.

The effective operation of a PTW system requires involvement and cooperation from a number of persons. The procedure for issuing a PTW should be written and adhered to.

Hot work

Work that involves actual or potential sources of ignition and carried out in an area where there is a risk of fire or explosion (eg welding, flame cutting, grinding).

Audit

To conduct a systematic review to make sure standards and management systems are being followed.

Safety signs

The Health and Safety (Safety Signs and Signals) Regulations 1996 require employers to ensure that safety signs are provided (or are in place) and maintained in circumstances where risks to health and safety have not been avoided by other means, for example, safe systems of work. The range of safety signs are shown on the following pages.

Prohibition signs

These signs indicate an activity that must not be done. They are circular white signs with a red border and red cross bar.

Red	
• Prohibition sign • Danger alarm • Fire-fighting equipment	• Dangerous behaviour • Stop, shutdown, emergency cut-out devices • Evacuate identification and location

Some prohibition signs

Warning signs

These provide safety information and/or give warning of a hazard or danger. They are triangular yellow signs with a black border and symbol.

Yellow or amber	
• Warning sign	• Be careful, take precautions • Examine

Some warning signs

Mandatory signs

These signs give instructions that must be obeyed. They are circular blue signs with a white symbol.

Blue	
• Mandatory	• Specific behaviour or action • Wear personal protective equipment

Hand protection must be worn in this area

Eye protection must be worn

Safety helmets must be worn in this area

High visibility clothing must be worn in this area

Some mandatory signs

A typical construction site safety sign

SUGGESTED ACTIVITY

List four other warning and mandatory signs not shown here. You can find more signs on the internet or look around the site where you are based.

Advisory or safe condition signs

These give information about safety provision. They are square or rectangular signs with a white symbol.

Green	
• Emergency escape, first-aid sign • No danger	• Doors, exits, routes, equipment, facilities • Return to normal

Some safe condition signs

Symbols for hazardous substances

Hazardous substances are given a classification according to the severity and type of hazard they may present to people in the workplace. However, all over the world there are different laws on how to identify the hazardous properties of chemicals. The United Nations has therefore created the Globally Harmonised System of Classification and Labelling of Chemicals (GHS). The aim of the GHS is to have, worldwide, the same:

- criteria for classifying chemicals according to their health, environmental and physical hazards
- hazard communication requirements for labelling and safety data sheets.

The GHS is not a formal treaty, but is a non-legally binding international agreement. In Great Britain the existing legislation is the Chemicals (Hazard Information and Packaging for Supply) Regulations 2009 (CHIPS), but this will gradually be replaced by the European Classification, Labelling and Packaging of Substances and Mixtures Regulations between now and 2015.

Examples of the GHS labelling system are shown below:

The hazard signs shown indicate substances that are
(top row, left to right) flammable, explosive, oxidising, gas under pressure, **toxic**
and (bottom row, left to right) long-term health hazards (causes of cancer),
corrosive, require caution (irritants), dangerous to the environment

Toxic

Poisonous.

Common hazardous substances

Hazardous substances at work may include the substances used directly in the work process, such as glue, paints, thinners, solvents and cleaning materials, and those produced by different work activities, such as welding fumes. Health hazards are always present during building services activities due to the nature of the activities and may include other hazards, such as vibration, dust (possibly including asbestos), cement and solvents.

Hazardous substance

Something that can cause ill health to people.

The Control of Substances Hazardous to Health (COSHH) Regulations 2002 provide a framework that helps employers assess risk and monitor effective controls. A COSHH assessment is essential before work starts and should be updated as new substances are introduced.

SmartScreen Unit 201
Worksheet 3

Hazardous substances include:

- any substance that gives off fumes that may cause headaches or respiratory irritation
- acids that cause skin burns or respiratory irritation (eg battery acid, metal-cleaning materials)
- solvents, for example, for PVC tubes and fittings, that cause skin and respiratory irritation
- man-made fibres that cause eye or skin irritation (eg thermal insulation, optical fibres)
- cement and wood dust that may cause eye irritation and respiratory irritation
- fumes and gases that cause respiratory irritation (eg soldering, brazing and welding fumes, or overheating/burning PVC)
- asbestos.

The COSHH Regulations require employers to assess risk and ensure the prevention or adequate control of exposure to hazardous substances by measures other than the provision of personal protective equipment (PPE) 'so far as is reasonably practicable'.

> **SUGGESTED ACTIVITY**
>
> Think of some common electrical components that could contain substances hazardous to health.

Precautions to be taken with hazardous substances

There is an acknowledged 'hierarchy of control' list of measures designed to control risks. These are considered in order of importance, effectiveness or priority. The measures are listed as follows:

- eliminate the risk by designing out or changing the process
- reduce the risk by substituting less hazardous substances
- isolate the risk using enclosures, barriers or by moving workers away
- control the risk by the introduction of guarding or local exhaust systems
- management control such as safe systems of work, training, etc
- PPE such as eye protection or respiratory equipment.

> **ASSESSMENT GUIDANCE**
>
> Make sure you can recognise a hazard. Be able to explain what you can do to reduce the risk.

> **SUGGESTED ACTIVITY**
>
> How would you safely dispose of discharge lamps containing mercury?

Asbestos encountered in the workplace

Asbestos is the single greatest cause of work-related deaths in the UK. It is a naturally occurring substance, which is obtained from the ground as a rock-like ore, normally through open-pit mining.

SmartScreen Unit 201
Worksheet 5

Asbestos was used extensively as a building material in the UK from the 1950s through to the mid-1980s. It was used for a variety of purposes and was ideal for fireproofing and insulation. Any building built before the year 2,000 (houses, factories, offices, schools, hospitals, etc) may contain asbestos.

Asbestos materials in good condition are safe unless the asbestos fibres become airborne, which happens when materials are damaged due to demolition or remedial works on or in the vicinity of asbestos ceiling tiles, asbestos cement roofs and wall sheets, sprayed asbestos coatings on steel structures and lagging. In older buildings the presence of asbestos in and around boilers, hot water pipes and structural fire protection must always be anticipated when undertaking electrical work. It is difficult to identify asbestos by colour alone and laboratory tests are normally required for positive identification.

What to do if the presence of asbestos is suspected

If asbestos is discovered during a work activity, work should be stopped and the employer or duty holder informed immediately.

The Control of Asbestos Regulations 2012 affect anyone who owns, occupies, manages or otherwise has responsibilities for the maintenance and repair of buildings that may contain asbestos. The regulations cover the need for a risk assessment, the need for method statements for the removal and disposal of asbestos, air monitoring and the control measures required. These control measures include personal protective equipment and training.

Implications of being exposed to asbestos

Abrade

To scrape or wear away.

Asbestos commonly comes in the form of chrysotile (white asbestos), amosite (brown asbestos) and crocidolite (blue asbestos). Chrysotile is the common form of asbestos and accounts for 90% to 95% of all asbestos in circulation. When **abraded** or drilled, asbestos produces a fine dust with fibres small enough to be taken into the lungs. Asbestos fibres are very sharp and can lead to mesothelioma (cancer of the lining of the lung), lung cancer, asbestosis (scarring of the lung), diffuse pleural thickening (thickening and hardening of the lung wall) and death.

Personal protective equipment and its use

Virtually all **personal protective equipment (PPE)** is covered by the Personal Protective Equipment at Work Regulations 1992 (PPE Regulations). The exception is respiratory equipment, which is covered by specific regulations relating to specific substances (lead, gases, substances hazardous to health, etc).

PPE is defined in the PPE Regulations as 'all equipment (including clothing affording protection against the weather) which is intended to be worn or held by a person at work and which protects them against one or more risks to his health or safety'. Such equipment includes safety helmets, gloves, eye protection, high-visibility clothing, safety footwear and safety harnesses. Employers are responsible for providing, replacing and paying for PPE.

Hearing protection and respiratory protective equipment provided for most work situations are not covered by these regulations because other regulations apply to them. However, these items need to be compatible with any other PPE provided.

PPE should be used only when all other measures are inadequate to control exposure. It protects the wearer only while it is being worn and, if it fails, PPE offers no protection at all. The provision of PPE is only one part of the protection package; training, selection of the correct equipment in all work situations, good supervision, monitoring and supervision of its use, all play a part in the success of PPE as a control measure.

> **Personal protective equipment (PPE)**
>
> All equipment, including clothing for weather protection, worn or held by a person at work, which protects that person from risks to health and safety.

PPE for different types of protection

Protection for the eyes

Hazards – chemical or metal splash, dust, projectiles, gas and vapour, radiation.

PPE options – safety spectacles, goggles, face shields, visors.

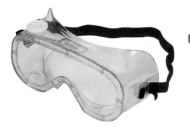

Different types of eye protection

Protection for the head

Hazards – impact from falling or flying objects, risk of head bumping or hair getting caught.

PPE options – a range of helmets and bump caps.

A safety helmet and bump cap

Protection for breathing

Hazards – dust, vapour, gas, oxygen-deficient atmospheres.

PPE options – disposable filtering face-piece or respirator, half or full-face respirators, air-fed helmets, breathing apparatus.

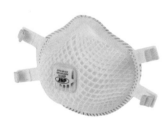

Examples of protective breathing equipment

Protection for the body

Hazards – temperature extremes, adverse weather, chemical or metal splash, spray from pressure leaks or spray guns, impact or penetration, contaminated dust, excessive wear or entanglement of own clothing.

PPE options – conventional or disposable overalls, boiler suits, specialist protective clothing such as chain-mail aprons, flame-retardant or high-visibility clothing.

Protection for hands and arms

Hazards – abrasion, temperature extremes, cuts and punctures, impact, chemicals, electric shock, skin infection, disease or contamination.

PPE options – gloves, gauntlets, mitts, wrist cuffs, armlets.

Riggers' gloves – for heavy duty/manual handling work, PVC gloves – for acids and oil, IEC 60903 gloves – made from insulating material, for live work

Protection for feet and legs

Hazards – spillages underfoot, electrostatic build-up, slipping, cuts and punctures, falling objects, metal and chemical splash, abrasion.

PPE options – safety boots and shoes with protective toecaps and penetration-resistant mid-sole, gaiters, leggings, spats.

It is important that employees know why PPE is needed and are trained to use it correctly, as otherwise it is unlikely to protect as required. The following points should be considered.

- Does it fit correctly?
- How does the wearer feel?
- Is it comfortable?
- Do all items of PPE work well together?
- Does PPE interfere with the job being done?
- Does PPE introduce another health risk, for example, of overheating or getting caught up in machinery?
- If PPE needs maintenance or cleaning, how is it done?

When employees find PPE comfortable they are far more likely to wear it.

ASSESSMENT GUIDANCE

Safety footwear is likely to protect against falling objects and against nails penetrating from below. Nails should be removed from wood that has been stripped out, but sometimes one will be missed. A nail in the foot is a very painful experience as well as carrying the risk of tetanus.

Procedures for manually handling heavy and bulky items

Manual handling is one of the most common causes of injury at work and causes over a third of all workplace injuries. Manual handling injuries can occur almost anywhere in the workplace. Heavy manual labour, awkward postures and previous or existing injury can increase the risk. Work-related manual handling injuries can have serious implications for both the employer and the person who has been injured.

Manual handling

The movement of items by lifting, lowering, carrying, pushing or pulling by human effort alone.

The introduction of the Manual Handling Operations Regulations 1992 saw a change from reliance on safe lifting techniques to an analysis, using risk assessment, of the need for manual handling. The regulations established a clear hierarchy of manual handling measures.

- Avoid manual handling operations as far as is reasonably practicable by re-engineering the task to avoid moving the load or by mechanising the operation.

- If manual handling cannot be avoided, a risk assessment should be made.

- Reduce the risk of injury as far as is reasonably practicable either by the use of mechanical aids or by making improvements to the task (eg using two persons), the load and the environment.

Even if mechanical handling methods are used to handle and transport equipment or materials, there may still be hazards. These hazards may still be present in the four elements that make up mechanical handling.

- Handling equipment – mechanical handling equipment must be suitable for the task, well maintained and inspected on a regular basis.

- The load – the load needs to be prepared in such a way as to minimise accidents, taking into account such things as security of the load, flammable materials and stability of the load.

- The workplace – if possible, the workplace should be designed to keep the workforce and the load apart.

- The human element – employees who use the equipment must be properly trained.

What to do in an accident or emergency

An accident is defined by the HSE as, 'any unplanned event that results in injury or ill health of people, or damage or loss to property, plant, materials or the environment, or a loss of a **business opportunity**'.

Emergencies

Emergency procedures are there to limit the damage to people and property caused by an incident. Although the most likely emergency to be dealt with is fire (see pages 25, 30 and 41–44), there are many more emergency situations that need to be considered, including the following.

Business opportunity

In this context, the opportunity to make profit from the work or contract.

Electrical fire or explosion

Fires involving electricity are often caused by lack of care in the maintenance and use of electrical equipment and installations. The use of electrical equipment should be avoided in potentially flammable atmospheres as far as is possible. However, if the use of electrical equipment in these areas cannot be avoided, then equipment purchased in accordance with the Equipment and Protective Systems Intended for Use in Potentially Explosive Atmospheres Regulations 1996 must be used.

Escape of toxic fumes or gases

Some gases are poisonous and can be dangerous to life at very low concentrations. Some toxic gases have strong smells such as the distinctive 'rotten eggs' smell of hydrogen sulphide (H_2S). The measurements most often used for the concentration of toxic gases are parts per million (ppm) and parts per billion (ppb). More people die from toxic gas exposure than from explosions caused by the ignition of flammable gas. With toxic substances, the main concern is the effect on workers of exposure to even very low concentrations. These could be inhaled, ingested (swallowed) or absorbed through the skin. Since adverse effects can often result from cumulative, long-term exposure, it is important not only to measure the concentration of gas, but also the total time of exposure.

Gas explosion

A gas explosion is an explosion resulting from a gas leak in the presence of an ignition source. The main explosive gases are natural gas, methane, propane and butane because they are widely used for heating purposes in temporary and permanent situations. However, many other gases, such as hydrogen, are **combustible** and have caused explosions in the past. The source of ignition can be anything from a naked flame to the electrical energy in a piece of equipment.

Industrial gas explosions can be prevented with the use of intrinsic safety barriers to prevent ignition. The principle behind intrinsic safety is to ensure that the electrical and thermal energy from any electrical equipment in a hazardous area is kept low enough to prevent the ignition of flammable gas. Items such as electric motors would not be permitted in a hazardous area.

Procedures for handling injuries sustained on site

The type of accident that can occur in the workplace is dependent on the work activity being undertaken but can range from a cut finger to a **fatality**, or from a vehicle collision to the collapse of a

> **ASSESSMENT GUIDANCE**
>
> When carrying out a practical assessment, always wear your PPE.

> **KEY POINT**
>
> Toxic substances can interfere with or stop one or more vital processes in the human body. Mercury, lead and carbon monoxide, for example, are all toxic.

Combustible
Able to catch fire and burn easily.

Fatality
Death.

structure. The person in control of the premises needs to be prepared to deal with all types of accidents to ensure that the injured person can be treated quickly and effectively and that all the legal obligations are met.

Having a well-established procedure that everyone on site is aware of and understands will enable the person in control of premises to cope calmly and effectively when dealing with an accident. Good management following an accident will ensure that appropriate records are made, the accident is reported correctly and any lessons to be learned from the accident are understood and communicated to the workforce.

The procedures to be followed in the event of any accident or incident should be clear and specific to the project or site and should detail the following as a minimum:

- name of the appointed person(s) who will take control when someone is injured or falls ill
- name of the person(s) who will administer first aid
- location of the first-aid boxes and name of the person(s) responsible for ensuring they are fully stocked
- course of action that must be followed by the appointed person who takes control in the event of an accident
- guidance on action to take after the accident
- how information should be recorded and by whom (F2508 RIDDOR Form, which can be found on the HSE website www.hse.gov.uk/forms/incident/).

Procedures for recording accidents and near misses at work

The HSE definition of an accident has been given in the section 'What to do in an accident or emergency' on page 24.

A **near miss** is an unplanned event that does not result in injury, illness or damage, but had the potential to do so. Normally only a fortunate break in the chain of events prevents an injury, fatality or damage taking place. So a near miss could be defined as any incident that could have resulted in an accident. The keeping of information on near misses is very important in helping to prevent accidents occurring. Research has shown that damage and near miss accidents occur much more frequently than injury accidents and therefore give an indication of hazards.

The Social Security Act 1975 specifically requires employers to keep information on accidents. This should be the Statutory Accident Book

Near miss

Any incident that could, but does not, result in an accident.

Employers are required to keep a record of all accidents in the Statutory Accident Book, or an equivalent

for all Employee Accidents (Form B1510, found on the HSE website www.hse.gov.uk/forms/incident/) or equivalent. Each entry should be made on a separate page and the completed page securely stored to protect personal data (under the Data Protection Act 2003). An entry may be made by the employee or by anyone acting on their behalf. This information should be kept for a period of not less than three years.

Reporting the incident

The reporting of certain types of injury and incidents to the enforcing authority (the HSE or the local authority) is a legal requirement under the Reporting of Injuries, Diseases and Dangerous Occurrences Regulations 2013 (RIDDOR). Failure to comply with these regulations is a criminal offence.

RIDDOR states that deaths, major injuries (specified in the regulations) and injuries resulting in absence from work for over three days, and dangerous occurrences (specified in the regulations) must be reported. It is the responsibility of employers or the person in control of premises to report these types of incidents. Reportable major injuries include those resulting from an electric shock or electrical burn leading to unconsciousness, resuscitation or admittance to hospital for more than 24 hours. Electrical burns include burns caused by arcing or arcing products. A dangerous occurrence is a near miss that could have led to serious injury or loss of life.

A death, reportable major injury or a dangerous occurrence must be reported immediately. (From October 2013, 'major injury' is likely to change to 'specified injury'.) Injuries resulting in absence from work for over seven days must be reported within 15 days.

The police and HSE have the right to investigate fatal accidents at work. Therefore all fatal accidents must also be notified to the police. The police will often notify the HSE, but it is always a sensible precaution to ensure that the HSE has been notified.

Investigating accidents

There is nearly always something to be learned following an accident and ideally the causes of all accidents should be established regardless of whether or not injury or damage resulted. The level and nature of an investigation should reflect the significance of the event being investigated. The results of the accident investigation may lead to a review, possible amendment to the risk assessment and appropriate action to prevent similar accidents from occurring.

SUGGESTED ACTIVITY

Would a small cut on a finger, which did not require time off work, be reportable under RIDDOR?

KEY POINT

Try to keep notes of what happened during any incident. People are very bad at remembering what they actually saw.

Keeping records

There are numerous records to keep following even a minor accident. Easily accessible records should be maintained for all accidents that have occurred. In addition to the legal requirements, accident information can help an organisation identify key risk areas within the business. The accident book must be kept for three years following the last entry.

- The F2508 for reportable incidents should be kept for a period of not less than three years from the date the accident occurred.

- Appointed trade union representatives and representatives of employee safety should be provided with a copy of the F2508 if it relates to the workplace or the group of employees represented.

- It is advisable to keep copies of any accident investigation reports for the same period as above (three years).

Know electrical safety requirements when working in the building services industry (LO3)

There are seven assessment criteria for this Outcome:

1 Identify the common electrical dangers to be aware of on site.
2 List different sources of electrical supply for tools and equipment.
3 Describe reasons for using reduced low-voltage electrical supplies for tools and equipment on site.
4 Identify how to conduct a visual inspection of portable electrical equipment for safe condition before use.
5 State actions to take when portable electrical equipment fails visual inspection.
6 Outline the Safe Isolation Procedure.
7 State the procedures for dealing with electric shocks.

Common electrical dangers on site

Modern living is shaped by electricity. It is a safe, clean and immensely powerful source of energy and is in use in every factory, office, workshop and home in the country. However this energy source also has the potential to be very hazardous, with a possibility of death, if it is not handled with care. The Electricity at Work Regulations 1989 were made under the Health and Safety at Work etc Act 1974 (HSW Act) and came into force on 1 April 1990.

SmartScreen Unit 201
Presentation 3

The 1989 Regulations are goal setting in that they specify the objectives concerning the design, specification and construction of, and work activities on, electrical systems, in order to prevent injury caused by electricity. They do not specify the means for achieving these objectives. The 1989 Regulations are supported by a Memorandum of Guidance, HSR25 (2nd edition 2007).

Electrical injuries can be caused by a wide range of voltages, and are dependent upon individual circumstances, but the risk of injury is generally greater with higher voltages. Alternating current (a.c.) and direct current (d.c.) electrical supplies can cause a range of injuries including:

- electric shock
- electrical burns
- loss of muscle control
- fires arising from electrical causes
- arcing and explosion.

Electric shock

Electric shocks may arise either by direct contact with a live part or indirectly by contact with an exposed conductive part (eg a metal equipment case) that has become live as a result of a fault condition. Faults can arise from a variety of sources:

- broken equipment case exposing internal bare live connections
- cracked equipment case causing 'tracking' from internal live parts to the external surface
- damaged supply cord insulation, exposing bare live conductors
- broken plug, exposing bare live connections.

The magnitude (size) and duration of the shock current are the two most significant factors determining the severity of an electric shock. The magnitude of the shock current will depend on the contact voltage and impedance (electrical resistance) of the shock path. A possible shock path always exists through ground contact (eg hand to feet). In this case the shock path impedance is the body impedance plus any external impedance. A more dangerous situation is a hand-to-hand shock path when one hand is in contact with an exposed conductive part (eg an earthed metal equipment case), while the other simultaneously touches a live part. In this case the current will be limited only by the body impedance.

As the voltage increases, so the body impedance decreases, which increases the shock current. When the voltage decreases, the body impedance increases, which reduces the shock current. This has important implications concerning the voltage levels that are used in

SUGGESTED ACTIVITY

Battery-powered equipment is much safer and more convenient than mains equipment, but generally lacks the constant power available with 110V equipment. Are there any circumstances when 230V equipment can be used on site?

work situations and highlights the advantage of working with reduced low-voltage (110V) systems or battery-operated hand tools.

At 230V, the average person has a body impedance of approximately 1,300Ω. At mains voltage and frequency (230V–50Hz), currents as low as 50 milliamps (0.05A) can prove fatal, particularly if flowing through the body for a few seconds.

Electrical burn

Burns may arise due to:

- the passage of shock current through the body, particularly if at high voltage
- exposure to high-frequency radiation (eg from radio transmission antennas).

Loss of muscle control

People who experience an electric shock often get painful muscle spasms that can be strong enough to break bones or dislocate joints. This loss of muscle control often means the person cannot 'let go' or escape the electric shock. The person may fall if they are working at height or be thrown into nearby machinery and structures.

Fire

Electricity is believed to be a factor in the cause of over 30,000 fires in domestic and commercial premises in Britain each year. One of the principal causes of such fires is wiring with defects such as insulation failure, the overloading of conductors, lack of electrical protection and poor connections.

Arcing

Arcing frequently occurs due to short-circuit flashover accidentally caused while working on live equipment (either intentionally or unintentionally). Arcing generates UV radiation, causing severe sunburn. Molten metal particles are also likely to be deposited on exposed skin surfaces.

Explosion

There are two main electrical causes of explosion: short circuit due to an equipment fault, and ignition of flammable vapours or liquids caused by sparks or high surface temperatures.

Controlling current flow

It is necessary to include devices in circuits to control current flow; that is, to switch the current on or off by making or breaking the circuit. This may be required:

- for functional purposes (to switch equipment on or off)
- for use in an emergency (switching in the event of an accident)

- so that equipment can be switched off to prevent its use and allow maintenance work to be done safely on the mechanical parts

- to isolate a circuit, installation or piece of equipment to prevent the risk of shock where exposure to electrical parts and connections is likely for maintenance purposes.

The preparation of electrical equipment for maintenance purposes requires effective disconnection from all live supplies and the means for securing that disconnection (by locking off).

There is an important distinction between switching and isolation. Switching is cutting off the supply. Isolation is the secure disconnection and separation from all sources of electrical energy.

A variety of control devices are available for switching, isolation or a combination of these functions, some incorporating protective devices. Before starting work on a piece of isolated equipment, checks should be made to ensure that the circuit is dead, using an approved testing device.

In the case of portable equipment connected via a supply cord and plug, removal of the plug from the socket provides a ready means of isolation.

SUGGESTED ACTIVITY

Why are light switches unsuitable as devices for isolation?

Electrical supply for tools and equipment

Portable electric tools can provide valuable assistance with much of the physical effort required in electrotechnical activities. These tools can use different sources of electrical supply (mains or battery) and different means of maintaining safety in relation to that electrical supply.

Basic equipment constructions, all aimed at preventing the risk of electric shock, are specified in BS 2754:1976 – Construction of Electrical Equipment for Protection against Electric Shock, as detailed below.

Class I

The basic insulation may be an air gap and/or some form of insulating material. External conductive parts (eg the metal case) must be earthed by providing the supply through a three-core supply lead incorporating a protective conductor. The most important aspect of any periodic inspection/testing of Class I equipment is to check the integrity of this protective conductor. There is no recognised symbol for Class I equipment, though some appliances may show the symbol on the right.

This symbol may appear on Class I items

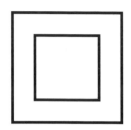

Class II symbol

Class III symbol

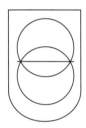

Class III supply symbol

Class II

Class II equipment has either no external conductive parts apart from fixing screws (insulation-encased equipment) or there is adequate insulation between any external conductive parts and the internal live parts to prevent the former becoming live as a result of an internal fault (metal-encased equipment). Periodic inspection or testing needs to focus on the integrity of the insulation. Class II equipment is identified by the symbol shown on the left.

Class III

This method of protection is not designed to prevent shock but to reduce its severity and therefore make the shock more survivable. The supply (no greater than 50V a.c.) must be provided from a separated extra-low-voltage (SELV) source such as a safety isolating transformer conforming to BS EN 61558, or a battery. Class III equipment is identified by the diamond symbol and the safety isolating symbol by two interlinked circles, as shown on the left.

Another way of reducing the risk of electric shock is by using a reduced low-voltage system. This is *not* a Class III system but is a safer arrangement than using mains-operated (230V) equipment because of the lower potential shock voltage. Supply is provided via a mains-powered (230V) step-down transformer with the centre point of the secondary winding connected to earth.

Reduced low-voltage electrical supplies for tools and equipment

The impedance of the human body is dependent on the touch voltage. Although 50V can be dangerous in certain circumstances, if the system voltage can be reduced to around this level, then the magnitude (size) of any current flow through the body will be significantly reduced.

The most common low-voltage system of this type in use in the UK uses a 230/110V double-wound step-down transformer with the secondary winding centre-tapped and connected to earth (CTE). While the supply voltage for equipment supplied from such a transformer is 110V, the maximum voltage to earth is 55V (63.5V for a 110V three-phase system). This is covered in BS 7671 (Regulation 411.8 Reduced low voltage systems). Overcurrent protection for the 110V supply may be provided by fuses at the transformer's 110V output terminals or by a thermal trip to detect excessive temperature rise in the secondary winding. The latter method is generally employed for portable units.

Such reduced voltage supplies may be provided:

- through fixed installations (workshops, plant rooms, lift rooms or other areas where portable electrical equipment is in frequent use)

- through small portable transformers designed to supply individual portable tools. See BS 4363:1998 and BS 7375:1996, which give a specification for the distribution of electricity on building and construction sites, based on the use of 230/110V CTE transformers.

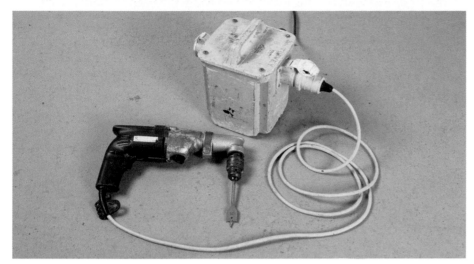

A portable CTE transformer and drill

One additional advantage of using a reduced low-voltage system is that this safeguard applies to all parts of the system on the load side of the transformer, including the flexible leads, as well as the tools, hand lamps, etc.

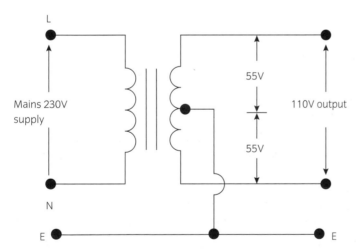

Schematic diagram of a 110/55V centre-tapped transformer in a reduced low voltage system

Alternatively, cordless or battery-powered tools offer a convenient way of providing a powered hand tool without the inconvenience of using a mains supply and without the hazard of trailing power leads.

How to conduct a visual inspection of portable electrical equipment

To maintain the safety and integrity of tools and equipment, regular in-service inspection and testing should be undertaken to confirm that the equipment remains in a safe condition. Portable appliance testing (PAT) is the term used to describe the examination of electrical appliances and equipment to ensure they are safe to use. There are three categories of in-service inspection or testing for portable tools and equipment:

- user checks (pre-use inspections) – users play an important role by checking equipment before use for signs of damage or obvious defects liable to affect safety in use
- periodic formal inspection or checks
- periodic combined inspection and testing.

The user checks are a visual check only, and should deal with the following inspection requirements separately for the equipment, supply lead and plug.

Equipment

- Equipment should be manufactured to relevant standards (BS or BS EN).
- The casing should have no visible damage and be free from dents and cracks.
- The switches should operate correctly.
- The supply lead should be secure and correctly connected.
- It should be suitable for the task.

Supply leads

- Supply leads should be manufactured to relevant standards (BS or BS EN).
- They should be suitable for the environment.
- They should be free from cuts or fraying.
- There should be no visible exposed conductor insulation (damaged sheath) or exposed live conductors.
- There should be no signs of damage to the cord sheath.
- There should be no joints evident.

Supply lead plugs

- Supply lead plugs should be manufactured to relevant standards (BS or BS EN).

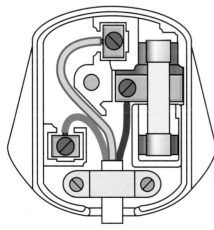

A correctly wired BS 1363 13A plug

- The insulation should have no damage.
- The cord/cable connections should be correct and secure.

The user checks must be backed up by periodic inspection and, where appropriate, testing. At that time, a thorough examination is undertaken by a nominated person, competent for that purpose. Schedules giving details of inspection and maintenance periods, together with records of the inspection, should form part of the procedure.

SUGGESTED ACTIVITY

What are the colours used to identify plugs and sockets for the following voltages?

- 110V single-phase
- 230V single-phase
- 400V three-phase

What to do when portable electrical equipment fails visual inspection

Most electrical safety defects can be found by visual examination, but some types of defect can only be found by testing. However, it is essential to understand that visual examination is an essential part of the process because some types of electrical safety defect cannot be detected by testing alone.

A relatively brief user check, based upon simple training and the use of a brief checklist, is a very useful part of any electrical maintenance regime. If the user checks detailed on page 34 and above are carried out, 95% of all faults will be identified and the appropriate action taken. No record is needed if there are no defects found. However, if equipment is found to be unsafe, it must be removed from service immediately. It must be labelled to show that it must not be used and the fault must be reported to a responsible person.

The Safe Isolation Procedure

Safe operating procedures for the isolation of plant and machinery during both electrical and mechanical maintenance must be prepared and followed. Wherever possible, electrical isolators should be fitted with a means by which the isolating mechanism can be locked in the open/off position. If this is not possible, an agreed procedure must be followed for the removal and storage of fuse links.

Regulation 12 of the Electricity at Work Regulations 1989 requires that, where necessary to prevent danger, suitable means (including, where appropriate, methods of identifying circuits) must be available for:

- cutting off the supply of electrical energy to any electrical equipment
- the isolation of any electrical equipment.

The aim of Regulation 12 is to ensure that work can be undertaken on an electrical system without danger, in compliance with Regulation 13 (work when equipment has been made dead).

Cutting off the supply – Depending on the equipment and the circumstances, this may be no more than normal functional switching (on/off) or emergency switching by means of a stop button or a trip switch.

Isolation – This means the disconnection and separation of the electrical equipment from every source of electrical energy in such a way that this disconnection and separation is secure.

From every source of electrical energy – Many accidents occur due to a failure to isolate all sources of supply to or within equipment (eg control and auxiliary supplies, uninterruptable power supply (UPS) systems or parallel circuit arrangements giving rise to back feeds).

Secure – Security can best be achieved by locking off with a safety lock (ie a lock with a unique key). The posting of a warning notice also serves to alert others to the isolation.

The steps below should be followed.

1 Select an approved voltage indicator to GS38 and confirm operation on a known source such as a proving unit.
2 Locate the correct source of supply and isolator for the section needing isolation.
3 Confirm that the device used for isolation is suitable and may be secured effectively.
4 Power down circuit loads if the isolator is not suitable for on-load switching.
5 Disconnect using the located isolator (from step 2).
6 Secure in the off position. If it is a lockable device,* keep the key on your person. Post warning signs.
7 Using the voltage indicator, confirm isolation by checking all combinations.
8 Prove the voltage indicator on a known source such as a proving unit.

Clear identification and labelling of circuits, switchgear and protective devices will minimise the risk of incorrect isolation.

How to deal with electric shocks

If all of the correct requirements, precautions and training of staff are taken, it is unlikely that an electrical accident will occur. However, procedures should be in place to deal with electric shock injury in

KEY POINT

GS38 is a document published by the HSE giving guidance on the safety of test instruments and leads.

SUGGESTED ACTIVITY

What type of device is it preferable to use when more than one person is working on an electrical circuit?

* If the device is a fuse or removable handle instead of a lockable device, keep this securely under supervision while work is undertaken.

the event of an accident. The recommended procedure for dealing with a person who has received a low-voltage shock is as follows.

- Raise the alarm (colleagues and a trained first-aider).

- Make sure the area is safe by switching off the electricity supply.

- Request colleagues to call an ambulance (999 or 112).

- If it is not possible to switch off the power supply, move the person away from the source of electricity by using a non-conductive item.

- Check whether the person is responsive, whether their airway is clear and whether they are breathing.

- If the person is unconscious but breathing, move them into the recovery position.

- If they are unconscious but not breathing, start to give cardiopulmonary resuscitation (CPR):

  - CPR is undertaken by interlocking the hands and giving 30 chest compressions in the centre of the chest, between the two pectoral muscles, at a rate of about 100 pulses per minute.

  - Tilt the casualty's head back gently, by placing one hand on the forehead and the other under the chin, to open the airway and give two mouth-to-mouth breaths.

  - Repeat the cycle of 30 compressions to two breaths until either help arrives or the patient recovers.

- Any minor burns should be treated by placing a sterile dressing over the burn and securing with a bandage.

SUGGESTED ACTIVITY

Find out the latest advice on giving CPR from the British Heart Foundation website at www.bhf.org.uk or the St John Ambulance website at www.sja.org.uk.

Know the safety requirements for working with gases and heat-producing equipment (LO4)

There are eight assessment criteria for this Outcome:

1 Identify different types of gases used on site.
2 Describe how bottled gases and equipment should be safely transported and stored.
3 Describe how to conduct a visual inspection on heat-producing equipment for safe condition.
4 Describe how combustion takes place.
5 State the dangers of working with heat-producing equipment.
6 State the procedures to follow on discovery of fires on site.
7 Identify different classifications of fires.
8 Identify types of fire extinguisher for different classifications of fires.

Different types of gases used on site

SmartScreen Unit 201

Presentation 4

Industrial gas makes up a group of gases that is used for a large range of applications in the building services sector. The gases may be organic or inorganic and are produced either by extracting them from air or making them synthetically from chemicals. They can take various forms, such as compressed, liquid or solid.

The most common industrial gases are stored under high pressure in metal containers (gas cylinders) and are used as a single gas or as a mixture (eg oxy-acetylene for welding). The table below lists the properties of the most common gases in use.

Gas	Colour	Chemical formula	Smell	Respiratory hazard	Flammability	Weight vs air
Acetylene	Maroon	C_2H_2	Pungent with hint of garlic	Asphyxiant	Highly flammable	Lighter
Argon	Dark green shoulder	Ar_2	None	Asphyxiant	Not flammable	Heavier
Butane	Blue or yellow	C_4H_{10}	Odourised	Asphyxiant	Highly flammable	Heavier
Carbon dioxide	Grey Shoulder	CO_2	None	Asphyxiant	Not flammable	Lighter
Chlorine	Yellow shoulder	Cl_2	Irritating, pungent odour	Toxic by inhalation	Can react violently with other substances	Heavier
Hydrogen	Red shoulder	H_2	None	None	Highly flammable	Lighter
Helium	Brown shoulder	He	None	Non-toxic	Not flammable	Lighter
Nitrogen	Black shoulder	N_2	None	Asphyxiant	Not flammable	Equal
Oxygen	White shoulder	O_2	None	None	Not flammable but supports combustion	Equal
Propane	Red shoulder	C_3H_8	Odourised	Asphyxiant	Highly flammable	Heavier

The design and manufacture of pressure equipment and assemblies must be in compliance with the Pressure Equipment Regulations 1999.

Gas cylinders are a convenient way to transport and store these gases and they may be used in many different applications:

- chemical processes
- soldering, welding and flame-cutting (oxy-acetylene)
- breathing (eg diving, emergency rescue)
- medical and laboratory uses
- dispensing drinks and beverages
- fuel for vehicles such as fork-lift trucks (propane)
- extinguishing fires (carbon dioxide)
- heating and cooking (butane, propane).

The main causes of accidents involving gas cylinders are:

- inadequate training and supervision
- poor installation
- poor examination and maintenance
- faulty equipment and/or design (eg badly fitted valves and regulators)
- poor handling
- poor storage
- inadequately ventilated working conditions
- incorrect filling procedures.

How to safely transport and store bottled gases and equipment

The most likely hazards that will be encountered due to the use of gas cylinders in the building services environment are in: handling, storage and transport.

Handling and use

- Gas cylinders should be used in a vertical position, unless specifically designed to be used otherwise.
- Cylinder colours are only a guide. Labels should be used as the primary means of identifying cylinder content.
- Valves, regulators and pipework must be suitable for the type of gas and pressure being used.
- Cylinders must not be used for any purpose other than the transport and storage of gas.
- Cylinders should not be dropped, rolled or dragged.

- Cylinder valves should be closed and dust caps should be replaced when a gas cylinder is not in use.
- Valves should be protected by a valve cap or collar.

Transport

- Stow gas cylinders securely when in transit.
- Gas cylinders should be contained within the body of the vehicle.
- Gas cylinders should be clearly marked to show their contents (including their United Nations unique number).
- The transport of gas cylinders is subject to the Carriage of Dangerous Goods and Use of Transportable Pressure Equipment Regulations 2009.

Storage

- Gas cylinders should not be stored for excessive periods of time.
- Stocks should be rotated to ensure that the first in is the first used.
- Gas cylinders should be stored in a dry, safe place on a flat surface in the open air or in an adequately ventilated building or part of a building specifically reserved for this purpose.
- Cylinders containing flammable gas should not be stored in part of a building used for other purposes.
- Gas cylinders should be protected from external heat sources that might stop them working properly.
- Gas cylinders should be stored away from sources of ignition and other flammable materials.
- Do not allow gas cylinders to be stored so that they stand or lie in water.
- Valves should be kept shut on empty cylinders to prevent contaminants getting in.
- Store gas cylinders securely when they are not in use. They should be properly fixed to prevent them falling over, unless designed to be freestanding.
- Store cylinders where they are not vulnerable to hazards caused by impact (eg from vehicles such as fork-lift trucks).

How to conduct a visual inspection on heat-producing equipment

An external visual inspection of the surfaces of gas cylinders and any attachments (valves, connecting hoses, flashback arresters and regulators) should identify the majority of in-service faults. The inspection is looking for signs of impact damage, scrapes, fire damage, corrosion, erosion and wear. These signs should be noted and any cylinder with defects that could cause a failure should be removed from service and returned to the supplier.

How combustion takes place

Most fires are preventable and, by adopting the right behaviours and procedures, prevention can easily be achieved.

A fire needs three elements to start: a source of ignition (heat), a source of fuel (something that burns) and oxygen. If any one of these elements is removed, a fire will not ignite or will cease to burn.

- Sources of ignition include heaters, lighting, naked flames, electrical equipment, smokers' materials (cigarettes, matches) and anything else that can get very hot or cause sparks.

- Sources of fuel include wood, paper, plastic, rubber or foam, loose packaging materials, waste rubbish, combustible liquids and furniture.

- Sources of oxygen include the air surrounding us.

A fire safety risk assessment using the same approach as used in the health and safety risk assessment should be used to determine the risks. Based on the findings of the assessment, employers must ensure that adequate and appropriate fire safety measures are in place to minimise the risk of injury or loss of life in the event of a fire. These findings must be kept up to date.

The fire triangle shows the three elements of a fire. If one element is not present, or removed, the fire will not ignite or will stop burning

Dangers of working with heat-producing equipment

The use of heat-producing equipment (hot work) is a common occurrence in building services work activities. Hot work is work that might generate sufficient heat, sparks or flame to cause a fire. It includes welding, flame cutting, soldering, brazing, grinding and other equipment that incorporates a flame, such as boilers for bitumastic materials.

The flames, sparks and heat produced during the hot work are ignition sources that can cause fires and explosions in many different situations. For example:

- Sparks produced during hot work can land on combustible materials and cause fires and explosions.

- Hot work performed on tanks and vessels with residual flammable substances and vapours can cause the tanks to explode.

To help prevent fire in the workplace, the risk assessment should identify what could cause a fire to start – the sources of ignition (heat or sparks), the substances that burn – and the people who may be at risk.

Once the risks have been identified, appropriate action can be taken to control them. Actions should be based on whether the risks can be avoided altogether or, if this is not possible, how they can be reduced. The checklist below will help with an appropriate action plan.

- Keep sources of ignition and flammable substances apart.
- Avoid accidental fires (eg make sure heaters cannot be knocked over).
- Ensure good housekeeping at all times; for example, avoid build-up of rubbish that could burn.
- Consider how to detect fires and how to warn people quickly if a fire starts; for example, install smoke alarms and fire alarms or bells.
- Have the correct fire-fighting equipment available.
- Keep fire exits and escape routes clearly marked and unobstructed at all times.
- Ensure workers receive appropriate training on procedures and fire drills.
- Review and update the risk assessment on a regular basis.

The Regulatory Reform (Fire Safety) Order 2005 covers general fire safety in England and Wales.

Procedures on discovery of fires on site

How people react in the event of fire depends on how well they have been prepared and trained for a fire emergency. It is therefore imperative that all employees (and visitors and contractors) are familiar with the company procedure to follow in the event of an emergency. A basic procedure is as follows.

- On discovery of a fire, raise the alarm immediately.
- If staff are trained and it is considered safe to do so, attempt to fight the fire using the equipment provided.
- If fire fighting fails, evacuate (leave the area) immediately.
- Ensure that no one is left in the fire area and close doors on exit in order to prevent the spread of fire.
- Go straight to the designated assembly point. These points are specially chosen as they are in locations of safety and where the emergency services are not likely to be obstructed on arrival.

KEY POINT

Remember never to ignore a fire alarm, even if it has gone off many times before that day. It may be the last one you ignore.

Different classifications of fires

All fires are grouped into classes, according to the type of materials that are burning.

The grid below is a guide to the different types of fire and the type of extinguisher that should be used.

Fire classification	Water	Foam	Powder	Carbon dioxide
Class A – Combustible materials such as paper, wood, cardboard and most plastics	✓	✓	✓	
Class B – Flammable or combustible liquids such as petrol, kerosene, paraffin, grease and oil		✓	✓	✓
Class C – Flammable gases, such as propane, butane and methane			✓	✓
Class D – Combustible metals, such as magnesium, titanium, potassium and sodium			Specialist dry powder	
Class F – Cooking oils and fats		Specialist wet chemical		

Fires involving equipment such as electrical circuits or electronic equipment are often referred to as Class E fires, although the category does not officially exist under the BS EN 3 rating system. This is because electrical equipment is often the cause of the fire, rather than the actual type of fire. Most modern fire extinguishers specify on the label whether they should be used on electrical equipment. Normally carbon dioxide or dry powder are suitable agents for putting out a fire involving electricity.

SmartScreen Unit 201
Worksheet 4

SUGGESTED ACTIVITY

Why should you not use water extinguishers on oil fires?

Know the safety requirements for using access equipment in the building services industry (LO5)

There are four assessment criteria for this Outcome:

1 Identify different types of access equipment.
2 Select suitable equipment for carrying out work at height, based on the work being carried out.
3 Describe the safety checks to be carried out on access equipment.
4 Describe safe erection methods for access equipment.

SmartScreen Unit 201
Presentation 5

Working at height remains one of the biggest causes of fatalities and major injuries within the construction industry with almost 50% of fatalities resulting from falls from ladders, stepladders and through fragile roofs. Work at height means work in any place, including at or below ground level (eg in underground workings), where a person could fall a distance liable to cause injury.

The Work at Height Regulations 2005 require duty holders to ensure that:

- all work at height is properly planned and organised
- those involved in work at height are competent
- the risks from work at height are assessed and appropriate work equipment is selected and used
- the risks of working on or near fragile surfaces are properly managed
- the equipment used for work at height is properly inspected and maintained.

Access equipment for different types of work

Many different types of access equipment are used in the building services industry, such as mobile elevated work platforms (MEWPs), ladders, mobile tower scaffolds, tube and fitting scaffolding and personal suspension equipment (harnesses).

MEWPs

There is a wide range of MEWPs (vertical scissor lift, self-propelled boom, vehicle-mounted boom and trailer-mounted boom) and if any of these are to be used it is important to consider:

- the height from the ground
- whether the MEWP is appropriate for the job
- the ground conditions
- training of operators
- overhead hazards such as trees, steelwork or overhead cables
- the use of a restraint or fall arrest system
- closeness to passing traffic.

A vertical scissor lift

Ladders

It is recommended that ladders are used only for low-risk, short-duration work (between 15 and 30 minutes depending upon the task). Common causes of falls from ladders are:

- overreaching – maintain three points of contact with the ladder
- slipping from the ladder – keep the rungs clean, wear non-slip footwear, maintain three points of contact with the ladder, make sure the rungs are horizontal
- the ladder slips from its position – position the ladder on a firm level surface, secure the ladder top and bottom, check the ladder daily
- the ladder breaks – position the ladder properly using the 1:4 rule (four units up for every one unit out), do not exceed the maximum weight limit of the ladder, only carry light materials or tools.

Stepladders

Many of the common causes of falls from ladders can also be applied to stepladders. If stepladders are used, ensure that:

- they are suitable, in good condition and inspected before use
- they are sited on stable ground
- they face onto the work activity
- knees are never above the top tread of the stepladder
- the stepladder is open to the maximum
- wooden stepladders (or ladders) are not painted as this may hide defects.

The correct angle for a ladder is 1:4

Scaffolding

Some work activities, such as painting, roof work, window replacement or brickwork, are almost certainly more conveniently undertaken from a fixed external scaffold. Tube and fitting scaffolding must only be erected by competent people who have attended recognised training courses. Any alterations to the scaffolding must also be carried out by a competent person. Regular inspections of the scaffold must be made and recorded.

Safety checks and safe erection methods for access equipment

The Work at Height Regulations 2005 require that all scaffolding and equipment that is used for working at height, where a person could fall 2 metres or more, is inspected on a regular basis, using both formal and pre-use inspections to ensure that it is fit for use.

A marking, coding or tagging system is a good method of indicating when the next formal inspection is due. However, regular pre-use checks must take place as well as formal inspections.

The following safety checks must be carried out.

MEWPs

These must only be operated by trained and competent persons, who must also be competent to carry out the following pre-use checks.

The ground conditions must be suitable for the MEWP, with no risk of the MEWP becoming unstable or overturning.

- Check for overhead hazards such as trees, steelwork and overhead cables.
- Guard rails and toe boards must be in place.
- Outriggers should be fully extended and locked in position.
- The tyres must be properly inflated and the wheels immobilised.
- Check the controls to make sure they work as expected.
- Check the fluid and/or battery charge levels.
- Check that the descent alarm and horn are working.
- Check that the emergency or ground controls are working properly.

Tube and fitting scaffolds

This equipment must only be erected by competent people who have attended recognised training courses. Any alterations to the scaffolding must also be carried out by a competent person. However, users should undertake the following fundamental checks to prevent accidents.

Tags to record that a ladder inspection has taken place

SUGGESTED ACTIVITY

What is the qualification that allows a person to use mobile elevating work platforms (MEWPs)?

- Toe boards, guard rails and intermediate rails must be in place to prevent people and materials from falling.
- The scaffold must be on a stable surface and the uprights must be fitted with base plates and sole plates.
- Safe access and exit (ladders) must be in place and secured.

Ladders and step ladders

Users must check that:

- ladders and stepladders are of the right classification (trade/industrial)
- the styles, rungs or steps are in good condition
- the feet are not missing, loose, damaged or worn
- rivets are in place and secure
- the locking bars are not bent or buckled.

Prefabricated mobile scaffold towers

Mobile scaffold towers are a convenient means of undertaking repetitive tasks in the building services industry. The erection and dismantling must only be undertaken by a competent person. The following pre-use checks will ensure safe use.

- The maximum height-to-base ratios must not be exceeded.
- Diagonal bracing and stabilisers must not be damaged or bent.
- The brace claws must work properly.
- Internal access ladders must be in place.
- Wheels must be locked when work is in progress.
- The working platform must be boarded, with guard rails and toe boards fitted.
- The towers must be tied in windy conditions.
- The platform trap door must be operating correctly.
- Rivets must be checked visually to ensure they are in place and not damaged.

Know the safety requirements for working safely in excavations and confined spaces in the building services industry (LO6)

There are five assessment criteria for this Outcome:

1 Identify the situations in which it may be necessary to work in excavations.
2 Describe how excavations should be prepared for safe working.

3 State precautions to be taken to make excavations safe.

4 Identify areas where working in confined space may be a consideration.

5 State safety considerations when working in confined spaces.

Where it may be necessary to work in excavations

SmartScreen Unit 201
Presentation 6

Every year people are killed or seriously injured by collapses and falling materials while working in excavations. These excavations may be required in the course of ground source heating projects or the installation of drains and soakaways, septic tanks, electrical distribution networks and retaining structures.

How to prepare excavations for safe working

The hazards associated with excavations are:

- excavations collapsing and burying or injuring people working in them
- material falling from the sides into any excavation
- people or plant falling into excavations
- contact with underground services (electricity, high pressure water and gas)
- undermining other structures
- exhaust fumes from petrol or diesel-engined equipment such as compressors or generators.

Planning is the key to the safety of any excavation. Before work commences a decision will be required on what temporary support will be needed and what precautions need to be taken. The equipment and precautions needed (trench sheets, props, baulks, etc) should be available on site before work starts.

The sides of the excavation must be prevented from collapsing either by battering at an angle of between 5° and 45° or by shoring the sides up with timber, sheeting or a proprietary support system. In **granular soils**, the angle of slope (**batter**) should be less than the natural angle of **repose** of the material being excavated. In wet ground, a considerably flatter slope will be required. Loose materials may fall from spoil heaps into the excavation; therefore edge protection should include toe boards or other means to protect against falling materials. Head protection should be worn in excavations.

ASSESSMENT GUIDANCE

Gas can accumulate in underground workings. Always beware of this and test the atmosphere in the underground workings first.

Granular soils

Gravel, sand or silt (coarse-grained soil) with little or no clay content. Although some moist granular soils exhibit apparent cohesion (grains sticking together forming a solid), they have no cohesive strength. Granular soil cannot be moulded when moist and crumbles easily when dry.

Batter or slope

The angle in relation to the horizontal surface, of the trench walls of an excavation, to prevent the walls collapsing.

Repose

The angle to the horizontal at which the material in the cut face is stable and does not fall away. Different materials have different angles of repose, for example, 90° for solid rock and 30° for sand.

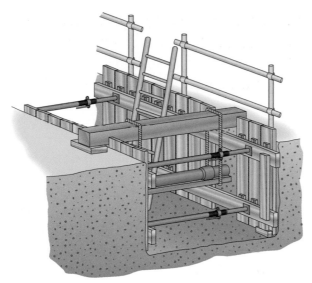

A timbered excavation with ladder access, barriers, wooden boards and supported services

---

Precautions to make excavations safe

To prevent people from falling into an excavation, the edges should be protected in one of the following ways:

- with guard rails and toe boards inserted into the ground immediately next to the supported excavation side
- by a fabricated guard rail assembly that connects to the sides of the trench box
- by using the trench support system itself (eg using trench box extensions or trench sheets longer than the trench depth).

A competent person who fully understands the dangers and necessary precautions should inspect the excavation at the start of each shift and after any event that may have affected their strength or stability, or after a fall of rock or earth.

A record of the inspections will be required and any faults that are found should be corrected immediately.

Areas where working in confined space may be a consideration

A confined space is a place which is substantially enclosed (though not always entirely) and where serious injury can occur from hazardous substances or conditions within the space or nearby (eg lack of oxygen). Examples include chambers, tanks, vessels, furnaces, boilers or cisterns, inspection chambers, pits, roof spaces and under suspended timber floors. Entry into a confined space requires the correct equipment, including PPE and a harness. Activities below ground require a descent control and rescue system.

> **KEY POINT**
>
> Never take a chance with an excavation. It only takes a second for an unsafe working area to collapse and suffocate you.

Safety considerations when working in confined spaces

The two main hazards associated with confined spaces are the presence of toxic or other dangerous substances and the absence of adequate oxygen.

The Confined Space Regulations 1997 detail the specific controls that are necessary when people enter confined spaces. These can be summarised as follows.

- Avoid entry to confined spaces (eg by doing the work from the outside).
- If entry to a confined space is unavoidable, follow a safe system of work which should include rigorous preparation, isolation, air testing and other precautions, and the use of a confined space entry permit.
- Put in place adequate emergency arrangements before the work starts.

Adequate ventilation, sufficient lighting, the provision of the correct PPE, emergency evacuation procedures and any medical conditions must all be catered for. Lone working should not be allowed.

Section 706 of BS 7671 gives the requirements for supplies to portable electrical equipment used in confined spaces that are restrictive and/or conducting.

Be able to apply safe working practice (LO7)

There are four assessment criteria for this Outcome:

1 Perform manual handling techniques.
2 Manually handle loads using mechanical lifting aids.
3 Demonstrate the safe method of assembly of access equipment.
4 Use access equipment safely.

This outcome is about your being able to demonstrate your understanding of health and safety by practical activity. The contents of this section have been covered in previous outcomes but are outlined again here to act as a reminder of the key points. Safe working practices are systems of work that are carried out in a safe manner for any given task.

Manual handling techniques

Manual handling is one of the most common causes of injury at work and causes over a third of all workplace injuries. Manual handling injuries can occur almost anywhere in the workplace and heavy manual labour, awkward postures and previous or existing injury can increase the risk. Work-related manual handling injuries can have serious implications for both the employer and the person who has been injured.

The introduction of the Manual Handling Operations Regulations 1992 saw a change from reliance on safe lifting techniques to an analysis, using risk assessment, of the need for manual handling. The regulations established a clear hierarchy of manual handling measures.

- Avoid manual handling operations as far as is reasonably practicable by re-engineering the task to avoid moving the load or by mechanising the operation.
- If manual handling cannot be avoided, a risk assessment should be made.
- Reduce the risk of injury as far as is reasonably practicable either by the use of mechanical aids or by making improvements to the task (eg using two persons), the load and the environment.

How to manually handle loads using mechanical lifting aids

Even if mechanical handling methods are used to handle and transport equipment or materials, hazards may still be present in the four elements that make up the mechanical handling: the handling equipment, the load, the workplace and the human element.

- Mechanical handling equipment must be suitable for the task, well maintained and inspected on a regular basis.
- The load needs to be prepared in such a way as to minimise accidents, taking into account such things as security of the load, flammable materials and stability of the load.
- If possible, the workplace should be designed to keep the workforce and the load apart.
- Employees who are to use the equipment must be properly trained.

The safe method of assembly of access equipment

The different types of access equipment likely to be used in electrotechnical work are:

- ladders
- fixed scaffolds
- mobile tower scaffolds
- mobile elevated work platforms (MEWPs).

The correct angle for a ladder is one unit out for every four units up

The main cause of accidents involving ladders is ladder movement while in use. This normally occurs when a ladder has not been secured to a fixed point. The ladder needs to be stable in use and this can be achieved by making sure the inclination is as near as possible to a 1 in 4 ratio of distance from the wall to the distance up the wall. The foot of the ladder should be tied to a rigid support and not used in high winds or heavy rain.

Fixed scaffolds are often more effective for long duration work than a ladder. Tube and fitting scaffolding must only be erected by competent people who have attended recognised training courses. Any alterations to the scaffolding must also be carried out by a competent person. Regular inspections of the scaffold must be made and recorded.

As scaffolding must only be erected by competent people, it will be difficult for non-competent persons to check the security of the scaffolding. However certain areas can be checked prior to use, as follows.

- Are there adequate toe boards, guard rails and intermediate rails to prevent people or materials from falling?
- Is the scaffold on a stable surface with base plates and timber sole plates?
- Is there safe access and exit?
- Is the working platform fully boarded?
- Are the lower-level uprights prominently marked?
- Is the scaffold braced or secured to the building?

Tube and fitting scaffold

Mobile scaffold towers are a convenient means of undertaking repetitive tasks in the building services industry. The following must be adhered to when mobile towers are used.

- Erection and dismantling must be done by competent persons only.

- The maximum height-to-base ratios must not be exceeded.
- Diagonal bracing and stabilisers must always be used.
- Internal access ladders must always be used.
- Wheels must be locked when work is in progress.
- The working platform must be boarded and guard rails and toe boards fitted.
- Towers must be tied in windy conditions.
- The manufacturer's instructions must be followed at all times.

The following factors must be considered when MEWPs are used:

- whether the MEWP is appropriate for the job
- whether the ground conditions affect stability
- operation only by trained and competent persons
- overhead hazards such as trees, steelwork or overhead cables
- the use of restraint or fall arrest systems
- closeness to passing traffic
- the height from the ground
- proper inflation of tyres and immobilisation of wheels.

How to use access equipment safely

Working at height remains one of the biggest causes of fatalities and major injuries within the construction industry with almost 50% of fatalities resulting from falls from ladders, stepladders and through fragile roofs. Work at height means work in any place, including at or below ground level (eg in underground workings), where a person could fall a distance liable to cause injury.

The Work at Height Regulations 2005 require duty holders to ensure that:

- all work at height is properly planned and organised
- those involved in work at height are competent
- the risks from work at height are assessed and appropriate work equipment is selected and used
- the risks of working on or near fragile surfaces are properly managed
- the equipment used for work at height is properly inspected and maintained.

Before deciding whether access equipment is necessary to work at height, it will be necessary to carry out a risk assessment. This will consist of:

- a careful examination of the work at height task to identify hazards

- consideration of whether the hazards pose a risk to people.

If it is absolutely necessary to work at height, the following points should be considered.

- Use an existing safe place of work to access work at height. If there is already a safe means of access such as a permanent stair and guard-railed platform, it should be used.
- Provide or use work equipment such as scaffolding, mobile access towers or mobile elevated work platforms (MEWPs) with guard rails around the working platform, to prevent falls.
- Minimise distance and consequences of a fall by using, for example, a properly set up stepladder or ladder within its limitations for low-level, short-duration work only.

Having decided that access equipment is necessary, the condition of all the equipment required will also need to be determined. To achieve this it may be necessary to find out how much equipment is in circulation. This may require an initial audit, giving each piece of equipment a thorough examination to establish whether it is safe to use and fit for purpose. A database can then be constructed recording the type, location and condition of each piece of access equipment to enable the equipment to be tracked and managed proactively.

For those employers who have no in-depth knowledge of working at height, the HSE has produced a Work at height Access equipment and Information Toolkit (WAIT). It gives practical advice on the factors to consider when selecting access equipment and how to work at height safely, as well as information on some of the different types of access equipment available.

Job-specific training will ensure that employees undertake their job in a safe manner. This must include training for the competent person in erecting and dismantling access equipment and also for the user of that equipment.

Always observe requirements of site safety signs

Test your knowledge questions

1 Under the Health and Safety at Work etc Act 1974 employers are responsible for:
   a) transport to and from work
   b) payment of trade union fees
   c) subsidised canteen and rest facilities
   d) a safe working environment with adequate welfare facilities.

2 A risk assessment should be completed at the start of a job and:
   a) filed away until the work is finished
   b) continuously reviewed as work progresses
   c) posted to the Health and Safety Executive
   d) fixed to the office notice board.

3 Which of the following is a non-statutory document?
   a) Personal Protective Equipment at Work Regulations 1992.
   b) Provision and Use of Work Equipment Regulations 1998.
   c) Manual Handling Operations Regulations 1992.
   d) BS 7671: IET Wiring Regulations.

4 Work at Height Regulations 2005 apply to all work at height:
   a) inside buildings only
   b) on roofs when fixing PV systems only
   c) where there is a risk of a fall liable to cause personal injury
   d) when there is no one around to hold the ladder.

5 Compliance with the IET Wiring Regulations will meet the requirements of:
   a) Health and Safety at Work etc Act 1974
   b) Electricity at Work Regulations 1989
   c) Control of Substances Hazardous to Health (COSHH) Regulations 2002 (as amended)
   d) Health and Safety (Display Screen Equipment) Regulations 1992.

6 A prohibition notice requires the duty holder to:
   a) stop an activity immediately
   b) stop an activity at the end of the day
   c) carry on but put up notices
   d) contact the Health and Safety Executive for advice.

7 The word that collectively means keeping a clean and tidy working environment is:
   a) maintenance
   b) tidysafe
   c) housekeeping
   d) sweepclear.

8 Scrap and rubbish should be:
   a) left on the stairs for collection later
   b) pushed into a corner out of the way
   c) placed in an appropriate container
   d) collected in a bin and burnt at night.

9 When working on **any** construction site, one essential item of PPE required would be:
   a) overalls
   b) earplugs
   c) respirator
   d) safety footwear.

10 A sign consists of a red circle with a red diagonal line. This type of sign is:
   a) a warning
   b) a prohibition
   c) mandatory
   d) advisory.

11 A sign consists of white symbols on a green background. This type of sign is:
   a) a warning
   b) a prohibition
   c) mandatory
   d) advisory.

12 Upon discovering asbestos on site, the action to take is:

a) continue working

b) cover with a damp cloth

c) stop work and inform duty holder

d) remove the asbestos at the end of work.

13 The safety wear shown is used for protecting.

a) ears

b) nose

c) face

d) eyes.

14 This type of protective equipment is known as a:

a) baseball cap

b) hard hat

c) bump cap

d) soft hat.

15 The minimum shock current that may be fatal is considered to be:

a) 1mA

b) 30mA

c) 50mA

d) 6A.

16 Class I power tools must have a protective conductor connection to the:

a) plastic parts of the casing

b) metallic parts of the casing

c) neutral conductor connection

d) line conductor connection.

17 What standard must approved voltage indicators comply with?

a) Guidance Note 3.

b) GS58.

c) BS38.

d) GS38.

18 Upon discovering an unconscious colleague who has received an electric shock and is still touching an electric cable, the first action to take is:

a) remove the person from the supply

b) apply CPR

c) shout for help

d) dial 999 or 112.

19 The three parts of a fire triangle are:

a) fuel, oxygen, heat

b) fuel, nitrogen, carbon

c) oxygen, helium, carbon

d) oxygen, carbon, fuel.

20 A Class I 'C' fire (flammable gases) can be fought using:

a) powder, carbon dioxide

b) water, carbon dioxide

c) water, foam

d) powder, foam.

Assessment checklist

What you now know (Learning Outcome)	What you can now do (Assessment criteria)	Where this is found (Page number)
1. Know health and safety legislation	1.1 State the aims of health and safety legislation	2
	1.2 Identify the responsibilities of individuals under health and safety legislation	3
	1.3 Identify statutory and non-statutory health and safety materials	5
	1.4 Identify the different roles of the Health and Safety Executive in enforcing health and safety legislation.	8
2. Know how to handle hazardous situations	2.1 Identify common hazardous situations found on site	10–14
	2.2 Describe safe systems at work	14
	2.3 Identify the categories of safety signs	16–17
	2.4 Identify symbols for hazardous substances	18
	2.5 List common hazardous substances used in the building services industry	18–19
	2.6 List precautions to be taken when working with hazardous substances	19
	2.7 Identify the types of asbestos that may be encountered in the workplace	20
	2.8 Identify the actions to be taken if the presence of asbestos is suspected	20
	2.9 Describe the implications of being exposed to asbestos	20
	2.10 State the application of different types of personal protective equipment	21–23
	2.11 Identify the procedures for manually handling heavy and bulky items	23–24
	2.12 Identify the actions that should be taken when an accident or emergency is discovered	24–25
	2.13 State procedures for handling injuries sustained on site	25–26
	2.14 State the procedures for recording accidents and near misses at work.	26–28
3. Know electrical safety requirements when working in the building services industry	3.1 Identify the common electrical dangers to be aware of on site	28–31
	3.2 List different sources of electrical supply for tools and equipment	31–32
	3.3 Describe reasons for using reduced low-voltage electrical supplies for tools and equipment on site	32
	3.4 Identify how to conduct a visual inspection of portable electrical equipment for safe condition before use	34–35

What you now know (Learning Outcome)	What you can now do (Assessment criteria)	Where this is found (Page number)
3. (cntd.) Know electrical safety requirements when working in the building services industry	3.5 State actions to take when portable electrical equipment fails visual inspection	35
	3.6 Outline the Safe Isolation Procedure	35–36
	3.7 State the procedures for dealing with electric shocks.	36–37
4. Know the safety requirements for working with gases and heat-producing equipment	4.1 Identify different types of gases used on site	38
	4.2 Describe how bottled gases and equipment should be safely transported and stored	39–40
	4.3 Describe how to conduct a visual inspection on heat-producing equipment for safe condition	40–41
	4.4 Describe how combustion takes place	41
	4.5 State the dangers of working with heat-producing equipment	41–42
	4.6 State the procedures to follow on discovery of fires on site	42
	4.7 Identify different classifications of fires	43
	4.8 Identify types of fire extinguisher for different classifications of fires.	43
5. Know the safety requirements for using access equipment in the building services industry	5.1 Identify different types of access equipment	44–46
	5.2 Select suitable equipment for carrying out work at height, based on the work being carried out	44–46
	5.3 Describe the safety checks to be carried out on access equipment	46–47
	5.4 Describe safe erection methods for access equipment.	46–47
6. Know the safety requirements for working safely in excavations and confined spaces in the building services industry	6.1 Identify the situations in which it may be necessary to work in excavations	48
	6.2 Describe how excavations should be prepared for safe working	48–49
	6.3 State precautions to be taken to make excavations safe	49
	6.4 Identify areas where working in confined space may be a consideration	49–50
	6.5 State safety considerations when working in confined spaces.	50
7. Be able to apply safe working practice	7.1 Perform manual handling techniques	50–51
	7.2 Manually handle loads using mechanical lifting aids	51
	7.3 Demonstrate the safe method of assembly of access equipment	51–52
	7.4 Use access equipment safely.	53–54

Assessment guidance

This unit is common to the Level 2 and Level 3 qualifications. You do not need to be reassessed on this unit if you have passed it at Level 2.

The assessment of this unit is in two parts.

Practical assessment (201)

Health & safety (Unit 201) task	Description	Assessment criteria	Recommended assignment time
H&S 1	Identify hazards	AC 2.1	1 hour, 30 minutes
H&S 2	Manual handling	AC 7.1	45 minutes
H&S 3	Manual handling using lifting aids	AC 7.2	45 minutes
H&S 4	Ladders	AC 7.3, 7.4	45 minutes
H&S 5	Stand steps	AC 7.3, 7.4	45 minutes
H&S 6	Scaffold	AC 7.3, 7.4	1 hour, 30 minutes
H&S 7	Safety signs	AC 2.3	1 hour

- The above table shows the practical assessments in which you are required to demonstrate the ability to carry out the tasks listed.
- Failure to work safely will cause the assessment to be terminated.
- Make sure you understand all of the verbal instructions given to you by your assessor before you begin.
- If you are not sure of anything, ask.
- Make sure that all personal protective equipment given to you fits properly before you begin.
- You will be given feedback at the end of the assessment.

Online multiple-choice assessment (501)

- This is a closed-book online e-volve multiple-choice assessment.
- Attempt all questions.
- Do not leave until you are confident that you have completed all questions.
- Keep an eye on the time as it moves quickly when you are concentrating.

- Make sure you read each question fully before answering.
- Ensure you know how the e-volve system works. Ask for a demonstration if you are not sure.
- Do not take any paperwork into the exam with you.
- If you need paper to work anything out, ask the invigilator to provide some.
- Make sure your mobile phone is switched *off* (not on silent) during the exam. You may be asked to give it to the invigilator.

Before the assessment

- You will find sample questions on SmartScreen and some questions in the section below to test your knowledge of the Learning Outcomes.
- Make sure you go over these questions in your own time.
- Spend time on revision in the run-up to the assessments.

Understand the fundamental principles and requirements of environmental technology systems

The UK government is committed to carbon dioxide (CO_2) reduction in order to address the issue of climate change. House building is an area that has been identified as a sector that can make a significant contribution to this goal of carbon dioxide reduction. It is estimated that 43% of CO_2 emissions in the UK are attributable to domestic properties. The government has introduced the Code for Sustainable Homes, which is the national standard for the sustainable design and construction of new homes. It aims to reduce CO_2 emissions and promote standards of sustainable design higher than the current minimum standards set out by the building regulations. The target is that all new homes will be carbon neutral by 2016. Because of this, any operative working within the construction industry is likely to come into contact with one or more forms of environmental technology systems. A knowledge of the working principles, the advantages and disadvantages, the requirements of the location and an overview of the planning and regulatory requirements for each environmental technology is therefore a prerequisite to working in the construction industry.

There are four Learning Outcomes to this unit. The learner will:

1 Know the fundamental working principles of micro-renewable energy and water conservation technologies.

2 Know the fundamental requirements of building location/building features for the potential to install micro-renewable energy and water conservation systems.

3 Know the fundamental regulatory requirements relating to micro-renewable energy and water conservation technologies.

4 Know the typical advantages and disadvantages associated with micro-renewable energy and water conservation technologies.

This unit will be assessed by:

■ an online multiple-choice test 2365-301.

How this unit is organised

SmartScreen Unit 301

Additional resources to support this unit are available on SmartScreen.

This unit of the book is divided up according to technology, rather than consecutively by outcome.

This has been done to enable the learner to appreciate fully the principles of each technology and how these impact on the installation requirements and the regulatory requirements. The regulatory requirements (Learning Outcome 3) are covered first, followed by a discussion of the advantages and disadvantages of each technology.

The environmental technologies that are discussed in this chapter are described below.

Heat-producing	Electricity-producing
• Solar thermal • Ground source heat pump • Air source heat pump • Biomass	• Solar photovoltaic • Micro-wind • Micro-hydro

Co-generation	Water conservation
• Micro-combined heat and power (heat-led)	• Rainwater harvesting • Greywater re-use

Environmental technology systems covered

Know the fundamental regulatory requirements relating to micro-renewable energy and water conservation technologies (LO3)

Learning Outcome 3 deals with planning requirements and building regulations for each technology. This section explains the terminology used, provides an insight into the workings of both planning and building regulations and explains the differences across the UK.

Planning and permitted development

In general, under the Town and Country Planning Act 1990, before any building work that increases the size of a building is carried out, a planning application must be submitted to the local authority. A certain amount of building work is, however, allowed without the need for a planning application. This is known as *permitted development*. Permitted development usually comes with criteria that must be met. When building an extension, for example, it may be possible to do so under permitted development, if the extension is under a certain size, is a certain distance away from the boundary of the property and is not at the front of the property. If the extension does not meet these criteria, then a full application must be made.

The permitted development is intended to ease the burden on local authorities and to smooth the process for the builder or installer. Permitted development exists for renewable technologies and these are outlined within each technology section.

Building regulations

The Climate Change and Sustainable Energy Act 2006 brought micro-generation under the requirements of the building regulations.

Even if a planning application is not required, because the installation meets the criteria for permitted development, there is still a requirement to comply with the relevant building regulations.

Local Authority Building Control (LABC) is the body responsible for checking that building regulations have been met. The person carrying out the work is responsible for ensuring that approval is obtained.

Building regulations are statutory instruments that seek to ensure that the policies and requirements of the relevant legislation are complied with.

The building regulations themselves are rather brief and are currently divided into 14 sections, each of which is accompanied by an approved document. The approved documents are non-statutory and give guidelines on how to comply with the statutory requirements.

The 14 parts of the Building Regulations in England and Wales are listed on the next page.

Part	Title
A	Structure
B	Fire safety
C	Site preparation and resistance to contaminants and moisture
D	Toxic substances
E	Resistance to the passage of sound
F	Ventilation
G	Sanitation, hot-water safety and water efficiency
H	Drainage and waste disposal
J	Combustion appliances and fuel-storage systems
K	Protection from falling, collision and impact
L	Conservation of fuel and power
M	Access to and use of building
N	Glazing – safety in relation to impact, opening and cleaning
P	Electrical safety – dwellings

There is a 15th Approved Document, which relates to Regulation 7 of the Building Regulations, entitled Approved Document 7 Materials and Workmanship.

Compliance with building regulations is required when installing renewable technologies but not all will be applicable and different technologies will have to comply with different building regulations. Building regulations applicable to each technology are indicated in each section.

Differences in building regulations across the UK

It should be noted, that due to devolution of government in the UK, each country's government takes responsibility for building regulations, so there are differences between the individual countries.

England and Wales

England and Wales currently follow the same legal structure when it comes to building regulations.

Primary legislation: Building Act 1984

Secondary legislation: Building Regulations 2010

Guidance: 15 Approved Codes of Practice

Although Wales follows the same model as England, the Welsh Government is now responsible for the majority of functions under the Building Act, including making Building Regulations in Wales. The functions that have remained with the UK Government are as set out in The Welsh Ministers (Transfer of Functions) (No. 2) Order 2009 (S.I. 2009/3019).

Scotland

Primary legislation: Building (Scotland) Act 1984

Secondary legislation: Building (Scotland) Regulations 2004 – Amended 2009

Non-statutory guidance: 2 Technical Guide Books – Dwellings, Non-dwellings

Northern Ireland

Primary legislation: Building Regulations (Northern Ireland) 1979 Order (Amended 2009)

Secondary legislation: The Building Regulations (Northern Ireland) 2012

Non-statutory guidance: 15 Technical Booklets

The technical booklets have similar content to the approved documents used in England and Wales but the order of the documents is different. A comparison is included overleaf, along with any differences in title.

England and Wales		Northern Ireland	
A	Structure	D	
B	Fire safety	E	
C	Site preparation and resistance to contaminants and moisture	C	
D	Toxic substances	No comparable document	
E	Resistance to the passage of sound	G	
F	Ventilation	K	

England and Wales		Northern Ireland	
G	Sanitation, hot-water safety and water efficiency	P	Sanitary appliances, unvented hot-water storage systems and reducing the risk of scalding
H	Drainage and waste disposal	J N	Solid waste in buildings Drainage
J	Combustion appliances and fuel storage systems	L	
K	Protection from falling, collision and impact	H	Stair ramps guarding and protection from impact
L	Conservation of fuel and power	F1 F2	Dwellings Non-dwellings
M	Access to and use of building	R	
N	Glazing – safety in relation to impact, opening and cleaning	V	
P	Electrical safety – dwellings	No comparable document	
7	Materials and workmanship	B	

Within this unit, reference to parts within the building regulations is applicable to the part designations used in the Building Regulations of England and Wales.

Heat-producing micro-renewable energy technologies (LO1-4)

Solar thermal (hot-water) systems

A solar thermal hot-water system uses solar radiation to heat water, directly or indirectly.

Working principles

The key components of a solar thermal hot water system are:

1 solar collector
2 differential temperature controller
3 circulating pump
4 hot-water storage cylinder
5 auxiliary heat source.

SUGGESTED ACTIVITY

Why is a circulating pump required with a solar thermal system?

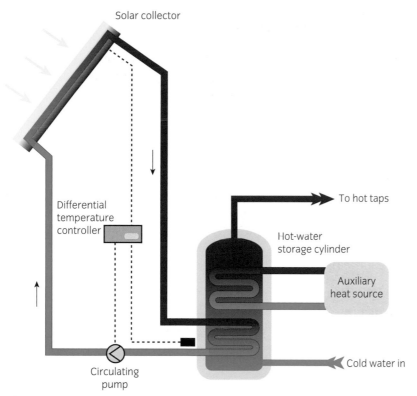

Solar collector

To hot taps

Differential temperature controller

Hot-water storage cylinder

Auxiliary heat source

Circulating pump

Cold water in

Solar thermal system components

1 Solar thermal collector

A solar thermal collector is designed to collect heat by absorbing heat radiation from the Sun. The heat energy from the Sun heats the heat-transfer fluid contained in the system.

Two types of solar collector are used.

Flat-plate collectors are less efficient but cheaper than evacuated tube collectors.

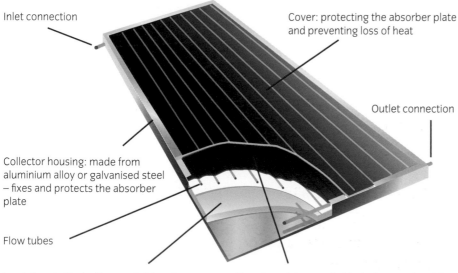

Inlet connection

Cover: protecting the absorber plate and preventing loss of heat

Outlet connection

Collector housing: made from aluminium alloy or galvanised steel – fixes and protects the absorber plate

Flow tubes

Insulation: to the bottom and sides of the collector to reduce the loss of heat

Absorber plate: usually black chrome absorbing coating to maximise heat-collecting efficiency

Cutaway diagram of a flat-plate collector

With this type of collector, the heat-transfer fluid circulates through the collectors and is directly heated by the Sun. The collectors need to be well insulated to avoid heat loss.

Evacuated-tube collectors are more efficient but more expensive than flat-plate collectors.

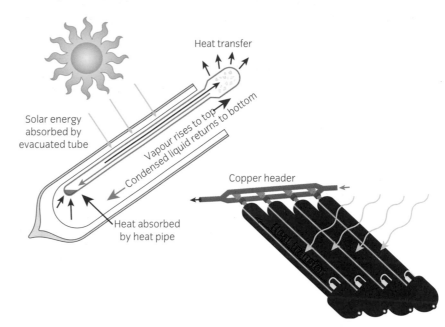

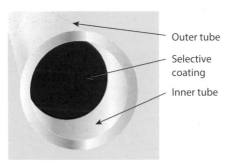

Evacuated-tube collector

An evacuated-tube collector consists of a specially coated, pressure-resistant, double-walled glass tube. The air is evacuated from the tube to aid the transfer of heat from the Sun to a heat pipe housed within the glass tube. The heat pipe contains a temperature-sensitive medium, such as methanol, that, when heated, vaporises. The warmed gas rises within the tube. A solar collector will contain a number of evacuated tubes in contact with a copper header tube that is part of the solar heating circuit. The heat tube is in contact with the header tube. The heat from the methanol vapour in the heat tubes is transferred by conduction to the heat-transfer fluid flowing through the solar heating circuit. This process cools the methanol vapour, which condenses and runs back down to the bottom of the heat tubes, ready for the process to start again. The collector must be mounted at a suitable angle to allow the vapour to rise and the condensed liquid to flow back down the heat pipes.

2 Differential temperature controller

The differential temperature controller (DTC) has sensors connected to the solar collector (high level) and the hot-water storage system (low level). It monitors the temperatures at the two points. The DTC turns the circulating pump on when there is enough solar energy available and there is a demand for water to be heated. Once the stored water reaches the required temperature, the DTC shuts off the circulating pump.

3 Circulating pump

The circulating pump is controlled by the DTC and circulates the system's heat-transfer fluid around the solar hot-water circuit. The circuit is a closed loop between the solar collector and the hot-water storage tank. The heat-transfer fluid is normally water-based but, depending on the system type, usually also contains glycol so that at night, or in periods of low temperatures, it does not freeze in the collectors.

4 Hot-water storage cylinder

The hot-water storage cylinder enables the transfer of heat from the solar collector circuit to the stored water. Several different types of cylinder or cylinder arrangement are possible.

Twin-coil cylinder

With this type of cylinder the lower coil is the solar heating circuit and the upper coil is the auxiliary heating circuit. Cold water enters at the base of the cylinder and is heated by the solar heating coil. If the solar heating circuit cannot meet the required demand, then the boiler will provide heat through the upper coil. Hot water is drawn off, by the taps, from the top of the cylinder.

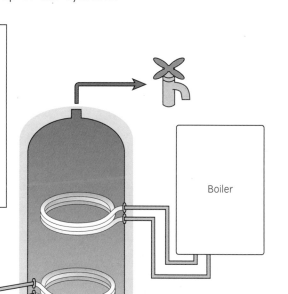

Twin-coil cylinder

Alternatives: One alternative arrangement is to use one cylinder as a solar preheat cylinder, the output of which feeds a hot-water cylinder. The auxiliary heating circuit is connected to the second cylinder.

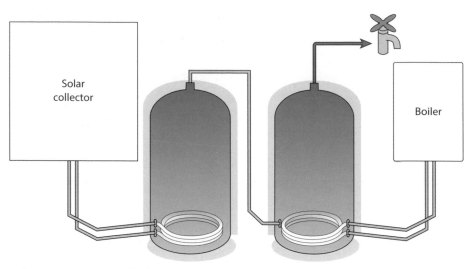

Using two separate cylinders

The two arrangements that have been described are indirect systems, with the solar heating circuit forming a closed loop.

Direct system: An alternative to the indirect system is the direct system, in which the domestic hot water that is stored in the cylinder is directly circulated through the solar collector and is the same water that is drawn off at the taps. Due to this fact, antifreeze (glycol) cannot be used in the system, so it is important to use freeze-tolerant collectors.

5 Auxiliary heat source

In the UK there will be times when there is insufficient solar energy available to provide adequate hot water. On these occasions an auxiliary heat source will be required. Where the premises have space-heating systems installed, the auxiliary heat source is usually this boiler. Where no suitable boiler exists, the auxiliary heat source will be an electric immersion heater.

Location and building requirements

When deciding whether or not a solar thermal hot-water system is suitable for particular premises, the following factors should be considered.

The orientation of the solar collectors

The optimum direction for the solar collectors to face is due south. However, as the Sun rises in the east and sets in the west, any location with a roof facing east, south or west is suitable for mounting a solar thermal system, although the efficiency of the system will be reduced for any system not facing due south.

The tilt of the solar collectors

During the year, the maximum elevation or height of the Sun, relative to the horizon, changes. It is lowest in December and highest in June. Ideally, solar collectors should always be perpendicular to the path of the Sun's rays. As it is generally not practical to change the tilt angle of a solar collector, a compromise angle has to be used. In the UK, the angle is 35°; however, the collectors will work, but less efficiently, from vertical through to horizontal.

Shading of the solar collectors

Any structure, tree, chimney, aerial or other object that stands between the collector and the Sun will block the Sun's energy. The Sun shines for a limited time and any reduction in the amount of heat energy reaching the collector will reduce its ability to provide hot water to meet the demand.

Shading	% of sky blocked by obstacles	Reduction in output
Heavy	> 80%	50%
Significant	> 60–80%	35%
Modest	> 20–60%	20%
None or very little	≤ 20%	No reduction

The suitability of the structure for mounting the solar collector

The structure has to be assessed as to its suitability for the chosen mounting system. Consideration needs to be given to the strength and condition of the structure and the suitability of fixings. The effect of wind must also be taken into account. The force exerted by the wind on the collectors, an upward force known as 'wind uplift', affects both the solar collector fixings and the fixings holding the roof members to the building structure.

In the case of roof-mounted systems on flats and other shared properties, the ownership of the structure on which the proposed system is to be installed must be considered.

The space needed to mount the collectors is dependent on demand for hot water. The number of people occupying premises determines the demand for hot water and, therefore, the number of collectors required and the space needed to mount them.

Compatibility with the existing hot-water system

Solar thermal systems provide stored hot water rather than instantaneous hot water.

- Premises using under-over-sink water heaters and electric showers will not be suitable for the installation of a solar thermal hot-water system.

1 person = 50 litres of hot water per day

50 litres of hot water = 1m² of solar collector

How much hot water is needed

■ Premises using a combination boiler to provide hot water will not be suitable for the installation of a solar thermal hot-water system unless substantial changes are made to the system.

Planning permission

Permitted development applies where a solar thermal system is installed:

■ on a dwelling house or block of flats

■ on a building within the grounds of a dwelling house or block of flats

■ as a stand-alone system in the grounds of a dwelling house or block of flats.

However there are criteria to be met in each case.

For building mounted systems:

■ the solar thermal system cannot protrude more than 200mm from the wall or the roof slope

■ the solar thermal system cannot protrude past the highest point of the roof (the ridgeline), excluding the chimney.

The criteria that must be met for stand-alone systems are that:

■ only one stand-alone system is allowed in the grounds

■ the array cannot exceed 4m in height

■ the array cannot be installed within 5m of the boundary of the grounds

■ the array cannot exceed $9m^2$ in area

■ no dimension of the array can exceed 3m in length.

For both stand-alone and building mounted systems:

■ the system cannot be installed in the grounds or on a building within the grounds of a listed building or a scheduled monument

■ if the dwelling is in a conservation area or a World Heritage Site, then the array cannot be closer to a highway than the house or block of flats.

In every other case, planning permission will be required.

Compliance with building regulations

The following building regulations will apply to solar thermal hot-water systems.

Part	Title	Relevance
A	Structure	Where solar collectors and other components can put extra load on the structure, particularly the roof structure, not only the additional downwards load but also the uplift caused by the wind must be considered
B	Fire safety	Where holes for pipes are made, this may reduce the fire resistance of the building fabric
C	Resistance to contaminants and moisture	Where holes for pipes and fixings for collectors are made, this may reduce the moisture resistance of the building and allow ingress of water
G	Sanitation, hot-water safety and water efficiency	Hot-water safety and water efficiency
L	Conservation of fuel and power	Energy efficiency of the system and the building as a whole
P	Electrical safety	The installation of electrical controls and components

Other regulatory requirements to consider

- BS 7671 Requirements for Electrical Installations
- Approved document Part G3: Unvented hot-water storage systems
- Water Regulations (WRAS)

Advantages

- It reduces CO_2 emissions.
- It reduces energy costs.
- It is low maintenance.
- It improves the energy rating of the building.

Disadvantages

- It may not be compatible with the existing hot-water system.
- It may not meet demand for hot water in the winter.
- There are high initial installation costs.
- It requires a linked auxiliary heat source.

Heat pumps

Working principles

A water pump moves water from a lower level to a higher level, through the application of energy. Pumping the handle draws water up from a lower level to a higher level through the application of kinetic energy.

As the name suggests, a heat pump moves heat energy from one location to another by the application of energy. In most cases, the applied energy is electrical energy.

Heat energy from the Sun exists in the air that surrounds us and in the ground beneath our feet. At *absolute zero* or 0K (kelvin), there is no heat in a system. This temperature is equivalent to −273°C so, even with outside temperatures of −10°C, there is a vast amount of free heat energy available.

A heat pump moves heat from one location to another, just as a water pump moves water from one location to another

| −273 | | −10 | 0 | 20 |

Heat energy exists down to absolute zero (0K ≈ −273°C)

Using a relatively small amount of energy, that stored heat energy in the air or in the ground can be extracted and put to use in heating our living accommodation.

Heat pumps extract heat from outside and transfer it inside, in much the same way that a refrigerator extracts heat from the inside of the refrigerator and releases it at the back of the refrigerator via the heat-exchange fins.

A basic rule of heat transfer is that heat moves from warmer spaces to colder spaces.

A heat pump contains a refrigerant. The external air or ground is the medium or heat source that gives up its heat energy. When the refrigerant is passed through this heat source the refrigerant is cooler than its surroundings and so absorbs heat. The compressor on the heat pump then compresses the refrigerant, causing the gas to heat up. When the refrigerant is passed to the interior, the refrigerant is now hotter than its surroundings and gives up its heat to the cooler surroundings. The refrigerant is then allowed to expand, where it once again turns into a liquid. As the refrigerant expands it cools and the cycle starts all over again.

ASSESSMENT GUIDANCE

Look at your refrigerator or freezer at home. The inside is cold because the heat energy has been removed from it. The tubes at the back are hot.

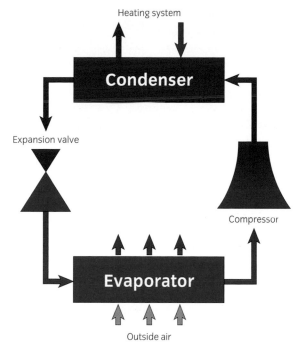

The refrigeration process

The only energy needed to drive the system is what is required by the compressor. The greater the difference in temperature between the refrigerant and the heat-source medium from where heat is being extracted, the greater the efficiency of the heat pump. If the heat-source medium is very cold then the refrigerant will need to be colder, to be able to absorb heat, so the harder the compressor must work and the more energy is needed to accomplish this.

Two main types of heat pump in common use are:

- ground source heat pumps (GSHP)
- air source heat pumps (ASHP).

Heat pumps extract heat energy from the air or the ground, but the energy extracted is replaced by the action of the Sun.

It is not uncommon for heat pumps to have efficiencies in the order of 300%; for an electrical input of 3kW, a heat output of 9kW is achievable. If we compare this with other heat appliances we can see where the savings are made.

Electricity input 1kW

Heat output 1kW

100% efficiency

The efficiency of an electric panel heater

Gas input 1kW

Heat output 0.95kW

95% efficiency

The efficiency of an A-rated condensing gas boiler

Electricity input 1kW

+2kW free heat
extracted from air

Heat output 3kW

equates to a 300%
efficiency

The efficiency of an air source heat pump

SUGGESTED ACTIVITY

Visit the website of a gas boiler supplier
and compare the efficiency of a
conventional boiler to that of a
condensing gas boiler.

The efficiency of a heat pump is measured in terms of the *coefficient of performance* (COP), which is the ratio between the heat delivered and the power input of the compressor.

$$COP = \frac{\text{heat delivered}}{\text{compressor power}}$$

The higher the COP value, the greater the efficiency. Higher COP values are achieved in mild weather than in cold weather because, in cold weather, the compressor has to work harder to extract heat.

Storing excess heat produced

Heat pumps are not able to provide instant heat and so therefore work best when run continuously. Stop–start operations will shorten the lifespan of a heat pump. A buffer tank, simply a large water-storage vessel, is incorporated into the circuit so that, when heat is not required within the premises, the heat pump can 'dump' heat to it and thus keep running. When there is a need for heat, this can be drawn from the buffer tank. A buffer tank can be used with both ground source and air source heat pumps.

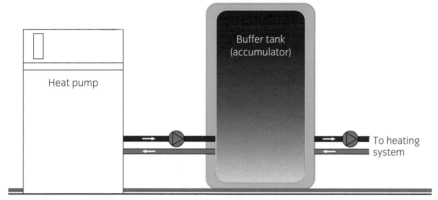

Storing heat in a buffer tank

Ground source heat pumps

Working principles

A ground source heat pump (GSHP) extracts low-temperature free heat from the ground, upgrades it to a higher temperature and releases it, where required, for space heating and water heating.

The key components of a GSHP are:

- heat-collection loops and a pump
- heat pump
- heating system.

The collection of heat from the ground is accomplished by means of pipes containing a mixture of water and antifreeze, which are buried in the ground. This type of system is known as a 'closed-loop' system. Three methods of burying the pipes are used, with each having advantages and disadvantages.

Horizontal loops

Piping is installed in horizontal trenches that are generally 1.5–2m deep. Horizontal loops require more piping than vertical loops – around 200m of piping for the average house.

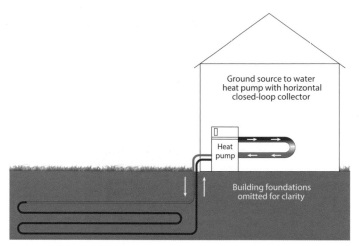

Horizontal ground loops

Vertical loops

Most commercial installations use vertical loops. Holes are bored to a depth of 15–60m, depending on soil conditions, and spaced approximately 5m apart. Pipe is then inserted into these boreholes. The advantage of this system is that less land is needed.

SUGGESTED ACTIVITY

What ground structure/material would make vertical loops impracticable or very expensive?

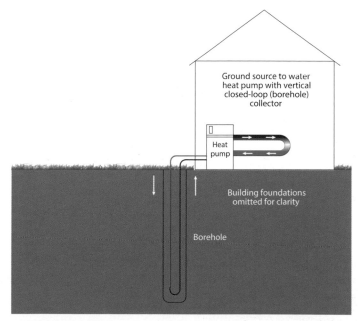

Vertical ground loops

Slinkies

Slinky coils are flattened, overlapping coils that are spread out and buried, either vertically or horizontally. They are able to concentrate the area of heat transfer into a small area of land. This reduces the length of trench needing to be excavated and therefore the amount of land required. Slinkies installed in a 10m long trench will yield around 1kW of heating load.

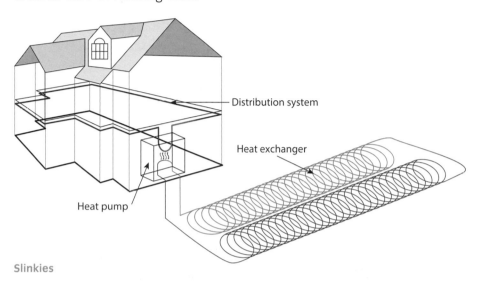

Slinkies

Slinkies being installed in the ground

ASSESSMENT GUIDANCE

Looking at this photograph, it is obvious that unless you want your garden to be given the 'Time Team' treatment it is best to have the ground source heat pump pipes installed during the original building stage.

The water–antifreeze mix is circulated around these ground pipes by means of a pump. The low-grade heat from the ground is passed over a heat exchanger, which transfers the heat from the ground to the refrigerant gas. The refrigerant gas is compressed and passed across a second heat exchanger, where the heat is transferred to a pumped heating loop that feeds either radiators or under-floor heating.

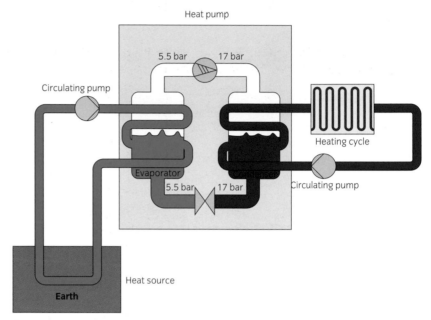

Ground source heat pump operating principle

Final heat output from the GSHP is at a lower temperature than would be obtained from a gas boiler. GSHP heat output is at 40°C, compared to a gas boiler at 60–80°C. For this reason, under-floor heating, which requires temperatures of 30–35°C, is the most suitable form of heating arrangement to use with a GSHP. Low-temperature or oversized radiators could also be used. A GSHP system in itself is unable to heat hot water directly to a suitable temperature. Hot water needs to be stored at a temperature of 60°C. An ancillary heating device will be required to reach the required temperatures.

SUGGESTED ACTIVITY

What is another name for a buffer tank?

A GSHP is unable to provide instant heat and, for maximum efficiency, should run all the time. In some cases it is beneficial to fit a buffer tank to the output so that any excess heat is stored, ready to be used when required.

By reversing the refrigeration process, a GSHP can also be used to provide cooling in summer.

Location and building requirements

For a GSHP system to work effectively, and due to the fact that the output temperature is low, the building has to be well insulated.

A suitable amount of land has to be available for trenches or, alternatively, land that is suitable for boreholes. In either case, access for machinery will be required.

Planning permission

The installation of a ground source heat pump is usually considered to be permitted development and will not require a planning application to be made.

If the building is a listed building or in a conservation area, the local area planning authority will be need to be consulted.

Compliance with building regulations

The following building regulations will apply.

Part	Title	Relevance
A	Structural safety	Where heat pumps and other components put additional load on the building structure or where openings are formed to pass from outside to inside
B	Fire safety	Where holes for pipes may reduce the fire-resistant integrity of the building structure
C	Resistance to contaminants and moisture	Where holes for pipes may reduce the moisture-resistant integrity of the building structure
E	Resistance to sound	Where holes for pipes may reduce the soundproof integrity of the building structure
G	Sanitation, hot-water safety and water efficiency	Hot-water safety and water efficiency Unvented hot-water system
L	Conservation of fuel and power	Energy efficiency of the system and the building as a whole
P	Electrical safety	The installation of electrical controls and components as well as the supply to the heat pump

ASSESSMENT GUIDANCE

It is not good practice to install a heat pump via a plug and socket. Any load over 3kW (13A) will need its own circuit wired back to the consumer unit by a competent electrician.

Other regulatory requirements to consider

- BS 7671 Requirements for Electrical Installations
- F (fluorinated) gas requirements if working on refrigeration pipework

Advantages

- There is high efficiency.
- There is a reduction in energy bills – cheaper to run than electric, gas or oil boilers.
- There is a reduction in CO_2 emissions.
- There are no CO_2 emissions on site.
- They are safe, as no combustion takes place and no emission of potentially dangerous gases.
- They are low maintenance, compared with combustion devices.
- They have a long lifespan.
- There is no requirement for fuel storage, so less installation space is required.
- They can be used to provide cooling in summer.
- They are more efficient than air source heat pumps.

Disadvantages

- The initial costs are high.
- They require large ground area or boreholes.
- The design and installation are complex tasks.
- They are unlikely to work efficiently with an existing heating system.
- They use refrigerants, which could be harmful to the environment.
- They are more expensive to install than air source heat pumps.

Air source heat pumps

Working principles

An air source heat pump (ASHP) extracts free heat from low-temperature air and releases it where required, for space heating and water heating.

The key components of an ASHP are:

- a heat pump containing a heat exchanger, a compressor and an expansion valve
- a heating system.

Air source heat pump

An ASHP works in a similar way to a refrigerator, but the cooled area becomes the outside world and the area where the heat is released is the inside of a building. The steps of the ASHP process are as follows.

- The pipes of the pump system contain refrigerant that can be a liquid or a gas, depending on the stage of the cycle. The refrigerant, as a gas, flows through a heat exchanger (evaporator), where low-temperature air from outside is drawn across the heat exchanger by means of the unit's internal fan. The heat from the air warms the refrigerant. Any liquid refrigerant boils to gas.

- The warmed refrigerant vapour then flows to a compressor, where it is compressed, causing its temperature to rise further.

- Following this pressurisation stage, the refrigerant gas passes through another heat exchanger (condenser), where it loses heat to the heating-system water, because it is hotter than the system water. At this stage, some of the refrigerant has condensed to a liquid. The heating system carries heat away to heat the building.

- The cooled refrigerant passes through an expansion valve, where its pressure drops suddenly and its temperature falls. The refrigerant flows once more to the evaporator heat exchanger, continuing the cycle.

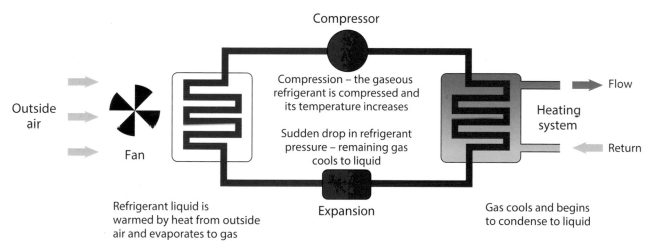

Air source heat pump operating principle

The two types of ASHP in common use are:

- air-to-water – the type described above, which can be used to provide both space heating and water heating
- air-to-air – this type is not suitable for providing water heating.

The output temperature of an ASHP will be lower than that of a gas-fired boiler. Ideally, the ASHP should be used in conjunction with an under-floor heating system. Alternatively, it could be used with low-temperature radiators.

SUGGESTED ACTIVITY

Name four different methods used to insulate a dwelling.

Location and building requirements

When deciding whether or not an ASHP is suitable for the premises, the following should be considered.

- The premises must be well insulated.
- There must be space to fit the unit on the ground outside the building, or mounted on a wall. There will also need to be clear space around the unit to allow an adequate airflow.
- The ideal heating system to couple to an ASHP is either under-floor heating or warm-air heating.
- An ASHP will pay for itself in a shorter period of time if it replaces an electric, coal or oil heating system than if it is replacing a gas-fired boiler.

Air sourced heat pumps are an ideal solution for new-build properties, where high levels of insulation and under-floor heating are to be installed.

Planning permission

Permitted development applies where an air source heat pump is installed:

- on a dwelling house or block of flats
- on a building within the grounds of a dwelling house or block of flats
- in the grounds of a dwelling house or block of flats.

There are, however, criteria to be met, mainly due to noise generation by the ASHP.

- The air source heat pump must comply with the MCS Planning Standards or equivalent.
- Only one ASHP may be installed on the building or within the grounds of the building.
- A wind turbine must not be installed on the building or within the grounds of the building.
- The volume of the outdoor unit's compressor must not exceed $0.6m^3$.
- It cannot be installed within 1m of the boundary.
- It cannot be installed on a pitched roof.
- If it is installed on a flat roof, it must not be within 1m of the roof edge.
- If installed on a wall that fronts a highway, it cannot be mounted above the level of the ground storey.
- It cannot be installed on a site designated as a monument.
- It cannot be installed on a building that is a listed building, or in its grounds.

- It cannot be installed on a roof or a wall that fronts a highway, or within a conservation area or World Heritage Site.
- If the dwelling is in a conservation area or a World Heritage Site, then the ASHP cannot be closer to a highway than the house or block of flats.

Compliance with building regulations

The following building regulations will apply.

Part	Title	Relevance
A	Structural safety	Where heat pumps and other components put additional load on the building structure, for instance, where the heat pump is installed on the roof or on a wall
B	Fire safety	Where holes for pipes may reduce the fire-resistant integrity of the building structure
C	Resistance from contaminants and moisture	Where holes for pipes may reduce the moisture-resistant integrity of the building structure
E	Resistance to sound	Where holes for pipes may reduce the soundproof integrity of the building structure
G	Sanitation, hot-water safety and water efficiency	Hot-water safety and water efficiency
L	Conservation of fuel and power	Energy efficiency of the system and the building as a whole
P	Electrical safety	The installation of electrical controls and components as well as the supply to the heat pump

Other regulatory requirements to consider

- BS 7671 Requirements for Electrical Installations.
- F (fluorinated) Gas Regulations if working on refrigeration pipework.

Advantages

- There is high efficiency.
- There is a reduction in energy bills – they are cheaper to run than

KEY POINT

An ASHP creates noise and therefore there are restrictions on the number that can be installed on a site.

SUGGESTED ACTIVITY

Why is it important to reduce CO_2 emissions?

electric, gas or oil boilers.

- There is a reduction in CO_2 emissions.
- There are no CO_2 emissions on site.
- They are safe, as no combustion takes place and there is no emission of potentially dangerous gases.
- They are low maintenance, compared with combustion devices.
- There is no requirement for fuel storage, so less installation space is required.
- They can be used to provide cooling in summer.
- They are cheaper and easier to install than GSHPs.

Disadvantages

- They are unlikely to work efficiently with an existing heating system.
- They are not as efficient as GSHPs.
- The initial cost is high.
- They are less efficient in winter.
- There is noise from fans.
- They have to incorporate a defrost cycle to stop the heat exchanger freezing in winter.

Biomass

What is biomass? Biomass is biological material from living or recently living organisms. Biomass fuels are usually derived from plant-based material but could be derived from animal material.

The major difference between biomass and fossil fuels, both of which are derived from the same source, is time. Fossil fuels, such as gas, oil and coal, have taken millions of years to form. Demand for these fuels is outstripping supply and replenishment. Biomass is derived from recently living organisms. As long as these organisms are replaced by replanting, and demand does not outstrip replacement time, the whole process is sustainable. Biomass is therefore rightly regarded as a renewable energy technology.

ASSESSMENT GUIDANCE

Biomass is in some ways a further development from the wood-burning back boiler, with more controls.

Both fossil fuels and biomass fuels are burnt to produce heat and both produce carbon dioxide. This is a greenhouse gas that has been linked to global warming. During their lives, plants and trees absorb carbon dioxide from the atmosphere, to enable growth to take place. When these plants are burnt, the carbon dioxide is released once again into the atmosphere.

So how does biomass have a carbon advantage over fossil fuels? The answer again is time. Fossil fuels absorbed carbon dioxide from the atmosphere millions of years ago and have trapped that carbon dioxide ever since. When fossil fuels are burnt, they release the carbon dioxide from all those millions of years ago and so add to the present-day atmospheric carbon dioxide level.

Biomass absorbs carbon dioxide when it grows, reducing current atmospheric carbon dioxide levels. When biomass is turned into fuel and burnt, it releases the carbon dioxide back into the atmosphere. The net result is that there is no overall increase in the amount of carbon dioxide in the atmosphere.

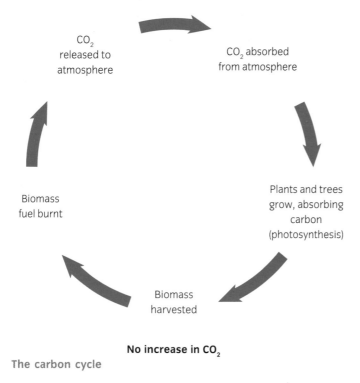

CO_2 released to atmosphere

CO_2 absorbed from atmosphere

Plants and trees grow, absorbing carbon (photosynthesis)

Biomass fuel burnt

Biomass harvested

No increase in CO_2

The carbon cycle

A disadvantage of biomass is that the material is less dense than fossil fuels so, to achieve the same heat output, a greater quantity of biomass than fossil fuel is required. However, with careful management, the use of biomass is sustainable, whereas the use of fossil fuels is not.

The classes of biomass raw material that can be turned into biomass fuels are:

- wood
- crops such as elephant grass, reed canary grass and oil-seed rape
- agricultural by-products such as straw, grain husks, forest product waste, animal waste such as chicken litter and slurry
- food waste – it is estimated that some 35% of food purchased in the UK ends up as waste
- industrial waste.

Woody biomass

For domestic use, wood-related products are the primary biomass fuels.

For wood to work as a sustainable material, the trees used need to be relatively fast growing, so short-rotation coppice woodlands containing willow, hazel and poplar are used on a 3–5-year rotation. Because of this, large logs are not available, neither can slow-growing timbers that would have a higher calorific value be used. Woody biomass as a fuel is generally supplied as:

- small logs
- wood chips – mechanically shredded trees, branches, etc
- wood pellets – formed from sawdust or shavings that are compressed to form pellets.

The *calorific value* (energy given off by burning) of woody biomass is generally low. The greener (wetter) the wood is, the lower the calorific value will be.

Woody biomass boilers

A biomass boiler can be as simple as a log-burner providing heat to a single room or may be a boiler heating a whole house.

Woody biomass boilers can be automated so that a constant supply of fuel is available. Wood pellets are transferred to a combustion chamber by means of an auger drive or, if the fuel storage is remote from the boiler, by a suction system. The combustion process is monitored via thermostats in the flue gases and adjustments are made to the fan speed, which controls air intake, and to the fuel-feed system, to control the feed of pellets. All of this is controlled by a microprocessor.

The hot flue gases are passed across a heat exchanger, where the heat is transferred to the water in the central-heating system. From this point the heated water is circulated around a standard central-heating system.

In automated biomass boilers, heat exchangers are self-cleaning and the amount of ash produced is relatively small. As a result, the boilers require little maintenance. The waste gases are taken away from the boiler by the flue and are then dispersed via the flue terminal.

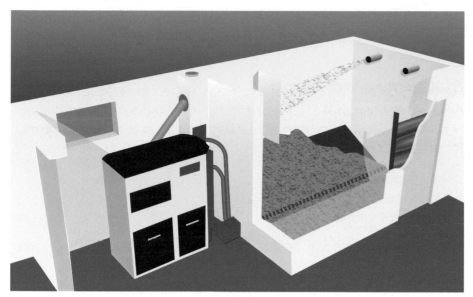

Biomass boiler with suction feed system

Location and building requirements

When considering the installation of a biomass boiler, the following considerations should be taken into account.

- Space will be required for storage of biomass fuel.
- Easy access will be required for delivery of biomass fuel.
- A biomass boiler may not be permitted in a designated smokeless zone.

Smoke-control areas and exempt appliances

In the past, when it was common to burn coal as a source of domestic heat or for the commercial generation of heat and power, many cities suffered from very poor air quality and *smogs*, which are a mixture of winter fog and smoke, were common. These smogs contained high levels of sulphur dioxide and smoke particles, both of which are harmful to humans. In December 1952, a period of windless conditions prevailed, resulting in a smog in London that lasted for five days. Apart from very poor visibility, it was estimated at the time that this smog resulted in some 4,000 premature deaths and another 100,000 people suffered smog-related illnesses. Recent reports have found that these figures were seriously underestimated and as many as 12,000 may have died. Not surprisingly, there was a massive public outcry, which lead to the government introducing the Clean Air Act 1956 and local authorities declaring areas as 'smoke-control areas'.

The Clean Air Act of 1956 was replaced in 1993. Under the Clean Air Act 1993 it is an offence to sell or burn an *unauthorised fuel* in a smoke-control area unless it is burned in what is known as an 'exempt appliance'. These appliances are able to burn fuels that would normally be 'smoky', without emitting smoke to the atmosphere. Each appliance is designed to burn a specific fuel.

SUGGESTED ACTIVITY

An existing building is to be converted to be heated by biomass. The fuel is to be delivered by a heavy goods vehicle via country lanes. Identify two problems that may have to be considered.

Lists of authorised fuels and exempt appliances can be found on the Department for the Environment Food and Rural Affairs (Defra) website.

Planning permission

Planning permission will not normally be required for the installation of a biomass boiler in a domestic dwelling if all of the work is internal to the building.

If the installation requires an external flue to be installed, it will normally be classed as permitted development as long as the following criterion is met.

- The flue is to the rear or side elevation and does not extend more than 1m above the highest part of the roof.

Listed building or buildings in a designated area

Check with the local planning authority for both internal work and external flues.

Buildings in a conservation area or in a World Heritage Site

Flues should not be fitted on the principal or side elevation if they would be visible from a highway.

If the project includes the construction of buildings for storage of the biofuels, or to house the boiler, then the same planning requirements as for extensions and garden outbuildings will apply.

Compliance with building regulations

The following building regulations will apply.

Part	Title	Relevance
A	Structural safety	Where the biomass appliance and other components put load on the structure
B	Fire safety	Where holes for pipes, etc may reduce the fire-resisting integrity of the building structure
C	Resistance to contaminants and moisture	Where holes for pipes, etc may reduce the moisture-resisting integrity of the building structure
E	Resistance to sound	Where holes for pipes, etc may reduce the soundproof integrity of the building structure

Part	Title	Relevance
G	Sanitation, hot-water safety, and water efficiency	Hot-water safety and water efficiency
J	Heat-producing appliances	Biomass boilers produce heat and therefore must be installed correctly
L	Conservation of fuel and power	Energy efficiency of the installed system and the building
P	Electrical safety	Safe installation of the electrical supplies and any controls

Advantages

- It is carbon neutral.
- It is a sustainable fuel source.
- When it is burnt, the waste gases are low in nitrous oxide, with no sulphur dioxide – both are greenhouse gases.

Disadvantages

- Transportation costs are high – wood pellets or chips will need to be delivered in bulk to make delivery costs viable.
- Storage space is needed for the fuel. As woody biomass has a low calorific value, a large quantity of fuel will be required. Consideration must be given to whether or not adequate storage space is available.
- Control – when a solid fuel is burnt, it is not possible to have instant control of heat, as would be the case with a gas boiler. The fuel source cannot be instantly removed to stop combustion.
- It requires a suitable flue system.

Electricity-producing micro-renewable energy technologies (LO1-4)

The electricity-producing micro-renewable energy technologies that will be discussed in this section are:

- solar photovoltaic
- micro-wind
- micro-hydro.

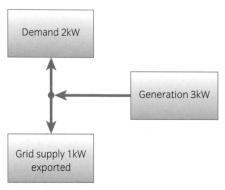

Generation exceeds demand

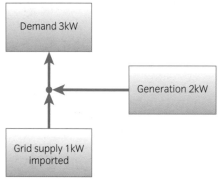

Demand exceeds generation

The major advantages of these technologies are that they do not use any of the planet's dwindling fossil fuel resources. They also do not produce any carbon dioxide (CO_2) when running.

With each of the electricity-producing micro-renewable energy technologies, two types of connection exist:

1 on-grid or grid tied – where the system is connected in parallel with the grid-supplied electricity
2 off-grid – where the system is not connected to the grid but supplies electricity directly to current-using equipment or is used to charge batteries and then supplies electrical equipment via an inverter.

The batteries required for off-grid systems need to be deep-cycle type batteries, which are expensive to purchase. The other downside of using batteries to store electricity is that the batteries' life span may be as short as five years, after which the battery bank will require replacing.

With on-grid systems, any excess electricity generated is exported back to the grid. At times when the generation output is not sufficient to meet the demand, electricity is imported from the grid.

While the following sections will be focused primarily on on-grid or grid-tied systems, which are the most common type in use, an overview of the components required for off-grid systems is included to provide a complete explanation of the technology.

Solar photovoltaic (PV)

Solar photovoltaic (PV) is the conversion of light into electricity. Light is electromagnetic energy and, in the case of visible light, is electromagnetic energy that is visible to the human eye. The electromagnetic energy released by the Sun consists of a wide spectrum, most of which is not visible to the human eye and cannot be converted into electricity by PV modules.

Working principles

The basic element of photovoltaic energy production is the PV cell, which is made from semiconductor material. A semiconductor is a material with resistivity that sits between that of an insulator and a conductor. Whilst various semiconductor materials can be used in the making of PV cells, the most common material is silicon. Adding a small quantity of a different element (an impurity) to the silicon, a process known as 'doping', produces n-type or p-type semiconductor material. Whether it is n-type (negative) or p-type (positive) semiconductor material is dependent on the element used to dope the silicon. Placing an n-type and a p-type semiconductor material together creates a p-n junction. This forms the basis of all semiconductors used in electronics.

When *photons*, which are particles of energy from the Sun, hit the surface of the PV cell they are absorbed by the p-type material. The additional energy provided by these photons allows electrons to overcome the bonds holding them and move within the semiconductor material, thus creating a potential difference or – in other words – generating a voltage.

Photovoltaic cells have an output voltage of 0.5V, so a number of these are linked together to form modules with resulting higher voltage and power outputs. Modules are connected together in series to increase voltage. These are known as 'strings'. All the modules together are known as an 'array'. An array therefore can comprise a single string or multiple strings. The connection arrangements are determined by the size of the system and the choice of inverter. It should be noted that PV arrays can attain d.c. voltages of many hundreds of volts.

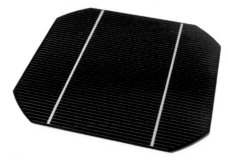

PV cell

There are many arrangements for PV systems but they can be divided into two categories:

- off-grid systems, where the PV modules are used to charge batteries
- on-grid systems, where the PV modules are connected to the grid supply via an inverter.

The key components of an off-grid PV system are:

- PV modules
- a PV module mounting system
- d.c. cabling
- a charge controller
- a deep-discharge battery bank
- an inverter.

Other components, such as isolators, will also be required.

Off-grid system components

Off-grid systems are ideal where no mains supply exists and there is a relatively small demand for power. Deep-discharge batteries are expensive and will need replacing within 5–10 years, depending on use.

On-grid systems where the PV modules are connected to the grid supply via an inverter

The key components of an on-grid PV system are:

- PV modules
- a PV module mounting system
- d.c. cabling
- an inverter
- a.c. cabling
- metering
- a connection to the grid.

Other components such as isolators will also be required.

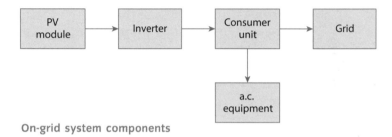

On-grid system components

PV modules

A range of different types of module, of various efficiencies, is available. The performance of a PV module is expressed as an *efficiency percentage*: the higher the percentage the greater the efficiency.

- Monocrystalline modules range in efficiency, from 15% to 20%.
- Polycrystalline modules range in efficiency, from 13% to 16%, but are cheaper to purchase than monocrystalline modules.
- Amorphous film ranges in efficiency, from 5% to 7%. Amorphous film is low efficiency but is flexible, so it can be formed into curves and is ideal for surfaces that are not flat.

Whilst efficiencies may appear low, the maximum theoretical efficiency that can be obtained with a single junction silicon cell is only 34%.

PV module mounting system

Photovoltaic modules can be fitted as on-roof systems, in-roof systems or ground-mount systems.

- On-roof systems are the common method employed for retrofit systems. Various different mounting systems exist for securing the modules to the roof structure. Most consist of aluminium rails, which are fixed to the roof structure by means of roof hooks. Mounting systems also exist for fitting PV modules to flat roofs.

Checks will need to be made to ensure that the existing roof structure can withstand the additional weight and also the uplift forces that will be exerted on the PV array by the wind.

- In in-roof systems the modules replace the roof tiles. The modules used are specially designed to interlock, to ensure that the roof structure is watertight. The modules are fixed directly to the roof structure. Several different systems are on the market, from single-tile size to large panels that replace a whole section of roof tiles. In-roof systems cost more than on-roof systems but are more aesthetically pleasing. In-roof systems are generally only suitable for new-build projects or where the roof is to be retiled.

On-roof mounting system

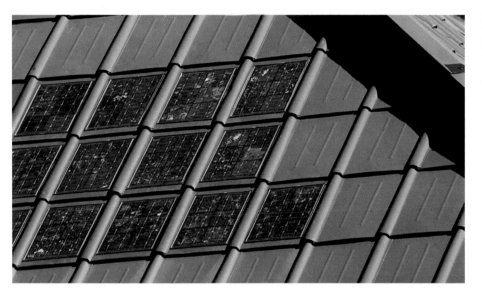

In-roof mounting system

- Ground-mount systems and pole-mount systems are available for free-standing PV arrays.

ASSESSMENT GUIDANCE

It is essential that the integrity of the roof is maintained and that no leaks occur due to the installation. Only approved contractors with the relevant insurance should be employed.

SUGGESTED ACTIVITY

With regards to a PV installation, who would normally be responsible for mounting the roof brackets and panels and testing and connecting the electrical system?

Ground-mount system

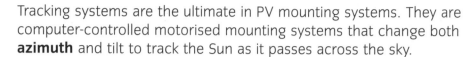

Azimuth

Ideally the modules should face due south, but any direction between east and west will give acceptable outputs. Azimuth refers to the angle that the panel direction diverges from facing due south

PV inverter

SUGGESTED ACTIVITY

What is the purpose of an inverter?

Tracking systems are the ultimate in PV mounting systems. They are computer-controlled motorised mounting systems that change both **azimuth** and tilt to track the Sun as it passes across the sky.

Inverter

The inverter's primary function is to convert the d.c. input to a 230V a.c. 50Hz output, and synchronise it with the mains supply frequency. The inverter also ensures that, in the event of mains supply failure, the PV system does not create a danger by continuing to feed power onto the grid. The inverter must be matched to the PV array with regard to power and d.c. input voltage, to avoid damage to the inverter and to ensure that it works efficiently. Both d.c. and a.c. isolators will be fitted to the inverter, to allow it to be isolated for maintenance purposes.

Metering

A generation meter is installed on the system to record the number of units generated, so that the feed-in tariff can be claimed.

Connection to the grid

Connection to the grid within domestic premises is made via a spare way in the consumer unit and a 16A overcurrent protective device. An isolator is fitted at the intake position to provide emergency switching, so that the PV system can easily be isolated from the grid.

Location and building requirements

When deciding on the suitability of a location or building for the installation of PV, the following considerations should be taken into account.

Adequate roof space available

The roof space available determines the maximum size of PV array that can be installed. In the UK, all calculations are based on 1,000Wp (watts peak) of the Sun's radiation on 1m^2 so, if the array uses modules with a 15% efficiency, each 1kWp of array will require approximately 7m^2 of roof space. The greater the efficiency of the modules, the less roof space that is required.

The orientation (azimuth) of the PV array

The optimum direction for the solar collectors to face is due south; however, as the Sun rises in the east and sets in the west, any location with a roof facing east, south or west is suitable for mounting a PV array, but the efficiency of the system will be reduced for any system not facing due south.

The tilt of the PV array

Throughout the year, the height of the Sun relative to the horizon changes from its lowest in December through to its highest in June. As it is generally not practical to vary the tilt angle throughout the year, the optimum tilt for the PV array in the UK is between 30° and 40°;

Ideal orientation is south

however, the modules will work outside the optimum tilt range and will even work if vertical or horizontal, but they will be less efficient.

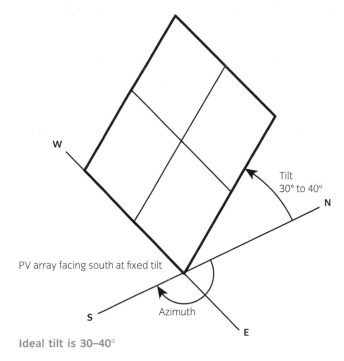

W

Tilt
30° to 40°

N

PV array facing south at fixed tilt

S

Azimuth

E

Ideal tilt is 30–40°

Shading of the PV array

Any structure, tree, chimney, aerial or other object that stands between the PV array (collector) and the Sun will prevent some of the Sun's energy from reaching the collector. The Sun shines for a limited time and any reduction in the amount of sunlight landing on the collector will reduces its ability to produce electricity.

Location within the UK

The location within the UK will determine how much sunshine will fall, annually, on the PV array and, in turn, this will determine the amount of electricity that can be generated. For example, a location in Brighton will generate more electricity than one in Newcastle, purely because Brighton receives more sunshine.

Shading of PV system

The suitability of the structure for mounting the solar collector

The structure has to be assessed for its suitability for fixing the chosen mounting system. Consideration needs to be given to the strength of the structure, the suitability of fixings and the condition of the structure. Consideration also needs to be given to the effect known as 'wind uplift', an upward force exerted by the wind on the module and mounting system. The strength of the PV array fixings and the fixings holding the roof members to the building structure must be great enough to allow for wind uplift.

In the case of roof-mounted systems on flats and other shared properties, consideration must also be given to the ownership of the structure on which the proposed system is to be installed.

A suitable place to mount the inverter

The inverter is usually mounted either in the loft space or at the mains position.

Connection to the grid

A spare way within the consumer unit will need to be available for connection of the PV system. If one is not available then the consumer unit may need to be changed.

Planning permission

Permitted development applies where a PV system is installed:

- on a dwelling house or block of flats
- on a building within the grounds of a dwelling house or block of flats
- as a stand-alone system in the grounds of a dwelling house or block of flats.

However, there are criteria to be met in each case.

For building-mounted systems:

- the PV system must not protrude more than 200mm from the wall or the roof slope
- the PV system must not protrude past the highest point of the roof (the ridgeline), excluding the chimney.

For stand-alone systems the following criteria must be met.

- Only one stand-alone system is allowed in the grounds.
- The array must not exceed 4m in height.
- The array must not be installed within 5m of the boundary of the grounds.
- The array must not exceed 9m^2 in area.
- No dimension of the array may exceed 3m in length.

For both stand-alone and building-mounted systems the following criteria must be met.

- The system must not be installed in the grounds or on a building within the grounds of a listed building or a scheduled monument.
- If the dwelling is in a conservation area or a World Heritage Site, then the array must not be closer to a highway than the house or block of flats.
- In every other case, planning permission will be required.

Compliance with building regulations

The following building regulations will apply.

Part	Title	Relevance
A	Structural safety	The PV modules will impose both downward force and wind uplift stresses on the roof structure
B	Fire safety	The passage of cables through the building fabric could reduce the fire-resisting integrity of the structure
C	Resistance to contaminants and moisture	The fixing brackets for on-roof systems and the passage of cables through the building fabric could reduce the moisture-resisting integrity of the structure
E	Resistance to sound	The passage of cables through the building fabric could reduce the sound-resisting properties of the structure
L	Conservation of fuel and power	The efficiency of the system and the building overall
P	Electrical safety	The installation of the components and wiring system

SUGGESTED ACTIVITY

What documentation should be completed by the electrical installer after testing the new PV installation?

Other regulatory requirements

- BS 7671 Wiring Regulations will apply to the PV installation.
- G83 requirements will apply to on-grid systems up to 3.68kW per phase; above this size the requirements of G59 will need to be complied with.
- Micro Generation Certification Scheme requirements will apply.

Advantages

- They can be fitted to most buildings.
- There is a feed-in tariff available for electricity generated, regardless of whether it is used on site or exported to the grid.
- Excess electricity can be sold back to the distribution network operator (DNO).
- There is a reduction in electricity imported.
- It uses zero carbon technology.
- It improves energy performance certificate ratings.
- There is a reasonable payback period on the initial investment.

ASSESSMENT GUIDANCE

All systems will require some penetration of the building fabric, be it the roof or walls, depending on the building type and construction. You should be able to describe the methods of making good for all building fabrics.

Disadvantages

- Initial cost is high.

- The system size is dependent on available, suitable roof area.

- It requires a relatively large array to offset installation costs.

- It gives variable output that is dependent on the amount of sunshine available. Lowest output is at times of greatest requirement, such as at night and in the winter. Savings need to be considered over the whole year.

- There is an aesthetic impact (on the appearance of the building).

Micro-wind

Wind turbines harness energy from the wind and turn it into electricity. The UK is an ideal location for the installation of wind turbines, as about 40% of Europe's wind energy passes over the UK. A micro-wind turbine installed on a suitable site could easily generate more power than would be consumed on site.

Working principles

The wind passing the rotor blades of a turbine causes it to turn. The hub is connected by a low-speed shaft to a gearbox. The gearbox output is connected to a high-speed shaft that drives a generator which, in turn, produces electricity. Turbines are available as either horizontal-axis wind turbines (HAWT) or vertical-axis wind turbines (VAWT).

A HAWT has a tailfin to turn the turbine so that it is facing in the correct direction to make the most of the available wind. The gearbox and generator will also be mounted in the horizontal plane.

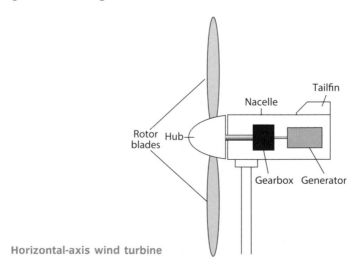

Horizontal-axis wind turbine

Vertical-axis wind turbines, of which there are many different designs, will work with wind blowing from any direction and therefore do not require a tailfin. A VAWT also has a gearbox and generator.

The two types of micro-wind turbines suitable for domestic installation are:

- pole-mounted, freestanding wind turbines
- building-mounted wind turbines, which are generally smaller than pole-mounted turbines.

Micro-wind generation systems fall into two basic categories:

- on-grid (grid-tied), which is connected in parallel with the grid supply via an inverter
- off-grid, which charge batteries to store electricity for later use.

The output from a micro-wind turbine is *wild* alternating current (a.c.). 'Wild' refers to the fact that the output varies in both voltage and frequency. The output is connected to a system controller, which rectifies the output to d.c.

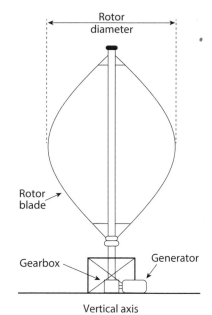

Vertical-axis wind turbine

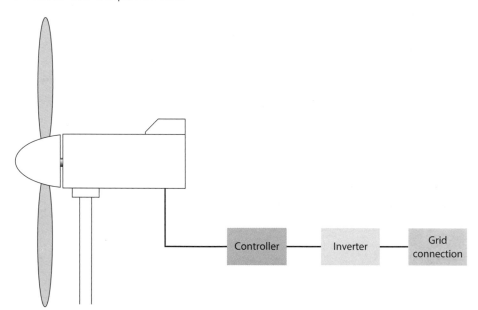

Block diagram of an on-grid micro-wind system

In the case of an on-grid system the d.c. output from the system controller is connected to an inverter which converts d.c. to a.c. at 230V 50Hz, for connection to the grid supply via a generation meter and the consumer unit.

SUGGESTED ACTIVITY

The first wind turbine is believed to have been installed in Scotland in July 1887 by James Blythe. Find out what he used it for.

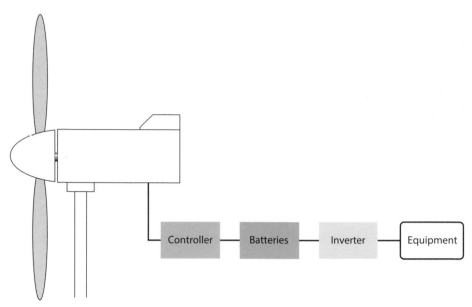

Block diagram of an off-grid micro-wind system

With off-grid systems the output from the controller is used to charge batteries so that the output is stored for when it is needed. The output from the batteries then feeds an inverter so that 230V a.c. equipment can be connected.

Location and building requirements

When considering the installation of a micro-wind turbine, it is important to consider the location or building requirements, including:

- the average wind speed on the site
- any obstructions and turbulence
- the height at which the turbine can be mounted
- turbine noise, vibration, flicker.

Wind speed

Wind is not constant, so the average wind speed on a site, measured in metres per second (m/s), is a prime consideration when deciding on a location's suitability for the installation of a micro-wind turbine.

Wind speed needs to be a minimum of 5m/s for a wind turbine to generate electricity. Manufacturers of wind turbines provide power curves for their turbines, which show the output of the turbine at different wind speeds. Most micro-wind turbines will achieve their maximum output when the wind speed is around 10m/s.

Obstructions and turbulence

For a wind turbine to work efficiently, a smooth flow of air needs to pass across the turbine blades.

The ideal site for a wind turbine would be at the top of a gentle slope. As the wind passes up the slope it gains speed, resulting in a higher output from the turbine.

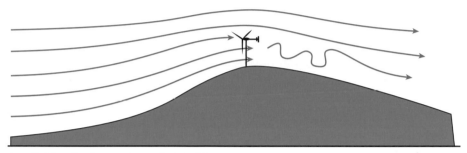

A suitable site for a micro-wind turbine

The diagram below illustrates the effect on the wind when a wind turbine is poorly sited. The wind passing over the turbine blades is disturbed and thus the efficiency is reduced.

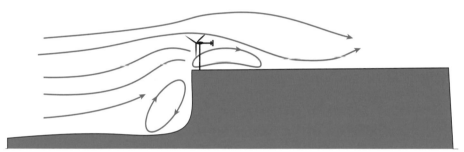

An unsuitable site for a micro-wind turbine

Any obstacles, such as trees or tall buildings, will affect the wind passing over the turbine blades.

Where an obstacle is upwind of the wind turbine, in the direction of the prevailing wind, the wind turbine should be sited at a minimum distance of 10 × the height of the obstacle away from the obstacle. In the case of an obstacle that is 10m in height, this would mean that the wind turbine should be sited a minimum of 10 × 10m away, which is 100m from the obstacle.

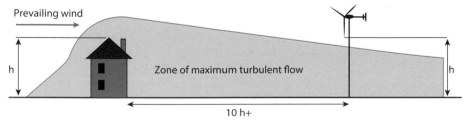

Placement of micro-wind turbine to avoid obstacles

Turbine mounting height

Generally, the higher a wind turbine is mounted, the better. The minimum recommended height is 6–7m but, ideally, it should be mounted at a height of 9–12m. As a wind turbine has moving parts, consideration needs to be given to access for maintenance. Where an obstacle lies upwind of the turbine, the bottom edge of the blade should be above the height of the obstacle.

Turbine noise

Consideration needs to be given to buildings sited close to the wind turbine as the turbine will generate noise in use.

Turbine vibration

Consideration needs to be given to vibration when the wind turbine is building mounted. It may be necessary to consult a structural engineer.

Shadow flicker

Shadow flicker is the result of the rotating blades of a turbine passing between a viewer and the Sun. It is important to ensure that shadow flicker does not unduly affect a building sited in the shadow-flicker zone of the wind turbine.

The distance of the shadow-flicker zone from the turbine will be at its greatest when the Sun is at its lowest in the sky.

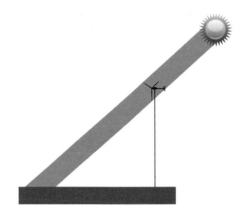

The area affected by shadow flicker

Planning permission

Whilst permitted development exists for the installation of wind turbines, it is severely restricted, so, in the majority of installations, a planning application will be required. The permitted development criteria are detailed below.

Permitted development applies where a wind turbine is installed:

- on a detached dwelling house
- on a detached building within the grounds of a dwelling house or block of flats
- as a stand-alone system in the grounds of a dwelling house or block of flats.

It is important to note that permitted development for building-mounted wind turbines only applies to detached premises. It does not apply to semi-detached houses or flats.

Even with detached buildings or stand-alone turbines there are criteria to be met.

- The wind turbine must comply with the MCS planning standards, or equivalent.

- Only one wind turbine may be installed on the building or within the grounds of the building.

- An air source heat pump may not be installed on the building or within the grounds of the building.

- The highest part of the wind turbine (normally the blades) must not protrude more than 3m above the ridge line of the building or be more than 15m in height.

- The lowest part of the blades of the wind turbine must be a minimum of 5m from ground level.

- The wind turbine must be a minimum of 5m from the boundary of the premises.

- The wind turbine cannot be installed on or within:

  - land that is safeguarded land (usually designated for military or aeronautical reasons)

  - a site that is designated as a scheduled monument

  - a listed building

  - the grounds of a listed building

  - land within a national park

  - an area of outstanding natural beauty

  - the Broads (wetlands and inland waterways in Norfolk and Suffolk).

- The wind turbine cannot be installed on the roof or wall of a building that fronts a highway, if that building is within a conservation area.

The following conditions also apply.

- The blades must be made of non-reflective material.

- The wind turbine should be sited so as to minimise its effect on the external appearance of the building.

Compliance with building regulations

The following building regulations will apply.

Part	Title	Relevance
A	Structural safety	A wind turbine mounted on a building will exert additional structural load, as well as forces, due to its operation
B	Fire safety	Cable entries and fixings may reduce the fire-resisting integrity of the building structure
C	Resistance to contaminants and moisture	Cable entries and fixings may reduce the moisture-resisting integrity of the building fabric

SUGGESTED ACTIVITY

It is possible that wind turbines may have an adverse effect on wildlife. Identify which groups could be affected.

KEY POINT

Rats, mice and insects can enter a building through surprisingly small holes. A wasps' nest can cause considerable distress to people.

Part	Title	Relevance
E	Resistance to sound	Cable entries may reduce the sound-resisting integrity of the building fabric
L	Conservation of fuel and power	The efficiency of the system and the building
P	Electrical safety	Installation of wiring and components

Other regulatory requirements

- For on-grid systems, the requirements of the Distribution Network Operator (DNO) will apply.
- Wiring Regulations BS 7671 will apply to the installation of micro-wind turbines.

Advantages

- They can be very effective on a suitable site as the UK has 40% of Europe's wind resources.
- There are no carbon dioxide emissions.
- They produce most energy in winter, when consumer demand is at its maximum.
- A feed-in tariff is available.
- This can be a very effective technology where mains electricity does not exist.

Disadvantages

- Initial costs are high.
- The requirements of the site are onerous.
- Planning can be onerous.
- Performance is variable and is dependent on wind availability.
- Micro-wind turbines cause noise, vibration and flicker.

Micro-hydro-electric

All rivers flow downhill. This movement of water from a higher level to a lower level is a source of free kinetic energy that hydro-electric generation harnesses. Water passing across or through a turbine can be used to turn a generator and thus produce electricity. Given the right location, micro-hydro-electric is the most constant and reliable source of all the micro-generation technologies and is the most likely of the technologies to meet all of the energy needs of the consumer.

As with the other micro-generation technologies, there are two possible system arrangements for micro-hydro schemes: on-grid and off-grid systems.

SUGGESTED ACTIVITY

What is meant by a 'feed-in tariff'?

Working principles

Whilst it is possible to place generators directly into the water stream, it is more likely that the water will be diverted from the main stream or river, through the turbine, and back into the stream or river at a lower level. Apart from the work involved with the turbines and generators, there is also a large amount of civil engineering and construction work to be carried out to route the water to where it is needed.

The main components of the water course construction are:

- intake – the point where a portion of the river's water is diverted from the main stream
- the canal that connects the intake to the forebay
- the forebay, which holds a reservoir of water that ensures that the penstock is pressurised at all times and allows surges in demand to be catered for
- the penstock, which is pipework taking water from the forebay to the turbines
- the powerhouse, which is the building housing the turbine and the generator
- the tailrace, which is the outlet that takes the water exiting the turbines and returns it to the main stream of the river.

See the diagram below.

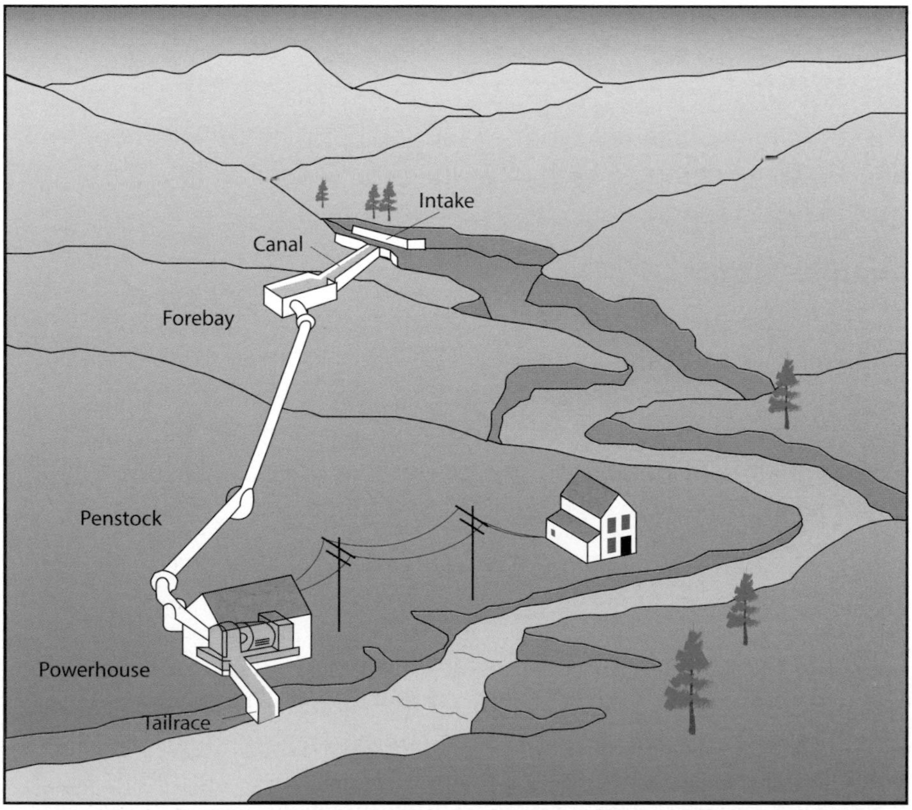

The component parts of a micro-hydro system

SUGGESTED ACTIVITY

Which types of fish are most likely to be adversely affected by the installation of turbines in rivers?

ASSESSMENT GUIDANCE

The idea of water turbines sounds great but of course few people live near a water course that is suitable for this application. PV and solar thermal systems are more adaptable for different locations.

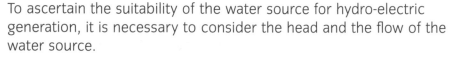

To ascertain the suitability of the water source for hydro-electric generation, it is necessary to consider the head and the flow of the water source.

Head

The head is the vertical height difference between the proposed inlet position and the proposed outlet. This measurement is known as 'gross head'.

Head height is generally classified as:

- low head – below 10m
- medium head – 10–50m
- high head – above 50m.

There is no absolute definition for each classification. The Environment Agency, for example, classifies low head as below 4m. Some manufacturers specify high head as above 300m.

Net head

This is used in calculations of potential power generation and takes into account losses due to friction, as the water passes through the penstock.

Flow

This is the amount of water flowing through the water course and is measured in cubic metres per second (m^3/s).

Turbines

There are many different types of turbine but they fall into two primary design groups, each of which is better suited to a particular type of water supply.

Impulse turbine

In an impulse turbine, the turbine wheel or *runner* operates in air, with water jets driving the runner. The water from the penstock is focused on the blades by means of a nozzle. The velocity of the water is increased but the water pressure remains the same so there is no requirement to enclose the runner in a pressure casing. Impulse turbines are used with high-head water sources.

Examples of impulse turbines are described below.

Pelton

This consists of a wheel with bucket-type vanes set around the rim. The water jet hits the vane and turns the runner. The water gives up most of its energy and falls into a discharge channel below. A multi-jet Pelton turbine is also available. This type of turbine is used with water sources with medium or high heads of water.

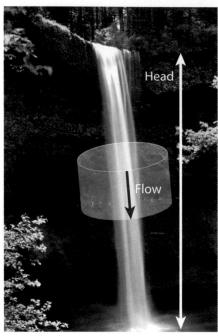

The meanings of 'head' and 'flow'

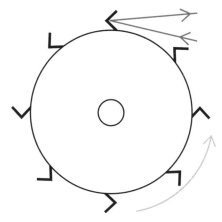

Impulse turbine

Turgo

This is similar to the Pelton but the water jet is designed to hit the runner at an angle and from one side of the turbine. The water enters at one side of the runner and exits at the other, allowing the Turgo turbine to be smaller than the Pelton for the same power output. This type of turbine is used with water sources with medium or high heads of water.

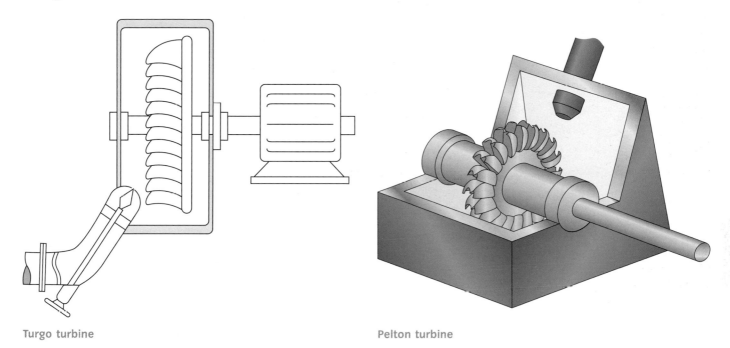

Turgo turbine   Pelton turbine

Cross-flow or Banki

With this type of turbine the runner consists of two end-plates with slats, set at an angle, joining the two discs, much like a water wheel. Water passes through the slats, turning the runner and then exiting from below. This type of turbine is used with water sources with low or medium heads of water.

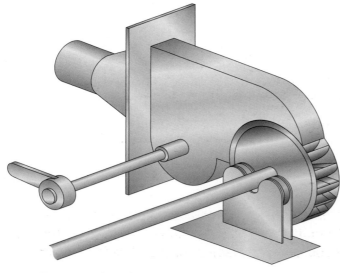

Cross-flow or Banki turbine

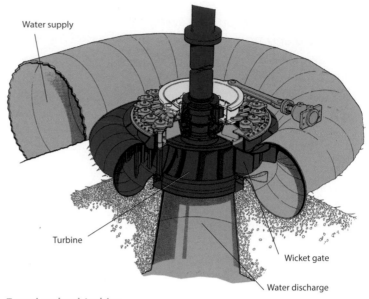

Francis wheel turbine

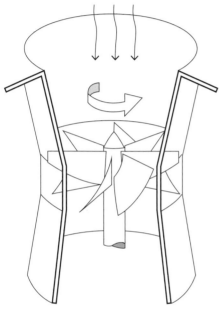

Reaction turbine

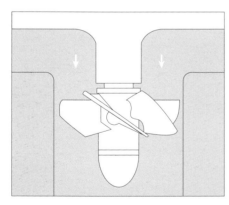

Kaplan or propeller turbine

Reaction turbine

In the reaction turbine, the runners are fully immersed in water and are enclosed in a pressure casing. Water passes through the turbine, causing the runner blades to turn or react.

Examples of reaction turbines are described below.

Francis wheel

Water enters the turbine housing and passes through the runner, causing it to turn. This type of turbine is used with water sources with low heads of water.

Kaplan (propeller)

This works like a boat propeller in reverse. Water passing the angled blades turns the runner. This type of turbine is used with water sources with low heads of water.

Reverse Archimedes' screw

The Archimedes' screw consists of a helical screw thread, which was originally designed so that turning the screw – usually by hand – would draw water up the thread to a higher level. In the case of hydro-electric turbines, water flows down the screw, hence *reverse*, turning the screw, which is connected to the generator. This type of turbine is particularly suited to low-head operations but its major feature is that, due to its design, it is 'fish-friendly' and fish are able to pass through it, so it may be the only option if a hydro-electric generator is to be fitted on a river that is environmentally sensitive.

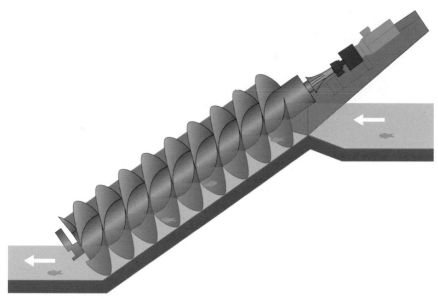

Reverse Archimedes' screw

Location building requirements

When considering the installation of a micro-hydro turbine, the following location or building requirements should be taken into account.

- The location will require a suitable water source with:
  - a minimum head of 1.5m
  - a minimum flow rate of 100 litres/second.

The water source should not be subject to seasonal variation that will take the water supply outside of the above parameters.

- The location has to be suitable to allow construction of:
  - the water inlet
  - the turbine/generator building
  - the water outlet or tailrace.

Planning permission

Planning permission will be required.

A micro-hydro scheme will have an impact on:

- the landscape and visual amenity
- nature conservation
- the water regime.

The planning application will need to be accompanied by an environmental statement detailing any environmental impact and what measures will be taken to minimise these. The environmental statement typically covers:

ASSESSMENT GUIDANCE

You should be able to draw the basic layout of a micro-hydro system.

- flora and fauna
- noise levels
- traffic
- land use
- archaeology
- recreation
- landscape
- air and water quality.

Compliance with building regulations

The following building regulations will apply.

Part	Title	Relevance
A	Structural safety	If any part of the system is housed in or connected to the building, then structural considerations will need to be taken into account
B	Fire safety	Where cables pass through the building fabric they may reduce the fire-resisting properties of the building fabric
C	Resistance to contaminants and moisture	Where cables pass through the building fabric they may reduce the moisture-resisting properties of the building fabric
E	Resistance to sound	Where cables pass through the building fabric they may reduce the sound-resisting properties of the building fabric
P	Electrical safety	Installation of components and cables

SUGGESTED ACTIVITY

What are the requirements of BS 7671 regarding repairing holes made in walls for the passage of cables?

Other regulatory requirements

- BS 7671 Wiring Regulations
- G83 requirements for grid-tied systems
- Micro Generation Certification Scheme requirements
- Environment Agency requirements

In England and Wales, all waterways of any size are controlled by the Environment Agency. To remove water from these waterways, even though it may be returned – as in the case of a hydro-electric system – will usually require permission and a licence.

There are three types of licence that may apply to a hydro-electric system.

- An **abstraction licence** will be required if water is diverted away from the main water course. The major concern will be the impact that the project has on fish migration, as the majority of turbines are not fish-friendly. This requirement may affect the choice of turbine (*see* Reverse Archimedes' screw, page 114). It may mean that fish screens are required over water inlets or, where the turbine is in the main channel of water, a fish pass around the turbine may need to be constructed.

- An **impoundment licence** – an impoundment is any construction that changes the flow of water, so if changes or additions are made to sluices, weirs, etc that control the flow within the main stream of water, an impoundment licence will be required.

- A **land drainage licence** will be required for any changes made to the main channel of water.

An Environment Site Audit (ESA) will be required as part of the initial assessment process. The ESA covers:

- water resources
- conservation
- chemical and physical water quality
- biological water quality
- fisheries
- managing flood risk
- navigation of the waterway.

Advantages

- There are no on-site carbon emissions.
- Large amounts of electricity are output, usually more than required for a single dwelling. The surplus can be sold.
- A feed-in tariff is available.
- There is a reasonable payback period.
- It is an excellent system where no mains electricity exists.
- It is not dependent on weather conditions or building orientation.

Disadvantages

- It requires a high head or fast flow of water on the property.
- It requires planning permission, which can be onerous.
- Environment Agency permission is required for water extraction.
- It may require strengthening of the grid for grid-tied systems.
- Initial costs are high.

ASSESSMENT GUIDANCE

Micro-hydro systems are no good on rivers that suffer seasonal droughts or are subject to winter flooding or freezing, unless preventative measures are taken to protect the generator and control gear.

Co-generation energy technologies (LO1-4)

Co-generation technologies – micro-combined heat and power (heat-led)

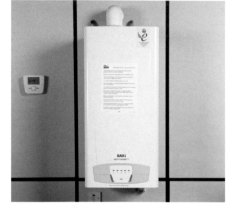

Baxi Ecogen domestic CHP boiler

In micro-combined heat and power (mCHP) technology, a fuel source is used to satisfy the demand for heat but, at the same time, generates electricity that can either be used or sold back to the supplier.

Currently, mCHP units used in domestic dwellings are powered by means of natural gas or liquid propane gas (LPG), but could be fuelled by using biomass, liquid propane gas (LPG) or other fuels.

The diagram below represents, from left to right, an old, inefficient gas boiler, a modern condensing boiler and an mCHP unit.

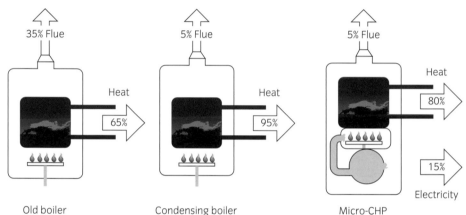

The efficiency of different boilers

With the old, inefficient boiler, 65% of the input energy is used to provide heating for the premises, 35% is lost up the flue. With the condensing boiler, this lost heat is re-used so that the output to the heating is 95%. The mCHP unit will achieve the same efficiencies as the modern condensing boiler but 80% of the input is used to provide heat and 15% is used to power a generator to provide power.

There are obvious savings to be made in replacing an old, inefficient boiler with a mCHP unit, but could the same savings not be made by fitting a condensing boiler? At first sight this may appear to be feasible; however, on a unit-by-unit comparison, gas is cheaper than electricity, so any electricity generated by using gas means a proportionally greater financial saving over using electricity.

In addition to this saving, locally generated power reduces transmission losses and consequently creates less carbon dioxide (CO_2) than if the electricity were generated at a power station some distance away.

This type of generation, using a mCHP unit, is known as 'heat-led', as the primary function of the unit is to provide space heating, while the generation of electricity is secondary. The more heat that is produced, the more electricity is generated. The unit generates electricity only when there is a demand for heating. Most domestic mCHP units will generate between 1kW and 1.5kW of electricity. Micro-combined heat and power is a carbon-reduction technology rather than a carbon-free technology.

Working principles

Combined heat and power (CHP) units have been available for a number of years but it is only recently that domestic versions have become available. Domestic versions are usually gas-fired and use a Stirling engine to produce electricity, though other fuel sources, and types of generator combinations, are available.

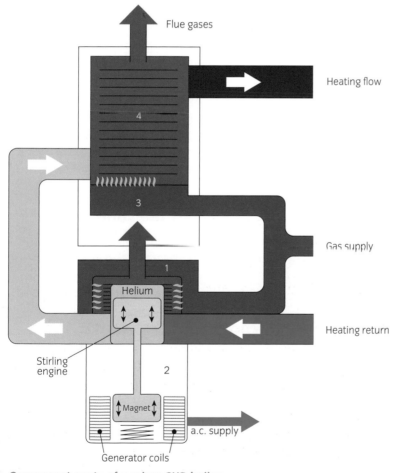

Component parts of a micro-CHP boiler

SUGGESTED ACTIVITY

Find out an alternative name for a Stirling engine.

The key components of the mCHP unit are:

1 the engine burner
2 the Stirling engine generator
3 the supplementary burner
4 the heat exchanger.

When there is a call for heat, the engine burner fires and starts the Stirling engine generator. The engine burner produces about 25% of the full heat output of the unit. The burner preheats the heating-system return water before passing it to the main heat exchanger. The hot flue gases from the engine burner are passed across the heat exchanger to heat the heating-system water further. If there is greater demand than is being supplied by the engine burner, then the supplementary burner operates to meet this demand.

How the Stirling engine generator works

The first Stirling engine was invented by Robert Stirling in 1816 and is very different from the internal combustion engine. The Stirling engine uses the expansion and contraction of internal gases, due to changes in temperature, to drive a piston. The gases within the engine do not leave the engine and no explosive combustion takes place, so the Stirling engine is very quiet in use.

In the case of the Stirling engine used in an mCHP unit, the gas contained within the engine is helium. When the engine burner fires, the helium expands, forcing the piston downwards. The return water from the heating system passing across the engine cools the gas, causing it to contract. A spring arrangement within the engine returns the piston to the top of the cylinder and the process starts all over again.

The piston is used to drive a magnet up and down between coils of wire, generating an electromotive force (emf) in the coils.

Connection of the mCHP unit to the supply

The preferred connection method between the mCHP unit and the supply is via a dedicated circuit, directly from the consumer unit. This method will allow for easy isolation of the generator from the incoming supply.

Location and building requirements

For mCHP to be viable, the following criteria must be met.

- The building should have a high demand for space heating. The larger the property the greater the carbon savings.
- A building that is well insulated will not usually be suitable, as a well-insulated building is unlikely to have a high demand for space heating.

If an mCHP unit is fitted to a building that is either too small or is well insulated, this will mean that the demand for heat will be small and the mCHP unit will cycle on and off, resulting in inefficient operation.

Planning permission

Planning permission will not normally be required for the installation of an mCHP unit in a domestic dwelling if all of the work is internal to

KEY POINT

Micro-CHP boilers are only suitable where there is a high demand for heat.

the building. If the installation requires an external flue to be installed, this will normally be classed as permitted development as long as the following criterion is met.

- Flues to the rear or side elevation do not extend more than 1m above the highest part of the roof.

Listed building or buildings in a designated area

Check with the local planning authority regarding both internal work and external flues.

Buildings in a conservation area or in a World Heritage Site

Flues should not be fitted on the principal or side elevation if they would be visible from a highway.

If the project includes the construction of buildings for fuel storage, or to house the mCHP unit, then the same planning requirements for extensions and garden outbuildings will apply.

Compliance with building regulations

The following building regulations will apply.

Part	Title	Relevance
A	Structural safety	Where the components increase load on the structure or where holes reduce the structural integrity of the building
B	Fire safety	Where installation of the system decreases the fire-resistant integrity of the structure, for example, where pipes or cables pass through fire compartments
C	Resistance to contaminants and moisture	Holes for pipes or cables could reduce the moisture-resisting integrity of the building
E	Resistance to sound	Holes for pipes or cables could reduce the sound-resisting qualities of the building
G	Sanitation, hot-water safety, and water efficiency	Hot-water safety and water efficiency
J	Heat-producing appliances	mCHP units are heat-producing systems

Part	Title	Relevance
L	Conservation of fuel and power	Energy efficiency of the system and the building
P	Electrical safety	Electrical installation of controls and supply

Other regulatory requirements

- Gas regulations will apply to the installation of the mCHP unit. The gas installation work will need to be carried out by an operative registered on the Gas Safe register.
- Water Regulations (WRAS) will apply to the water systems.
- BS 7617 Wiring Regulations will apply to the installation of control wiring and the wiring associated with the connection of the mCHP electrical generation output.
- G83 requirements will apply to the connection of the generator, although mCHP units do have a number of exemptions.
- Micro Generation Certification Scheme requirements.

Advantages

- The ability to generate electricity is not dependent on building direction or weather conditions.
- The system generates electricity while there is a need for heat.
- A feed-in tariff is available but is limited to generator outputs of less than 2kW and is only applicable to the first 30,000 units.
- Saves carbon over centrally generated electricity.
- Reduces the building's carbon footprint.

Disadvantages

- The initial cost of an mCHP is high, when compared with an efficient gas boiler.
- It is not suitable for properties with low demand for heat – small or very well insulated properties.
- There is limited capacity for generation of electricity.

Water conservation technologies (LO1-4)

SUGGESTED ACTIVITY

Water conservation has been used for hundreds of years, especially by gardeners. How was this achieved?

Many people regard the climate of the United Kingdom as wet. It is a common perception that the UK has a lot of rain and, in some locations, this is true, especially towards the west, where average annual rainfall is in excess of 1,000mm, but along the east coast the average is less than half of this.

The population of the UK is expanding and the demands on the water supply systems are ever increasing. Hose-pipe bans in many parts of the country are a regular feature of the summer months. In the UK, unlike many of our European neighbours, the water supplied is suitable for consumption straight from the tap, but we use it not only for drinking, but for bathing, washing clothes, watering gardens and washing cars.

Even in the UK, clean, fresh water is a limited resource. With growing demand, the pressure on this vital resource is increasing. Water conservation is one way of ensuring that demand does not outstrip supply and that shortages are avoided.

The two methods of water conservation covered in this unit are:

- rainwater harvesting
- re-use of greywater.

Water conservation is one way of reducing water bills. Whether calculated as measured (metered) or unmeasured, water bills, both contain two charges:

- charges for fresh water supplied
- charges for sewage or waste water taken away.

The amount of water taken away, which includes surface water (rainwater), is assumed to be 95% of the water supplied.

By conserving water, the waste-water charge, as well as the charge for fresh water supplied, can be reduced. Besides producing a reduction in household bills, water conservation will also help to relieve the pressure on a vital resource.

The Code for Sustainable Homes sets a target for reducing average drinking water consumption from 150 litres per person per day to an optimum 80 litres. The adopted target is currently 103 litres. Part G of the Building Regulations sets the level at 125 litres. Whichever target is used, the conclusion to be drawn is that a reduction in consumption is vital.

ASSESSMENT GUIDANCE

Given the demands for water and the problems water extraction from rivers may cause, it is essential that better use is made of our available water resources. The historic use of wholesome water for flushing toilets has to change.

SUGGESTED ACTIVITY

Water is a scarce resource that we cannot live without. Most of the water on Earth is seawater which is not fit for human consumption. Find out where your water supply comes from.

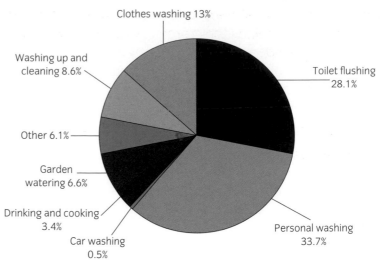

Clothes washing 13%

Washing up and cleaning 8.6%

Toilet flushing 28.1%

Other 6.1%

Garden watering 6.6%

Drinking and cooking 3.4%

Car washing 0.5%

Personal washing 33.7%

How water is used

Within the average home the amount of water used for drinking or food preparation is estimated at only 3.4% of total consumption, but there are obvious opportunities elsewhere for water savings.

The technologies to be covered are not concerned with direct carbon reduction or with financial savings, but are concerned with the reduction in consumption of a valuable resource.

Terminology used	Meaning
Wholesome water	Water that is palatable and suitable for human consumption. The water that is obtained from the utility company supply and is also known as 'wholesome', 'potable' and 'white water'
Rainwater	Water captured from rainwater gutters and downpipes
Greywater	Waste water from wash basins, showers, baths, sinks and washing machines
Black water	Sewage

Rainwater harvesting

Rainwater harvesting refers to the process of capturing and storing rainwater from the surface it falls on, rather than letting it run off into drains or allowing it to evaporate. Re-use of rainwater can result in sizeable reductions in wholesome water usage and thus monetary savings as well as carbon reductions.

If harvested rainwater is filtered, stored correctly and used regularly, so that it does not remain in the storage tanks for an excessive period of time, it can be used for:

- flushing toilets
- car washing
- garden watering
- supplying a washing machine.

Harvested rainwater cannot be used for:

- drinking water
- washing of dishes
- food preparation or washing of food
- personal hygiene, ie washing, bathing or showering.

Rainwater is classified as 'fluid category 5' risk, which is the highest risk category.

Working principles

The process steps in re-using rainwater are:

- collection
- filtration
- storage
- re-use.

Collection or capture of rainwater

Rainwater can be captured from roofs or hard standings. In the case of roofs the water is captured by means of gutters and flows to the water-harvesting tank via the property's rainwater downpipes. The amount of water that can be collected will be governed by:

- the size of the capture area
- the annual rainfall in the area.

Not all of the water that falls on the surface can be captured. During periods of very heavy rainfall, water may overflow gutters or merely bounce off the roof surface and avoid the guttering system completely.

Water collected from roofs covered in asbestos, copper, lead or bitumen may not be suitable for re-use and may pose a health risk, or result in discolouration of the water or odour problems. Water collected from hard standings such as driveways may be contaminated with oil or faecal matter.

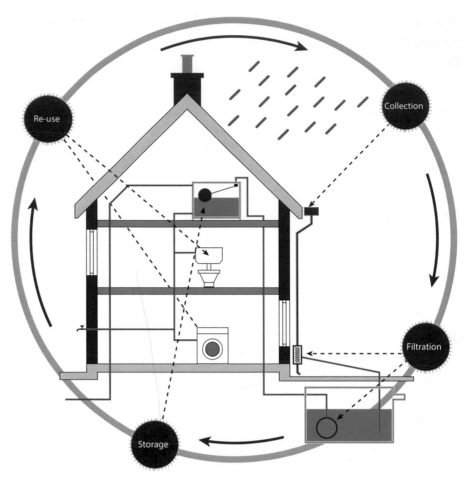

Rainwater-harvesting cycle

Filtration

As the rainwater passes from the rainwater-capture system to the storage tanks, it passes through an in-line filter to remove debris such as leaves. These are flushed out. The efficiency of this filter will determine how much of the captured water ends up in the storage tank. Manufacturers usually quote figures in excess of 90% efficiency.

Storage of rainwater

Rainwater storage tanks can be either above-ground or below-ground types and can vary in size from a small tank next to a house, to a buried tank that is able to hold many thousands of litres of water. Below-ground tanks will require excavation works, whilst above-ground tanks will need a suitably sized space to be available. Whichever type of tank is used, it will need to be protected against freezing, heating from direct sunlight and contamination.

The size of the tank will be determined by the rainwater available and the annual demand. It is common practice to size a tank to be 5% of annual rainwater supply or the anticipated annual demand.

A submersible pump is used to transport water from the storage tank to the point of demand.

> **KEY POINT**
>
> Most people with a garden will have a water butt to collect rainwater. Special adaptors are available that can be cut into the drainpipe to divert the water into the water butt.

The tank will incorporate an overflow pipe connected to the drainage system of the property, for times when the harvested rainwater exceeds the capacity of the storage tank.

Re-use of stored rainwater

With rainwater harvesting, two system options are available for the re-use of the collected and stored water: indirect and direct distribution.

Indirect distribution system

With indirect distribution systems, water is pumped from the storage tank to a supplementary storage tank or header tank located within the premises. This, in turn, feeds the water outlets via pipework separated from the wholesome water supply pipes.

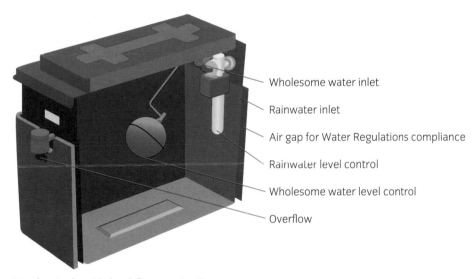

Wholesome water inlet

Rainwater inlet

Air gap for Water Regulations compliance

Rainwater level control

Wholesome water level control

Overflow

Header tank with backflow protection

The header tank will incorporate a backflow prevention air gap to meet the Water Regulation requirements. The arrangement of water control and overflow pipes will ensure that the air gap is maintained. The rainwater level control connects to a control unit that operates the submersible pump, so that water is drawn from the main storage tank when required. At times when there is not enough rainwater to meet the demand, fresh water is introduced into the system via the wholesome water inlet, which is controlled by means of the wholesome water level control.

Direct distribution system

In direct distribution systems, the control unit pumps rainwater directly to the outlets on demand. At times of low rainwater availability, the control unit will provide water from the wholesome supply to the outlets. The backflow prevention methods to meet the requirements of the Water Regulations will be incorporated into the control unit. This type of system uses more energy than the indirect distribution system.

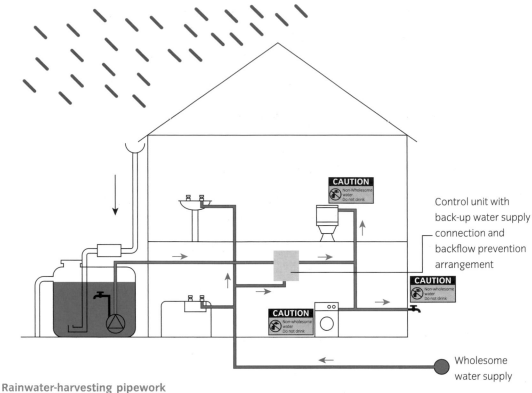

Control unit with back-up water supply connection and backflow prevention arrangement

CAUTION
Non-Wholesome water
Do not drink

CAUTION
Non-wholesome water
Do not drink

CAUTION
Non-wholesome water
Do not drink

Wholesome water supply

Rainwater-harvesting pipework

Location and building requirements

When deciding on the suitability of a location for the installation of a water-harvesting system, the following points will need to be taken into account.

- Is there suitable supply of rainwater to meet the demand? This is determined by finding the rainfall available and the level of water use by the occupants.

- A suitable supply of wholesome water will be required to provide back-up at times of drought.

- For above-ground storage tanks, the chosen location must avoid the risk of freezing, or the warming effects of sunlight, which may encourage algal growth.

- For below-ground tanks, consideration will need to be given to access for excavation equipment.

Planning permission

In principle, planning permission is not normally required for the installation of a rainwater-harvesting system if it does not alter the outside appearance of the property. It is, however, always worth enquiring, especially if the system is installed above ground, the building is in a designated area or the building is listed.

Compliance with building regulations

The following building regulations will apply.

Part	Title	Relevance
A	Structural safety	Where the components affect the loadings placed on the structure of the building or excavations are in close proximity to the building
B	Fire safety	Where holes for pipework reduce the fire-resisting integrity of the structure
C	Resistance from contaminants and moisture	Where holes for pipework reduce the moisture-resisting integrity of the structure, for example, pipework passing through vapour barriers
E	Resistance to sound	Where holes for pipework reduce the sound-resisting integrity of the structure
G	Sanitation, hot-water safety, and water efficiency	Water efficiency
H	Drainage and waste disposal	Where gutters and rainwater pipes are connected to the system
P	Electrical safety	Installation of supply and control wiring for the system

Other regulatory requirements

- The Water Supply (Water Fittings) Regulations 1999 apply to rainwater-harvesting systems. The key area of concern will be the avoidance of cross-contamination between rainwater and wholesome water. This is known as 'backflow prevention' and, as rainwater is classified as a category 5 risk, the usual method of providing backflow prevention is with a type AA air gap between the wholesome water and the rainwater.

Any pipework used to supply outlets with the rainwater will need to be labelled to distinguish it from wholesome water. Outlets will also need to be labelled to indicate that the water supplied is not suitable for drinking.

Pipework labels

- Wiring Regulations BS 7671 will apply to the installation of supplies and control systems for the rainwater-harvesting system.
- BS8515:2009 Rainwater Harvesting Systems – Code of Practice

Advantages

- There is a reduction in use of wholesome water.
- Water bills are reduced if the supply is metered.
- Water does not require any further treatment before use.
- The system is less complicated than greywater re-use systems.

Disadvantages

- The quantity of available water is limited by roof area. It may not meet the demand in dry periods.
- Initial costs are high.
- A water meter should be fitted.

Greywater re-use

Greywater is the waste water from baths, showers, hand basins, kitchen sinks and washing machines. It gets its name from its cloudy, grey appearance. Capturing and re-using the water for permitted uses reduces the consumption of wholesome (drinking) water.

Greywater collected from wash basins, showers and baths will often be contaminated with human intestinal bacteria and viruses, as well as organic material such as skin particles and hair. As well as these contaminants, it will also contain soap, detergents and cosmetic products, which are ideal nutrients for bacteria growth. Add to this the relatively high temperature of the greywater and the ideal conditions exist to encourage the growth of bacteria.

For these reasons, untreated greywater cannot be stored for more than a few hours. The less polluted water from wash basins, showers and baths is usually used in greywater re-use systems. This is known as 'bathroom greywater'. Where a greater supply of greywater is required, washing-machine waste water is collected.

KEY POINT

Whilst kitchen sink waste is classified as greywater, it is not usually collected and recycled. This is because the FOGs (fats, oils and greases) contained within it will emulsify as the water cools and will not be kind to the filtration systems.

ASSESSMENT GUIDANCE

Combined systems divert the surface water (rainwater) into the main drainage system. This can lead to flooding of premises in times of heavy rain.

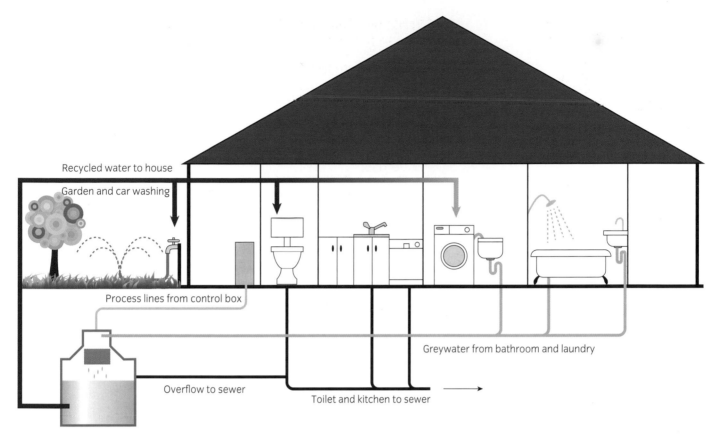

Recycled water to house

Garden and car washing

Process lines from control box

Greywater from bathroom and laundry

Overflow to sewer

Toilet and kitchen to sewer

Greywater re-use system

Greywater is classified as fluid category 5 risk (the highest) under the Water Supply (Water Fittings) Regulations 1999. Greywater can pose a serious health risk, due to its potential pathogen content. Untreated greywater deteriorates rapidly when stored, so all systems that store greywater will need to incorporate an appropriate level of treatment.

If greywater is filtered and stored correctly then it can be used for:

- flushing toilets
- car washing
- garden watering
- washing clothes (after additional processing).

Greywater cannot be used for:

- drinking water
- washing of dishes
- food preparation or washing of food
- personal hygiene – washing, bathing or showering.

SUGGESTED ACTIVITY

Currently, greywater is normally discharged directly into the drainage system of the house and then to the main drainage system. Think of the changes that would have to be made to direct all the greywater to a central collection point.

Working principles

Several types of greywater re-use systems exist but, apart from the direct re-use system, they all have similar common features:

- a tank for storage of the treated water
- a pump
- a distribution system for moving the water from storage to where it is to be used
- some form of treatment.

Direct re-use

Greywater is collected from appliances and directly re-used, without treatment or storage, and can be used for such things as watering the garden. Even so, greywater is not considered suitable for watering fruit or vegetable crops.

Short retention system

Greywater from baths and showers is collected in a cleaning tank, where it is treated, by means such as surface skimming, to remove debris such as soap, hair and foam. Heavier particles are allowed to settle to the bottom, where they are flushed away as waste. The remaining water is then transferred to a storage tank, ready for use. The storage tanks are usually relatively small, at around 100 litres, which is enough for 18–20 toilet flushes. If the water is not used within a short time, generally 24 hours, the stored greywater is purged, the system is cleaned and a small amount of flush water is introduced to allow toilet flushing. This avoids the greywater deteriorating, and beginning to smell, at times when the premises are unoccupied for a lengthy period of time. This type of system can result in water savings of 30%.

This type of system would be ideal for installation in a new-build project but would be more difficult to retrofit. It is usually fitted in the same room as the source of greywater.

Physical and chemical system

This system uses a filter to remove debris from the collected water (physical cleaning). After the greywater has been filtered, chemical disinfectants such as chlorine or bromine are added, to inhibit bacterial growth during storage.

Biomechanical system

This type of system is the most advanced of the greywater re-use systems. It uses both biological and physical methods to treat the collected greywater. An example of such a system is the German

Ecoplay greywater system

1 Cleaning tank
2 Storage tank

system AquaCycle® 900, which comprises an indoor unit about the size of a large refrigerator.

Greywater enters the system and passes through the filtering unit (1), where particles such as hair and textiles debris are filtered out. The filtering unit is electronically controlled to provide automatic flushing of the filter.

Water enters the main recycling chamber (2), where organic matter is decomposed by bio-cultures. The water remains in this chamber for 3hours before being pumped to the secondary recycling chamber (3), for further biological treatment. Biological sediment settles to the bottom of each chamber (4), where it is sucked out and transferred to a drain.

After a further period of 3 hours, the water passes through a UV filter (5), to the final storage chamber (7), where it is ready for use. When there is demand for the treated water this is pumped (8) to the point of demand. At times when treated water availability is low, fresh water (6) can be introduced to the system.

Water from this unit can be used for washing clothes as well as the other uses previously stated.

SUGGESTED ACTIVITY

Go to Water Works UK www.wwuk.co.uk or alternative websites and see the range of systems available.

AquaCycle® 900 system

Biological system

This type of system uses some of the principles employed by sewage treatment works. In this case, bacterial growth is encouraged rather than inhibited, by the introduction of oxygen to the waste water. Oxygen can be introduced by means of pumps pushing air through the storage tanks. Bacteria then 'digest' the organic matter contained within the greywater.

A more 'natural' method of oxygenating the water is by the use of reed beds. In nature, reeds, which thrive in waterlogged conditions, transfer oxygen to their roots. The greywater is allowed to infiltrate reed beds. The added oxygen and naturally occurring bacteria will remove any organic matter contained in the waste water. The disadvantages of using reed beds are the land area required for the reed beds and the expertise required to maintain them.

Alternative Water Solutions produces a system based on these principles, but on a smaller scale, called the Green Roof Recycling System (GROW). This uses a system of tiered gravel-filled troughs planted with native plants, the roots of which can perform the same function of filtering as a reed bed would.

GROW system

Location and building requirements

When considering the installation of a greywater re-use system, the following location or building requirements should be taken into account.

- There needs to be a suitable supply of greywater to meet the demand. Premises with a low volume of greywater are not suitable.

- Suitability of the location and availability of space to store enough greywater to meet the demand of the premises must be assessed.

- Storage tanks need to be located away from heat, including direct sunshine, to avoid the growth of algae. They need to be located so they are not subject to freezing in cold weather. There needs to be a wholesome water supply.

- Where greywater tanks are retrofitted, access for excavation equipment will need to be considered.

- A water meter will need to be fitted on the water supply to maximise the benefits.

Planning permission

In principle, planning permission is not normally required for the installation of a greywater re-use system, if the system does not alter the outside appearance of the property. It is, however, always worth enquiring, especially if the system is installed above ground, the building is in a designated area or the building is listed.

If a building is required to house the greywater storage system, then a planning application will need to be submitted.

Compliance with building regulations

The following building regulations will apply.

Part	Title	Relevance
A	Structural safety	Where the components affect the loadings placed on the structure of the building or excavations are in close proximity to the building
B	Fire safety	Where holes for pipework reduce the fire-resisting integrity of the structure
C	Resistance to contaminants and moisture	Where holes for pipework reduce the moisture-resisting integrity of the structure, for example, pipework passing through vapour barriers
E	Resistance to sound	Where holes for pipework reduce the sound-resisting integrity of the structure
G	Sanitation, hot-water safety, and water efficiency	Water efficiency
H	Drainage and waste disposal	Where waste pipes are connected to the system
P	Electrical safety	Installation of supply and control wiring for the system

Other regulatory requirements

- The Water Supply (Water Fittings) Regulations 1999 apply to greywater recycling installations. The key area of concern will be the avoidance of cross-contamination between greywater and wholesome water. This is known as 'backflow prevention' and, as greywater is classified as a category 5 risk, the usual method of providing backflow prevention is with an air gap between the wholesome water and the greywater.

Greywater warning label

■ Any pipework used to supply outlets with the treated greywater will need to be labelled to distinguish it from pipework for wholesome water. Outlets will need to be clearly identified by means such as labelling. Outlets will also need to be labelled to indicate that the water supplied is not suitable for drinking.

■ The local water authority must be notified when a greywater re-use system is to be installed.

■ Wiring Regulations BS 7671 will apply to the installation of supplies and control systems for the greywater re-use system.

Advantages

■ There will be a reduction in water bills if the supply is metered.

■ It reduces demands on the wholesome water supply.

■ A wide range of system options exists.

■ It has the potential to provide more re-usable water than a rainwater harvesting system.

Disadvantages

■ There are long payback periods.

■ It can be difficult to integrate into an existing system.

■ Only certain types of appliance or outlet can be connected. This causes additional plumbing work.

■ Cross-contamination can be a problem.

■ A water meter will need to be fitted to make maximum gains.

■ The need for filtering and pumping may actually increase rather than decrease the carbon footprint.

Test your knowledge questions

1 A system that uses a fluid to capture heat from the Sun is known as:
   a) photovoltaic
   b) solarvoltaic
   c) solar thermal
   d) solar thermostatic.

2 A buffer tank can be used for storing:
   a) cold water
   b) rainwater
   c) electricity
   d) hot water.

3 A system that uses sunlight to generate electricity is known as:
   a) photovoltaic
   b) photosynthetics
   c) solar thermal
   d) solar chemical.

4 Excess electricity can be sold back to the grid using:
   a) an off-peak tariff
   b) a feed-in tariff
   c) a fccd out tariff
   d) a daytime tariff.

5 One disadvantage of ground source heat pumps using horizontal pipework is:
   a) the need for unshaded ground
   b) the need for large ground area
   c) the need for deep boreholes
   d) that it is less than 100% efficient.

6 The air source heat pump works on the principle of:
   a) cold exchange
   b) electromagnetic induction
   c) the refrigeration cycle
   d) the Otto cycle.

7 Which one of the following fuels is commonly used in biomass systems?
   a) Coal.
   b) Oil.
   c) Shale.
   d) Wood chips.

8 Rainwater harvesting should not be used for:
   a) car washing
   b) drinking water
   c) toilet flushing
   d) garden watering.

9 A micro-hydro system relies upon:
   a) a constant wind speed
   b) reliable water flow
   c) a deep well
   d) sufficient sunlight.

10 A micro-wind system generates d.c. that is changed into a.c. by the use of:
   a) a rectifier
   b) an inverter
   c) a commutator
   d) slip rings.

11 Which orientation is NOT suitable for mounting a photovoltaic panel?
   a) North.
   b) South.
   c) East.
   d) West.

12 The maximum distance a solar thermal panel may extend above the roof surface is:
   a) 50mm
   b) 150mm
   c) 200mm
   d) 250mm.

13 A ground source heat pump installed on a building with limited grounds will require:

a) deep boreholes

b) long pipe trenches

c) an extra low-voltage supply

d) shallow trenches.

14 An air source heat pump has an electrical input of 2.5kW and output of 7.5kW. The coefficient of performance will be:

a) 0.33

b) 3

c) 5

d) 0.

15 The position of the biomass flue should be:

a) at the front of the building

b) at the side or rear of a building

c) roof-mounted at least 2m above roof level

d) roof-mounted at least 3m above roof level.

16 An 'on-grid' generating system is one where the system:

a) is connected in a 4 × 4 grid

b) is connected to a battery bank only

c) can be moved within an installed grid

d) is connected to the national grid.

17 Installation of greywater systems would depend upon:

a) total black and greywater available

b) adequate greywater availability

c) sufficient rainwater only

d) direct mixing with wholesome water.

18 One disadvantage associated with biomass systems is:

a) water produced by burning fuel

b) inability to control heat output

c) need for large storage area

d) large amount of ash produced.

19 One disadvantage of photovoltaic systems is:

a) they cannot be installed facing south

b) there is little or no output at night

c) they will always require roof strengthening

d) they cannot be exposed to snow or ice.

20 The minimum recommended height of wind turbine blades from the ground is:

a) 2m

b) 3m

c) 4m

d) 5m.

Assessment checklist

What you now know (Learning Outcome)	What you can now do (Assessment criteria)	Where this is found (page number)
1. Know the fundamental working principles of micro-renewable energy and water conservation technologies	1.1 Identify the fundamental working principles for each of the following heat-producing micro-renewable energy technologies:	
	■ solar thermal (hot water)	66–74
	■ ground source heat pump	77–82
	■ air source heat pump	82–86
	■ biomass	86–91
	1.2 Identify the fundamental working principles for each of the following electricity-producing micro-renewable energy technologies:	
	■ solar photovoltaic	92–100
	■ micro-wind	100–106
	■ micro-hydro	106–113
	1.3 Identify the fundamental working principles of the following co-generation technologies:	
	■ micro-combined heat and power (heat-led)	114–118
	1.4 Identify the fundamental working principles for each of the following water conservation technologies:	118
	■ rainwater harvesting	120–126
	■ greywater re-use.	126–132
2. Know the fundamental requirements of building location/building features for the potential to install micro-renewable energy and water conservation systems	2.1 Clarify the fundamental requirements for the potential to install a solar water heating system	66–74
	2.2 Clarify the fundamental requirements for the potential to install a solar photovoltaic system	92–100
	2.3 Clarify the fundamental requirements for the potential to install a ground source heat pump system	77–82
	2.4 Clarify the fundamental requirements for the potential to install an air source heat pump system	82–86
	2.5 Clarify the fundamental requirements for the potential to install a biomass system	86–91
	2.6 Clarify the fundamental requirements for the potential to install a micro-wind system	100–106

What you now know (Learning Outcome)	What you can now do (Assessment criteria)	Where this is found (page number)
2. (cntd). Know the fundamental requirements of building location/building features for the potential to install micro-renewable energy and water conservation systems	2.7 Clarify the fundamental requirements for the potential to install a micro-hydro system	106–113
	2.8 Clarify the fundamental requirements for the potential to install a micro-combined heat and power (heat-led) system	114–118
	2.9 Clarify the fundamental requirements for the potential to install a rainwater harvesting/greywater re-use system.	120–132
3. Know the fundamental regulatory requirements relating to micro-renewable energy and water conservation technologies	3.1 Confirm what would be typically classified as 'permitted development' under town and country planning regulations in relation to the deployment of the following technologies: ■ solar thermal (hot water) ■ solar photovoltaic ■ ground source heat pump ■ air source heat pump ■ micro-wind ■ biomass ■ micro-hydro ■ micro-combined heat and power (heat-led) ■ rainwater harvesting ■ greywater re-use	63–66
	3.2 Confirm which sections of the current building regulations/building standards apply in relation to the deployment of the following technologies: ■ solar thermal (hot water) ■ solar photovoltaic ■ ground source heat pump ■ air source heat pump ■ micro-wind ■ biomass ■ micro-hydro ■ micro-combined heat and power (heat-led) ■ rainwater harvesting ■ greywater re-use.	63–66

What you now know (Learning Outcome)	What you can now do (Assessment criteria)	Where this is found (page number)
4. Know the typical advantages and disadvantages associated with micro-renewable energy and water conservation technologies	4.1 Identify typical advantages associated with each of the following technologies:	
	■ solar thermal (hot water)	66–74
	■ solar photovoltaic	92–100
	■ ground source heat pump	77–82
	■ air source heat pump	82–86
	■ micro-wind	100–106
	■ biomass	86–91
	■ micro-hydro	106–113
	■ micro-combined heat and power (heat-led)	114–118
	■ rainwater harvesting	120–126
	■ greywater re-use	126
	4.2 Identify typical disadvantages associated with each of the following technologies:	
	■ solar thermal (hot water)	66–74
	■ solar photovoltaic	92–100
	■ ground source heat pump	77–82
	■ air source heat pump	82–86
	■ micro-wind	100–106
	■ biomass	106–113
	■ micro-hydro	
	■ micro-combined heat and power (heat-led)	114–118
	■ rainwater harvesting	120–126
	■ greywater re-use.	126–132

Assessment guidance

- The assessment for this unit is by a multiple-choice exam.
- You will find sample questions in this book and on SmartScreen.
- Make sure you go over these questions in your own time.
- Spend time on revision in the run-up to the exam.
- Ensure you know how the e-volve system works. Ask for a demonstration if you are not sure.
- It is better to have time left over at the end rather than have to rush to finish.
- Make sure you read every question carefully.
- If you need paper ask the invigilator to provide some.
- Make sure you have a scientific (non-programmable) calculator.
- Do not take any paperwork into the exam with you.
- Make sure your mobile phone is switched off during the exam. You may be asked to give it to the invigilator.

UNIT 303
Complex cold water systems

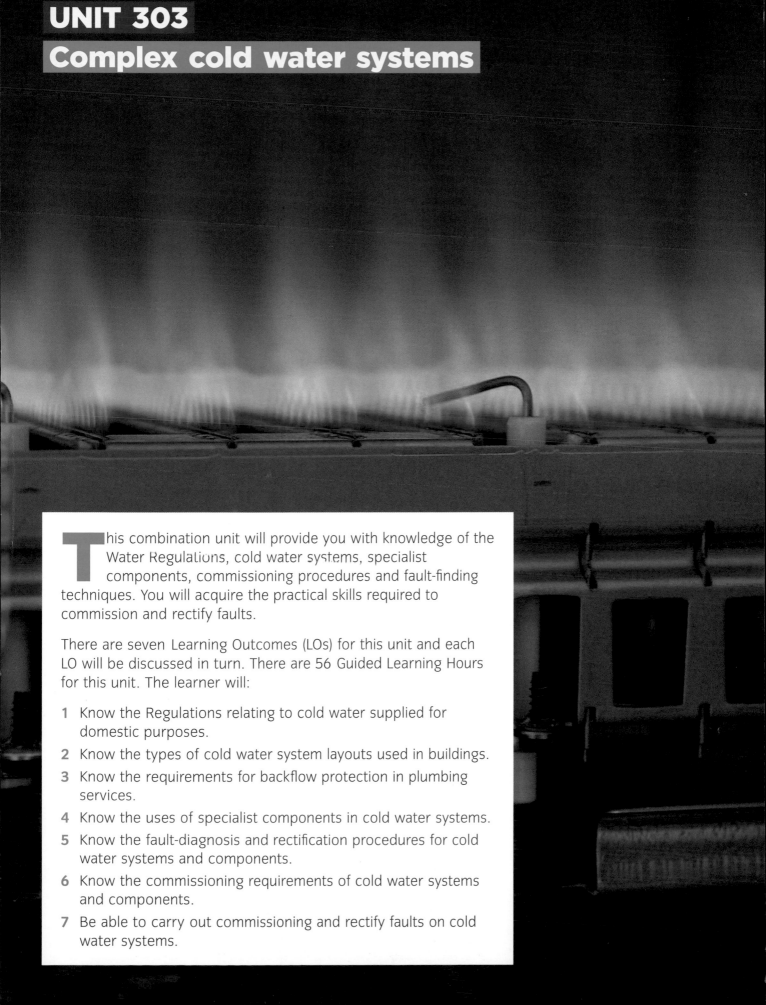

This combination unit will provide you with knowledge of the Water Regulations, cold water systems, specialist components, commissioning procedures and fault-finding techniques. You will acquire the practical skills required to commission and rectify faults.

There are seven Learning Outcomes (LOs) for this unit and each LO will be discussed in turn. There are 56 Guided Learning Hours for this unit. The learner will:

1 Know the Regulations relating to cold water supplied for domestic purposes.
2 Know the types of cold water system layouts used in buildings.
3 Know the requirements for backflow protection in plumbing services.
4 Know the uses of specialist components in cold water systems.
5 Know the fault-diagnosis and rectification procedures for cold water systems and components.
6 Know the commissioning requirements of cold water systems and components.
7 Be able to carry out commissioning and rectify faults on cold water systems.

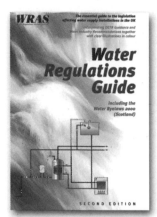

Water Regulations Guide

SmartScreen Unit 303
Presentation 1 and Handout 1

Regulations

Rules, procedures and administrative codes set by authorities or governmental agencies to achieve an objective. They are legally enforceable and must be followed to avoid prosecution.

Know the Regulations relating to cold water supplied for domestic purposes (LO1)

There are five assessment criteria for this Outcome:

1 Describe the purpose of the Water Supply (Water Fittings) Regulations 1999.

2 Describe how the Water Regulations impact on the installation and use of water systems.

3 Explain the requirements for advance notification of work.

4 Differentiate between the installer and user responsibilities under the Water Regulations.

5 Define the legal requirements for drawing water from a water undertaker's water main using a pump or booster.

	Range
Purpose	To protect against waste, misuse, undue consumption, contamination of drinking water and erroneous measurement of water
Water systems	Supplied by a water undertaker, private source
Requirements	Refer to the Water Regulations Advisory Scheme (WRAS) on notifiable works, notify the water undertaker
Legal requirements	Drawing more than 12 litres per minute

The water industry in England and Wales is regulated by the Water Industry Act 1991, as amended by the Water Act 2003 (Commencement No. 11) Order 2012, the Private Water Supplies **Regulations** 2009 and the Water Supply (Water Fittings) Regulations 1999. These three documents have specific roles to play within the plumbing industry. We will concentrate on the Water Supply (Water Fittings) Regulations as they apply to 95% of the water supplied to buildings in the UK.

The purpose of the Water Regulations

Before 1999, each water authority had its own set of water byelaws, based on the 101 Model Water Byelaws issued by the UK government in 1986. The problem was that each water undertaker had local variations, which caused much confusion, as there was no 'common' standard throughout the UK.

On 1 July 1999, the office of the Deputy Prime Minister issued the first ever Water Regulations to be enforced in the UK. They are known as the Water Supply (Water Fittings) Regulations 1999, and they offer a common standard of practice throughout the UK. They are linked to a British Standard, **BS EN 806 – Specification for installations inside buildings conveying water for human consumption**.

The Water Supply (Water Fittings) Regulations 1999 were made under Section 74 of the Water Industry Act 1991 to ensure that the plumbing systems we install and maintain prevent the following:

- contamination of water
- wastage of water
- misuse of water
- undue consumption of water
- **erroneous** metering of water.

An important factor is that these Regulations only cover installations in which the water is supplied from a water undertaker's water main, and are enforced by the water undertaker in your area. They are not enforceable where the water is supplied from a private water source. Private water supplies are regulated by the Private Water Supplies Regulations 2009 and are not covered here.

A free copy of the Water Supply (Water Fittings) Regulations 1999 can be downloaded from www.opsi.gov.uk/si/si1999/19991148.htm.

> **KEY POINT**
>
> **BS EN 806 – Specification for installations inside buildings conveying water for human consumption**
>
> This came into force in 2011. It is divided into five parts:
> 1. General recommendations
> 2. Design
> 3. Pipe sizing
> 4. Installation
> 5. Operation and maintenance.
> Linked with this is BS EN 8558 – Guide to the design, installation, testing and maintenance of services supplying water for domestic use within buildings and their curtilages. These two publications replace the old British Standard, BS 6700 – Design, installation, testing and maintenance of services supplying water for domestic use within buildings and their curtilages. However, parts of BS 6700 that are not covered in either BS EN 806 or BS EN 8558 will be retained. For more information, see www.wras.co.uk.

Erroneous

Incorrect.

The Water Regulations' impact on the installation and use of water systems

For a plumbing system to comply with the Water Regulations, it must be designed and installed in such a way that the Regulations are satisfied. There are several points to note.

- Plumbing materials and fittings must be of a suitable standard. It is not illegal to sell unsuitable materials but it is illegal to install such materials and fittings. Materials and fittings must be listed in the Water Fittings and Materials Directory.

- Plumbing systems must be designed, installed and maintained to ensure that:
  - the quality of the water is not compromised
  - the safety of the user is maintained
  - the system pipework, fittings and components are easily accessible
  - leaks are easily detectable
  - the system can be easily maintained

- protection against frost damage is adequate.

Systems should be designed and installed in accordance with BS EN 806 – Specification for installations inside buildings conveying water for human consumption (see page 141).

Plumbers should consider applying to join an Approved Plumber Scheme. These will be discussed later in this unit.

The requirements for advance notification of work

This part of the unit investigates the requirements for notification and consent with regard to cold water installation work. There are four questions to consider.

- Who needs to notify?
- Which types of installations require notification and consent?
- Why do you need to notify?
- Who should you notify?

The Water Supply (Water Fittings) Regulations 1999 requirements

Regulation 5 of the Water Supply (Water Fittings) Regulations 1999 requires that the water undertaker is notified before work starts on most types of plumbing installations. Anyone installing or using the installation without the water undertaker's written consent could be committing a criminal offence.

Notification and consent are also legally required where water fittings are to be installed on any water or waste water plants. This requirement applies irrespective of whether the plant is owned/operated by the same organisation as the enforcing water undertaker.

Consent is necessary for the installation of fittings in new buildings and dwellings, for extensions and alterations to water systems in existing non-domestic premises, where there is a material change of use of a building and for the installation of certain specified items. These include:

- a bidet with an ascending spray or flexible hose
- a bath larger than 230 litres (measured to the centre of the overflow)*
- a shower unit of a type specified by the Regulator (but none is currently specified)
- a pump or booster drawing more than 12 litres per minute

Flexible bidet hose

- a **reverse osmosis** unit

- a water treatment unit producing a waste water discharge or requiring water for regeneration or cleaning

- a reduced pressure zone (RPZ) valve or other mechanical device for protection against backflow in fluid category 4 or 5

- a garden watering system, unless it is designed to be hand operated*

- any water system laid outside a building and either less than 750mm or more than 1,350mm below ground level

- construction of an automatically replenished pond or swimming pool of more than 10,000 litres.*

Reverse osmosis

A water filtration process whereby a membrane filters unwanted chemicals, particles and contaminants out of the water.

When notifying the water undertaker, the following information must be sent:

- the name and address of the person giving notice and, if different, of the person to whom the consent should be sent

- a description of the proposed work and any related change of use of the premises

- the location of the premises and its use or intended use

- except for items marked (*) above, a plan of that part of the premises which relates to the proposed work and a diagram showing the pipework and fittings to be installed

- the plumbing contractor's name and address, if an approved plumber is to do the work.

Consent cannot be withheld unless there are reasonable grounds to do so, and may be granted subject to conditions, which must be followed. If written approval is not given within 10 working days it can be assumed that consent has been granted, but this does not alter the obligation on the installer and the owner or occupier to ensure that the Regulations have been complied with.

Notification and consent requirements: the Building Regulations Document G requirements

On 6 April 2010 an updated and extended version of Approved Document G (Part G) was implemented, bringing a number of new areas under Building Regulations control.

As a result the Local Authority Building Control must be notified of the installation of the systems mentioned in the bullet list on the next page. In general terms, notification should take place before the work starts and definitely within five working days of the work being completed.

Building Regulations Part G is broken down into six parts.

- **G1: Cold water supply** – there are new requirements for the supply of wholesome water for the purposes of drinking, washing or food preparation, and also for the provision of water of a suitable quality to sanitary conveniences fitted with a flushing device.

- **G2: Water efficiency** – G2 and Regulation 17K of the Building Regulations 2000 set out new requirements for water efficiency in dwellings.

- **G3: Hot water supply and systems** – this sets out enhanced and amended provisions for hot water supply and safety, applying safety provisions to all types of hot water systems and a new provision for scalding prevention.

- **G4: Sanitary conveniences and washing facilities** – this sets out requirements for sanitary conveniences and hand washing facilities.

- **G5: Bathrooms** – this sets out requirements for bathrooms, which apply to dwellings and to buildings containing one or more rooms for residential purposes.

- **G6: Kitchens and food preparation areas** – this contains a new provision requiring sinks to be provided in areas where food is prepared.

A free copy of the Building Regulations Part G can be downloaded from www.planningportal.gov.uk/buildingregulations/approveddocuments/partg/.

Certification of plumbing installations

As a direct result of the introduction of the Water Supply (Water Fittings) Regulations in 1999, water undertakers began to recognise Competent Persons Schemes, Approved Plumber Schemes and self-certification of certain plumbing installations. These are administered and regulated by the local water undertaker.

Under Water Supply Regulation 5, if a person proposes to carry out plumbing work on any building other than a house or dwelling, they can:

- notify their local water undertaker

- or use an approved plumber/contractor who will issue a Water Regulation Compliance Certificate. This certificate confirms that the work has been carried out in accordance with the regulations and that the fittings and materials used meet the strict requirements of the regulations.

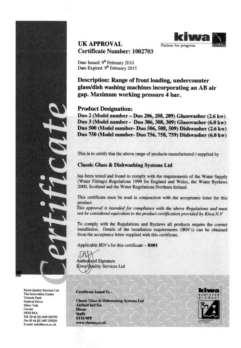

Competent Persons Schemes and Approved Plumber Schemes

A plumber who can prove, via an assessment, a substantial knowledge of the Water Supply (Water Fittings) Regulations 1999 can join a **Competent Persons Scheme** (CPS). The main benefit for members of these schemes is that they can self-certify plumbing supply work (domestic, commercial and industrial) that would otherwise have to wait for water company approval before work could commence.

The first of these schemes to emerge was the Water Industry Approved Plumber Scheme (WIAPS), adopted by most of the water undertakers in England and Wales.

Competent Persons Scheme

Members of CPS must follow certain rules to ensure that their work complies with Building Regulations.

Installer and user responsibilities under the Water Regulations

Plumbers who install plumbing systems and fittings, as well as users, owners and occupiers, have legal responsibilities under the Water Supply (Water Fittings) Regulations 1999 to ensure that any installation and the materials and fittings used comply with the regulations. In most cases, as you have seen, advance notification of the proposed installation must be given, so architects, building developers and plumbers have to abide by the regulations on behalf of their client. These regulations are not retrospective, so systems installed before the regulations came into force can still be used without the need to update them to current standards, even if that installation would not meet the current requirements. However, any alteration or extension completed on an existing installation must comply with the regulations in force at the time of the installation date. Regulation 3 makes this very clear.

1 No person shall:
- install a water fitting to convey or receive water supplied by a water undertaker, or alter, disconnect or use such a water fitting
- cause or permit such a water fitting to be installed, altered, disconnected or used
- in contravention of the following provisions of this Part.

2 No water fitting shall be installed, connected, arranged or used in such a manner that it causes or is likely to cause:
- waste, misuse, undue consumption or contamination of water supplied by a water undertaker
- the erroneous measurement of water supplied by a water undertaker.

3 No water fitting shall be installed, connected, arranged or used which by reason of being damaged, worn or otherwise faulty, causes or is likely to cause:

- waste, misuse, undue consumption or contamination of water supplied by a water undertaker
- the erroneous measurement of water supplied by a water undertaker.

The following points are worth remembering.

- Plumbers should obtain a copy of the regulations, any amendments and guidance notes from Her Majesty's Stationary Office (HMSO) or a copy of the Water Regulations Guide (ISBN-10: 0953970809) from WRAS to ensure that any plumbing work complies with the regulations. They must ensure that any fittings and materials are of a sufficient quality and that any installation is installed in a workmanlike manner to an approved installation standard requirement. Membership of a Competent Persons Scheme is advisable.

- Users, owners and occupiers must ensure that the person employed to undertake the proposed work is aware of the regulations and that any work completed is done so in accordance with the regulations. A Certificate of Compliance must be obtained for the work and retained for future reference by the user, owner or occupier. Regulation 3 also makes it clear that the user, owner or occupier is responsible for ensuring that waste, misuse, undue consumption, contamination or erroneous metering of the water supply does not occur during usage.

- The government requires the water undertaker to enforce the regulations within their area of supply. They will undertake inspections of new and existing installations to check that the regulations are being complied with. Where breaches of the regulations are found, they must be remedied as soon as practicable. Where breaches present a significant health risk, the water supply to the premises may be isolated or, in severe cases, disconnected immediately to protect the health of occupants and/or others fed from the same public supply. The government has deemed that it is a criminal offence to breach the regulations, and offenders – including users, owners and occupiers in cases where the original installer cannot be traced – may face prosecution.

The legal requirements for drawing water from a main using a pump or booster

In some instances, where the water main cannot provide sufficient water pressure or flow rate, a booster pump may be installed directly to the water undertaker's mains cold water supply. Under Regulation 5 of the Water Supply (Water Fittings) Regulations 1999, written consent is required before a booster pump supplying more than 12 litres per minute (0.2 litres per second) can be installed. This is to

boosting systems from a break cistern. Indirect systems are more common, because direct boosting systems are often forbidden by the water undertaker. They often reduce the mains pressure available to other consumers in the locality and can increase the risk of contamination by backflow. However, where insufficient water pressure exists and the demand is below 0.2 litres per second, drinking water may be boosted directly from the supply pipe provided that the water undertaker agrees. With indirect systems, a series of float switches in the break cistern start and stop the pumps depending on the water levels in the cistern.

Boosting pumps can create excessive aeration of the water, which, although causing no deterioration of water quality, may cause concern to the consumer because of the opaque, milky appearance of the water. There are several common examples of these systems:

- direct boosting systems
- direct boosting to a drinking water header and duplicate storage cisterns
- indirect boosting to a storage cistern
- indirect boosting with a pressure vessel.

Direct boosting systems

Where permission from the water undertaker has been granted, pumps can be directly fitted to the incoming supply pipe to enable the head of pressure to be increased.

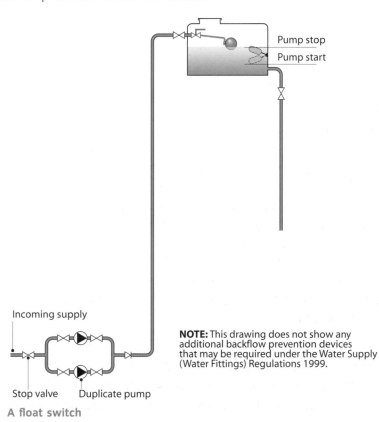

Incoming supply

Pump stop
Pump start

NOTE: This drawing does not show any additional backflow prevention devices that may be required under the Water Supply (Water Fittings) Regulations 1999.

Stop valve   Duplicate pump

A float switch

A float switch or some equally effective device situated inside the high-level cistern controls the pumps. The pumps either switch on or off depending on the water level in the cistern. The pumps are activated when the water drops to a depth normally equal to about half cistern capacity and switch off again when the water level reaches a depth approximately 50mm below the shut-off level of the float-operated valve (FOV).

If the cistern is to be used for drinking water, it must be the protected type.

Direct boosting to a drinking water header and duplicate storage cisterns

This system is mainly used for large and multi-storey installations. With this system, the cisterns at high level are for supplying non-drinking water only. A drinking water header sited on the boosted supply pipe provides limited storage of 5–7 litres of drinking water to sinks in each dwelling when the pump is not running. Excessive pressure should be avoided, as this can lead to an increase in the wastage of water at the sink taps, along with the nuisance of excessive splashing.

A pipeline switch on the header bypass starts the pumps when the water level falls to a **predetermined** level. The pumps can be time controlled or activated to shut down by a pressure switch. When filling the cisterns, the pumps should shut down when the water levels in the cisterns are approximately 50mm below the shut-off level of the float-operated valve.

Secondary backflow devices may be required at the drinking water outlets on each floor.

SmartScreen Unit 303
Handout 4

Predetermined

Decided in advance.

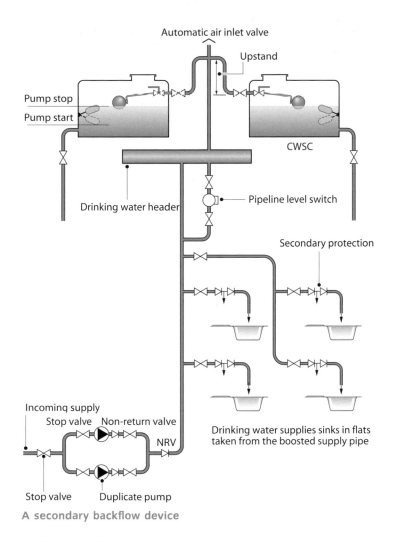

Automatic air inlet valve

Upstand

Pump stop

Pump start

CWSC

Drinking water header

Pipeline level switch

Secondary protection

Incoming supply

Stop valve   Non-return valve

NRV

Drinking water supplies sinks in flats
taken from the boosted supply pipe

Stop valve   Duplicate pump

A secondary backflow device

Indirect boosting to a storage cistern

This system incorporates a break cistern to store the water before it is pumped via a boosting pump (known as a booster set) to a storage cistern at high level. The pumps should be fitted to the outlet of the break cistern. The capacity of the break cistern needs careful consideration and will depend on the total water storage requirements and the cistern's location within the building, but it should not be less than 15 minutes of the pump's maximum output. However, the cistern must not be oversized either, as this may result in water **stagnation** inside it.

The water level in the storage cistern (or cisterns) is usually controlled by means of water level switches which control the pumps. When the water drops to a predetermined level the pumps start filling the storage cisterns. The pumps are then switched off when the water level reaches a point about 50mm from the shut-off level of the FOV. A water level switch should also be positioned in the break cistern to automatically shut off the pumps if the water level drops to within 225mm of the suction connection near the bottom of the break cistern. This is simply to ensure that the pumps do not run dry.

SmartScreen Unit 303
Handout 5

Stagnation
Where water has stopped flowing.

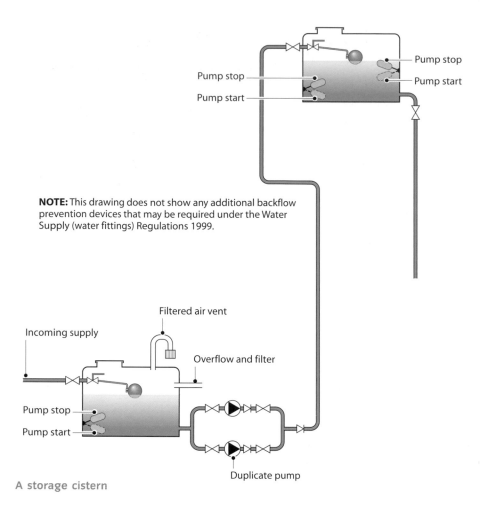

Pump stop

Pump start

Pump stop

Pump start

NOTE: This drawing does not show any additional backflow prevention devices that may be required under the Water Supply (water fittings) Regulations 1999.

Filtered air vent

Incoming supply

Overflow and filter

Pump stop

Pump start

Duplicate pump

A storage cistern

Indirect boosting with a pressure vessel

SmartScreen Unit 303

Handout 7

This rather complicated system is mainly used in buildings where a number of storage cisterns are fed at various floor levels, making it impractical to control pumps by water level switches. It uses a pneumatic pressure vessel to maintain the pressure boost to the higher levels of the building.

The pneumatic pressure vessel comprises a small water reservoir with a cushion of compressed air. The water pumps and the compressed air pump operate intermittently. The water pumps replenish the water level inside the vessel, which then maintains the system pressure with aid of the compressed air cushion. Because the system may be supplying drinking water, the water capacity is kept purposely low to ensure a rapid and regular turnover of water. The compressed air must be filtered to ensure that dust and insects are eliminated.

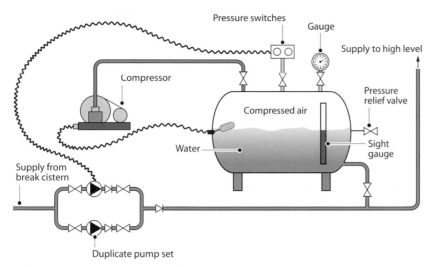

An auto-pneumatic pressure vessel

Normally the controls, including the pressure vessel, pumps, air compressor and control equipment, are purchased as a package, although self-assembled booster sets are also available.

As shown in the diagram, some of the floors below the limit of the mains cold water supply pressure are supplied un-boosted directly from the cold water main, with the floors above the mains pressure limit being supplied via the break cistern and booster set. Drinking water supplies must be from a protected cistern.

A typical booster set with pressure vessel and control boards

NOTE: This drawing does not show any additional backflow prevention devices that may be required under the Water Supply (Water Fittings) Regulations 1999.

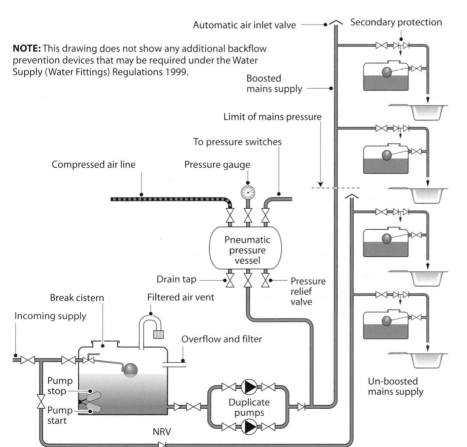

A protected cistern

The requirements for large-scale cisterns

The installation of large-scale cisterns differs somewhat from the cisterns you saw at Level 2. Large cisterns must be installed in accordance with the Water Supply (Water Fittings) Regulations 1999 (and the Scottish Water Byelaws 2004). Regulation 5 states that the water undertaker must be notified before the installation of large cisterns begins, and it is important to remember that the correct **backflow protection** must be present in relation to the fluid category of the contents of the cistern.

This section of the unit looks at the general requirements of large-scale cisterns.

Materials for large-scale cisterns

Large-scale cisterns can be made from several materials and may be either in one piece or sectional. Sectional cisterns are constructed, usually on site, from 1m^2 sections, which are bolted together and can be made to suit any capacity and tailored to fit any space. Sectional cisterns may be internally or externally flanged and are bolted together with stainless steel bolts.

The main materials are as follows.

- For one-piece cisterns:
  - glass reinforced plastic (GRP) – BS EN 13280:2001
  - plastic – BS 4213:2004 and BS EN 12573-1:2000 (polypropylene/PP, polyethylene/PE, polyvinyl chloride/PVC).

A selection of one-piece GRP cisterns

Backflow protection

Prevention of contamination of water through backflow or back siphonage.

- For sectional cisterns:
  - glass reinforced plastic (GRP) BS EN 13280:2001
  - steel to BS 1564:1975, with protection against corrosion and subsequent water contamination in the form of paint listed in the Water Fittings and Materials Directory, glass coating, galvanising, rubber lining, aluminium and rubber lining.

A sectional cistern

Overflow and warning pipe requirements of large-scale cisterns

Overflows for large-scale cisterns are quite different from those fitted to cisterns for domestic purposes. The objective is the same – to warn that the FOV is malfunctioning and to remove water that may otherwise damage the premises – but with larger cisterns, the potential for water wastage and water damage is far greater. Therefore, the layout is different.

The overflow/warning pipe on large-scale cisterns must:

- contain a vermin screen to prevent the ingress of insects and vermin
- be capable of draining the maximum inlet flow without compromising the inlet air gap
- contain an air break before connection to a drain
- not be of such a length that it will restrict the flow of water, causing the air gap to be compromised
- discharge in a visible, conspicuous position.

The warning pipe invert needs to be located a minimum of 25mm above the maximum water level of the cistern and the air gap should not be less than 20mm or twice the internal diameter of the inlet pipe, whichever is greater.

The general features of large-scale cisterns are as follows.

■ Cisterns with an actual capacity of 1,000–5,000 litres:

● The discharge level of the inlet device must be positioned at least twice the diameter of the inlet pipe above the top of the overflow pipe.

● The overflow pipe invert must be located at least 25mm above the invert of the warning pipe (or warning level if an alternative warning device is fitted).

● The warning pipe invert must be located at least 25mm above the water level in the cistern and must be at least 25mm in diameter.

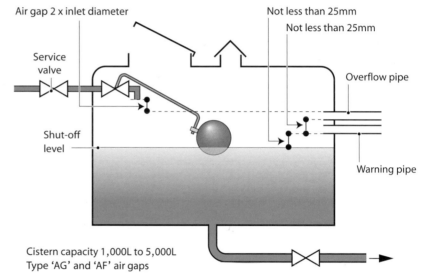

A cistern with a capacity of 1,000–5,000 litres

■ Cisterns with an actual capacity greater than 5,000 litres:

● The discharge level of the inlet device must be positioned at least twice the diameter of the inlet pipe above the top of the overflow pipe.

● The overflow pipe invert must be located at least 25mm above the invert of the warning pipe (or warning level if an alternative warning device is fitted).

● The warning pipe invert must be located at least 25mm above the water level in the cistern and must be at least 25mm in diameter.

● Alternatively, the warning pipe may be discarded provided a water level indicator with an audible or visual alarm is installed that operates when the water level reaches 25mm below the invert of the overflow pipe.

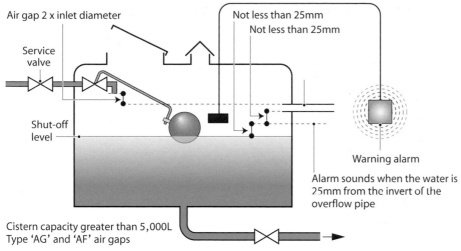

Air gap 2 x inlet diameter

Service valve

Shut-off level

Not less than 25mm
Not less than 25mm

Warning alarm

Alarm sounds when the water is 25mm from the invert of the overflow pipe

Cistern capacity greater than 5,000L
Type 'AG' and 'AF' air gaps

A cistern with a capacity greater than 5,000 litres

In both cases, the size of the overflow pipe will depend on the type of air gap incorporated into the cistern (air gaps and backflow protection are covered later in this unit) and this will depend on the fluid category of the cistern contents.

Remember the following.

- If a type AG air gap (fluid category 3) is fitted, the overflow diameter shall be a minimum of twice the inlet diameter.

- If a type AF air gap (fluid category 4) is fitted, the minimum cross-sectional area of the overflow pipe must be, throughout its entire length, four times the cross-sectional area of the inlet pipe.

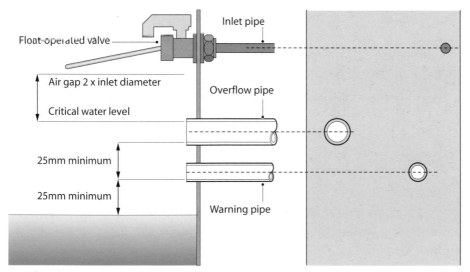

Float-operated valve

Inlet pipe

Air gap 2 x inlet diameter

Critical water level

Overflow pipe

25mm minimum

25mm minimum

Warning pipe

Relative positions of inlet, warning and overflow pipes

For all cisterns greater than 1,000 litres, the invert of the overflow must not be less than 50mm above the working level of the cistern.

Methods of filling large-scale cisterns

Section 7, Schedule 2: Paragraph 16 of the Water Supply (Water Fittings) Regulations states that:

'Every pipe supplying water connected to a storage cistern shall be fitted with an effective adjustable valve capable of shutting off the inflow of water at a suitable level below the overflowing level of the cistern.'

In domestic cistern installations up to 1,000 litres, this is usually a FOV to BS 1212. In large-scale cisterns, however, other means of filling the cistern are available:

- FOVs
- solenoid valves.

FOVs: BS 1212

There are four types of FOV that can be installed on large-scale cisterns. These are:

- BS 1212 Parts 1 and 2 FOVs
- equilibrium FOVs
- pressure-operated float valves
- Keraflo®-type delayed action float valves.

This section looks at the merits of each valve in turn.

- **BS 1212 Parts 1 and 2** – these are the most common types of FOV. The main problem with these types of FOV is that they are very restrictive to water flow and incur a much greater pressure (head) loss than other types of FOV, making cistern filling a long process. Wear on the washer and orifice can also be problematic when the valve is in constant use. They are, however, satisfactory when intermittent use is anticipated. These were looked at in detail at Level 2.

- **Equilibrium FOVs** – the equilibrium FOV offers a greater flow rate and lower pressure loss than the BS 1212 type of valve, and can be especially beneficial for a large-scale cistern with a high-pressure inlet cold water supply.

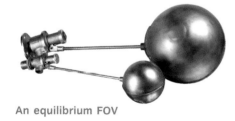

An equilibrium FOV

Unlike other FOVs, the equilibrium type of valve does not rely solely on the float to successfully close it. Instead, the closing operation is aided by the water pressure of the incoming mains cold water supply, allowing a smaller float to be used.

As can be seen in the diagram, the piston has a hole running through its length. This allows water to pass through to the back of the piston, which has the effect of pushing the piston towards valve shut-off while the water at the front of the piston tends to push it away, equalising the pressure on both sides of the barrel. The float just has to lift the arm to close the valve, greatly reducing the effort required to stop the flow of water.

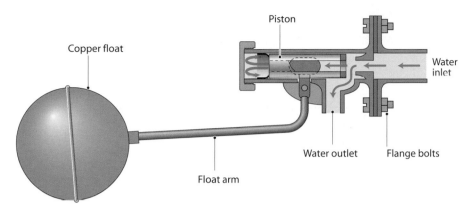

The operation of an equilibrium FOV

Equilibrium valves are an advantage where the water pressure is high and water hammer may be a concern. Almost all FOVs above 54mm are of this type.

■ **Pressure-operated float valves** – pressure-operated float valves use the pressure of the cold water main to assist the valve closure through the use of a pilot valve controlled by the lever and the float. Often called pilot-operated valves, the advantage here is that the variations in water tank levels between fully open and fully closed are greatly reduced. Although the head loss is greater than with the equilibrium type, pressure-operated valves are particularly suited to large-scale cisterns with a high-pressure supply.

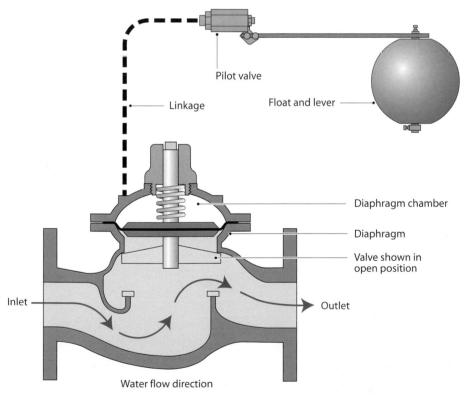

A pressure-operated valve

Pressure-operated valves work as follows. At zero pressure, the valve is closed. As water enters the valve inlet and the pressure increases, the valve opens to allow water to flow into the cistern. When the water has reached its shut-off level, the pilot valve, operated by the float and lever, closes. This causes the pressure within the diaphragm chamber to increase, thereby closing the water inlet and stopping the flow of water. As the water level drops, the float-operated pilot valve opens, releasing the pressure of the diaphragm chamber. Water pressure then re-opens the inlet valve and the cistern fills again to its shut-off level.

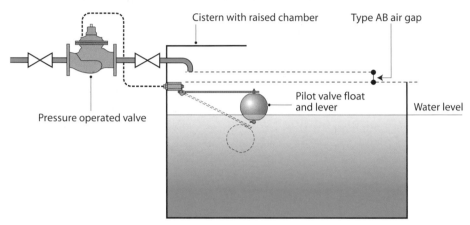

Pressure-operated valve cistern layout

Stratification

This describes how the temperature of the water varies with its depth. The nearer the water is to the top of the cistern, the warmer it will be. The deeper the water, the colder it will be. This tends to occur in layers, whereby there is a marked temperature difference from one layer to the next. The result is that water quality can vary, with the warmer water near the top being more susceptible to biological growth such as *Legionella pneumophila* (the bacterium causing Legionnaires' disease).

- **Keraflo®-type delayed action float valve** – an alternative approach to cistern filling is the Keraflo® delayed action float valve. This type of valve opens only when the water falls to a pre-set level in the cistern, opening fully to achieve a fast cistern fill. The benefits here are that not only does the cistern fill quickly, but also the velocity of the water entering the cistern means that it will mix with the cistern contents, preventing **stratification**.

The valve opens at a predetermined water level, opening and closing to fully eliminate water hammer and unwanted system noise. An adjustable water level enables the levels to be set based on water usage or, in the case of large domestic installations, occupancy levels.

The design of the valve means that boosting pumps are used less frequently. With a conventional FOV, pumps activating every few minutes waste energy and increase pump wear. A delayed action float valve eliminates this by allowing the pumps to be activated only once every few hours when the water level has fallen sufficiently.

Solenoid valves

A solenoid valve is an electromechanical valve that controls the flow of water into the cistern. The solenoid is an electromagnet that operates when an electrical current runs through the coil. When the coil is not energised, a spring keeps the valve shut.

Most solenoid valves used on large-scale cisterns are of the servo type (also called the pilot-type solenoid valve). With this type of valve, the electromagnet operates a plunger, which opens and closes

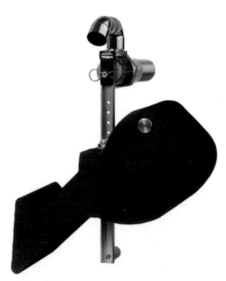

A Keraflo® 'Aylesbury'-type delayed action float valve

a pilot orifice. The incoming water pressure, which is fed through the pilot orifice, opens the valve seal and allows water to flow through the valve. As the pilot valve closes, the pressure on the valve seal decreases and a spring closes the valve.

Although very rarely used with modern systems, a solenoid valve discharging through an open-ended pipe is an acceptable alternative method of filling large-scale cisterns when used in conjunction with a float switch to activate it. Float switches are covered later in this unit – they are generally associated with boosted cold water systems.

A servo-type solenoid valve

Multiple-cistern installations

Where large quantities of water are required but space is limited, cisterns can be interlinked provided that they are the same size and capacity. Problems can occur if the cisterns are not linked correctly, especially where the cisterns are to supply drinking water. Stagnation of the water in some parts of the cistern may cause the quality of the water to deteriorate. The number of cisterns to be linked should be kept to a minimum.

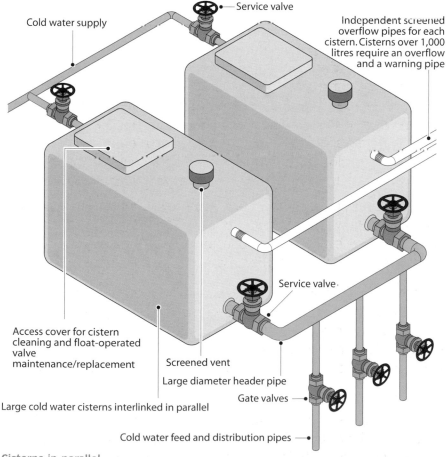

Cisterns in parallel

Stagnation can be avoided by following some basic rules. Connection must be arranged to encourage the flow of water through each cistern. This can be achieved by:

- keeping the cistern volumes to a minimum to ensure rapid turnover of water to prevent stagnation
- connecting the cisterns in parallel wherever possible
- connecting the inlets and the outlets at opposite ends of the cistern
- using delayed action FOVs to limit stratification.

Where it is not possible to connect cisterns in parallel, cisterns may be connected in series.

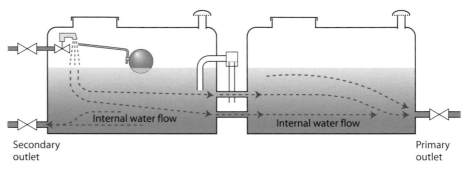

Secondary outlet

Primary outlet

Cisterns in series

In practice, cisterns in series should be interconnected to allow free movement of water from one cistern to another. They should be connected at the bottom and the middle so that water passes evenly through them. The primary outlet connection should be made on the opposite cistern to the FOV to encourage water movement, with the secondary connection being made on the cistern with the FOV installed. The overflow/warning pipe should be fitted on to the same cistern as the FOV. Both cisterns must be of the same size and capacity.

When connecting two or more cisterns together, you should ensure that the water movement is regular and even across all cisterns. In this situation, it is a good idea to install FOVs on all cisterns with appropriate service valves, as detailed in the Water Regulations: 'Every FOV must have a service valve fitted as close as is reasonably practicable.'

Wherever an FOV is fitted, an overflow/warning pipe must accompany it. These pipes should terminate in a conspicuous, visible position outside the building. On no account should they be coupled together.

There should be service/gate valves positioned to allow for isolation and maintenance of the cisterns without interrupting the supply. In the diagram on the next page, you will see that any two of the four cisterns can be de-commissioned, leaving two in operation. This ensures continuation of the supply.

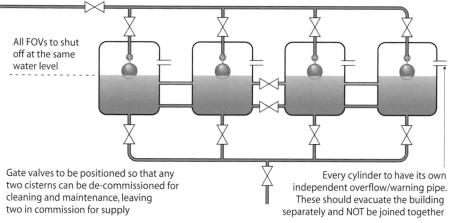

Every cylinder has a float-operated valve to allow movement of water in every cylinder
Each FOV is fitted with a service valve as detailed in the Water Supply (Water Fittings) Regulations

All FOVs to shut off at the same water level

Gate valves to be positioned so that any two cisterns can be de-commissioned for cleaning and maintenance, leaving two in commission for supply

Every cylinder to have its own independent overflow/warning pipe. These should evacuate the building separately and NOT be joined together

Installing three or more cisterns

Break cisterns

Break cisterns (often called break tanks) are used in large cold water installations in order to supply the system with water via a set of boosting pumps when the mains supply is insufficient. They provide a break between the mains supply and the installation. This has several advantages over pumping direct from the mains supply. Break cisterns ensure that:

- there is no surge on the mains supply when the boosting pumps either start or stop

- contamination of the mains cold water supply from multi-storey installations does not occur

- there is sufficient supply for the installation requirements at peak demand

- the water supply to other users is safeguarded by not drawing large amounts of water from the mains supply through the boosting pumps.

Break cisterns are often used in very tall buildings as intermediate cisterns on nominated service floors, dividing the system into a number of manageable pressure zones. The break cisterns provide water to both user outlets and other break cisterns higher up where the water is then boosted to other pressure zones further up the building.

As with all cistern installations, break cisterns must be fitted with an appropriate air gap that ensures zero backflow into any part of the system.

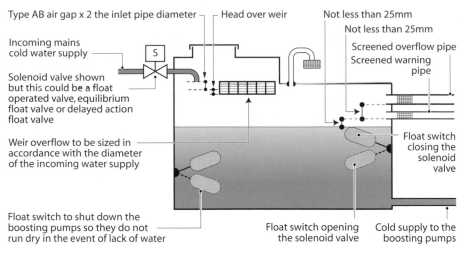

Type AB air gap x 2 the inlet pipe diameter — Head over weir

Incoming mains cold water supply

Solenoid valve shown but this could be a float operated valve, equilibrium float valve or delayed action float valve

Weir overflow to be sized in accordance with the diameter of the incoming water supply

Not less than 25mm
Not less than 25mm
Screened overflow pipe
Screened warning pipe

Float switch closing the solenoid valve

Float switch to shut down the boosting pumps so they do not run dry in the event of lack of water

Float switch opening the solenoid valve

Cold supply to the boosting pumps

The layout of a break cistern with a raised chamber

The function of components used in boosted cold water systems in multi-storey buildings

SmartScreen Unit 303

Worksheet 3

Boosted cold water systems contain various specialist components:

- booster pumps
- accumulators
- pressure switches
- float switches
- tranducers
- temperature sensors.

Booster pumps

Booster pumps for commercial buildings, often referred to as 'boosting sets', consist of duplicate vertical, multi-stage pumps that are used to pump the water to the required height within the building. Pumps have a common manifold on both the suction and discharge side of the pump, non-return valves to prevent water backflow through the system, shut-off valves and pressure switches for each pump. There are two types.

- 'Packaged' systems with all necessary integral controls – these will be delivered to site on a common base frame and will often contain the accumulator or pressure vessel (see below) as part of the unit. All that is required is connection to the system.

- Self-assembly sets – these are delivered as individual components that are assembled on site.

A booster set must be set up by a competent engineer. Manufacturers will often insist that commissioning is undertaken by a company commissioning engineer to ensure that the system is installed correctly.

A typical packaged boosting set

Accumulators

An accumulator is a pressurised vessel that holds a small amount of water for distribution within the installation. Accumulators are designed to maintain mains operating pressure when the pump is not working, and to reduce pump usage. Small accumulators can also be used to suppress water hammer.

Small domestic installations use bladder-type accumulators. These consist of a synthetic rubber bladder or bag within a coated steel cylinder or vessel.

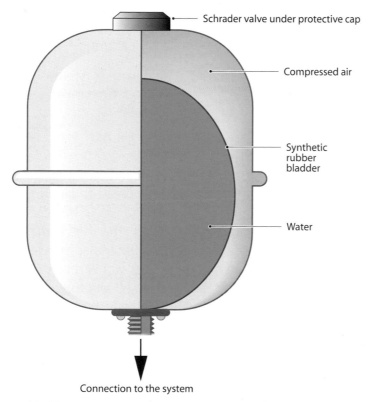

Schrader valve under protective cap

Compressed air

Synthetic rubber bladder

Water

Connection to the system

A bladder-type accumulator

The operation of an accumulator can be broken down into three stages.

1 When the pump operates it forces water into the accumulator bladder, compressing the air surrounding it to a pressure greater than the vessel's pre-charge pressure. This is the source of the stored energy.

2 When the bladder expands due to water being forced in by the pump, it deforms and the pressure within the accumulator increases. Bladder deformation stops when the water and the now-compressed air charge become balanced.

3 When a tap is opened, the pressure within the system drops and the compressed air forces the water out of the accumulator. When all the water inside the accumulator is used and the pressure falls to a predetermined level, the pump energises to recharge the accumulator water storage and pressure and the cycle begins again.

Probably the most important consideration when applying an accumulator is calculating the correct pre-charge pressure. You need to consider the following points:

- the type of accumulator being used
- the work to be done
- the system operating limits.

The pre-charge pressure is usually 80–90% of the minimum system cut-in pressure (the pressure at which the pump energises), to allow a small amount of water to remain in the vessel at all times. This prevents the bladder from collapsing totally. To calculate the pre-charge pressure, follow this simple procedure:

If the minimum working pressure of a cold water system is 2 bar, then:

$$2 \times 0.9\ (90\%) = 1.8\ \text{bar}$$

$$\text{Pre-charge pressure} = 1.8\ \text{bar}$$

The accumulator air charge must be lower than the mains pressure for water to enter the vessel, and on average, a pressure differential of around 1.5 bar lower than the supply pressure would be acceptable (but no more than 2 bar and no less than 0.8 bar). This means that if the supply pressure is 3.5 bar, the air charge within the accumulator must be around 2 bar. A supply pressure of 4.5 bar would require a 3 bar air charge, and so on. Air pressure can be checked and topped up as necessary at the Schrader valve situated at the top of the accumulator.

Auto-pneumatic pressure vessels

These were described on page 152 – Indirect boosting with a pressure vessel.

Float switches, transducers and temperature sensors

Float switches, transducers and temperature sensors play a vital part in modern boosted large-scale cold water systems. The problems encountered are not just related to how to install them but also where to install them. Installations of large-scale cisterns are often undertaken in tight and restricted spaces. Difficulties arise in positioning these components while providing access for maintenance and inspection.

Float switches

Float switches, often called level switches, detect water levels within the cistern to activate various other pieces of remote equipment such as start/stop functions on boosting pumps, open/close functions on solenoid valves, water level alarms and water level indicators.

There are many different types of float switches available, and these can vary in sophistication from simple magnetic toggle switches to ultrasonic and electronic types. Popular types include:

- magnetic toggle – a simple float switch that uses the opposing forces of magnets to activate a micro-switch
- sealed float
- pressure-activated diaphragm
- electronic
- ultrasonic.

A magnetic toggle float switch

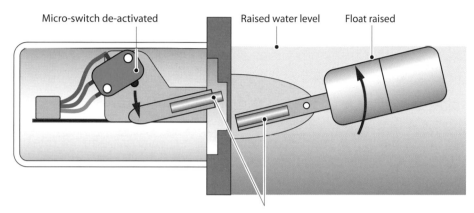

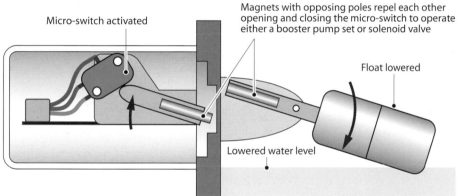

The magnetic toggle float switch – how it works

Transducers

A transducer is an electronic sensor that converts a signal from one form to another. In large-scale, multi-storey water systems it senses system pressure variations and converts a pre-set low pressure into voltage to activate either the boosting pumps or the compressor, feeding the pressure vessel to boost the pressure to normal operating pressure. It may also be used to sense over-pressurisation.

Temperature sensors

Temperature sensors are often used to monitor the temperature of large volumes of stored wholesome, **potable** water where the installation serves a large number of people, such as in a hospital, prison or any place where there is a 'duty of care'.

The Water Supply (Water Fittings) Regulations 1999 advise that stored wholesome water should not exceed 20°C in order to minimise the risk of microbacterial growth.

A water pressure transducer with pressure gauge

Potable

Potable is pronounced 'poe-table'. It comes from the French word *potable*, meaning drinkable.

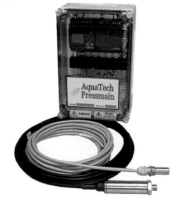

A water temperature sensor

Describe alternative water supplies to buildings

A private water supply is defined as any water supply that is not provided by a water undertaker. It is not connected to any part of the mains water network and therefore water rates are not charged – although the owner of any such supply may make a charge for water used. Private supplies are commonly used in rural areas where connection to the water mains is difficult. A private supply may serve one property or many properties on a private network. The water may be supplied from a borehole, spring, well, river, stream or pond.

Under the Water Act 2003, the local authority in the area where the private water supply is located is responsible for the inspection and testing of the water supply to ensure that it is maintained to a quality that is fit for human consumption. These inspections and tests are made in accordance with the Private Water Supplies Regulations 2009. Generally speaking, the more people that use the supply, the more detailed the tests and the more regular the inspections have to be. Supplies for commercial properties and activities or food production and preparation have to be tested more frequently and meet more stringent requirements than domestic supplies.

The Private Water Supplies Regulations 2009 stipulate that a risk assessment must be made for all private water supplies including the source, storage tanks, treatment systems and the premises using them.

A free copy of the Private Water Supplies Regulations 2009 can be downloaded from http://www.legislation.gov.uk/uksi/1999/1148/pdfs/uksi_19991148_en.pdf.

A private water supply can serve a single-occupancy dwelling providing less than 1,000 litres (1m^3) per day, or it may supply many properties or commercial/industrial buildings where the extraction rate may be in excess of 1,000,000 litres (1,000m^3) per day.

Sources of water vary, from boreholes, wells, springs and streams to rivers and lakes, and each one needs to be assessed for its water quality and suitability. The monitoring requirements of the Drinking Water Directive will vary according to the source and the size of the supply. In addition to this, the volume of water produced and the population it is serving should also be classified by the nature of the supply and whether the supply is:

- serving a single-occupancy dwelling
- for domestic use by the people living in the dwelling
- for commercial food production.

The issue of private water supply is an especially important part of a plumber's work, as the Water Supply (Water Fittings) Regulations 1999 do not apply in this instance. The special regulations mentioned previously (see page 168) need to be followed with respect to cleansing, sterilising and testing the water to ensure that it is fit for human consumption.

Methods of supplying water from a private water supply

There are several methods of water extraction that can be used to supply single-occupancy dwellings. This is the first of the assessment criteria for this section – here, the methods and equipment used for the following are investigated:

SmartScreen Unit 303
Handout 8

- pumping water from wells and boreholes
- collecting water from surface water sources such as rivers, streams and springs
- storing treated water.

Pumping water from wells and boreholes

Many small private drinking water supplies are extracted from boreholes and wells. Wells are usually large and circular of not less than 1m diameter, often dug by hand and occasionally by mechanical excavators. Boreholes are smaller in diameter and are drilled by specialist companies using a variety of methods including percussion and rotary drilling.

The quantity of available water will largely depend on the type of **aquifer** that the borehole is to access. Obviously, the bigger the aquifer, the more water will be available for extraction. Estimated amounts can be calculated by test pumping after the borehole has been sunk. Perched aquifers are the most unreliable as these may well dry up after long periods without rain.

A borehole

Aquifer

An aquifer is a type of rock that holds water like a sponge. There are three basic types.

- Confined aquifers – these have a confining layer between water level and ground level. A confining layer is a layer of material that is impermeable, with little or no porosity.
- Unconfined aquifers – these have no confining layers between water level and ground level.
- Perched aquifers – these have a confining impermeable layer below the water-bearing strata. They are situated above the main water table.

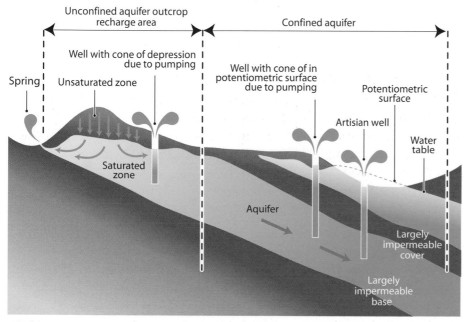

Aquifers

Normally, a well-sunk and properly constructed borehole will comfortably be able to supply a single-occupancy domestic dwelling with water pumped directly from the aquifer, after appropriate treatment, by either of the following:

Submersible pumps

- **A submersible pump** – this is a centrifugal pump that has a hermetically sealed motor and is therefore waterproof. The whole assembly is submerged in the water. The main advantage of this type of pump is that it prevents water cavitation (the formation of bubbles). Submersible pumps are designed to push water from the well or borehole rather than pull water from it. They are more efficient than surface pumps, which have to pull the water upwards from the borehole.

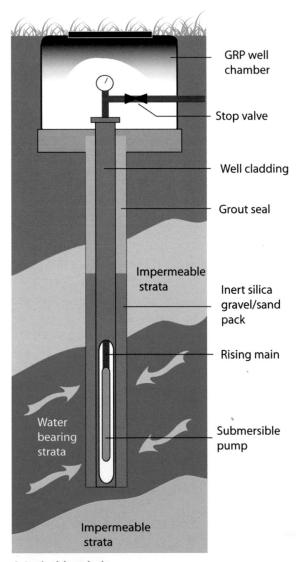

A typical borehole

- **A surface-mounted pump** – usually situated in a pump house, these pumps can either be horizontal single-stage or vertically mounted multi-stage centrifugal-type pumps. Multi-stage pumps are often accompanied by pressure vessels to aid boosting, especially where the water supply is unreliable or inefficient.

THE CITY & GUILDS TEXTBOOK

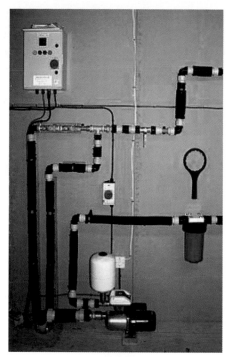

A typical surface pumping set using multi-stage centrifugal pumps and pressure vessels

A typical surface pumping set using a horizontal single-stage centrifugal pump

Water taken from deep wells and boreholes may have travelled from catchments several kilometres away. If the water supply is extracted from a sand and gravel aquifer it will be very clean, having gone through many stages of filtration in the sand/gravel geology. If the aquifer is predominantly limestone, the water will have travelled through fissures in the rock and will generally not be as clean as water from sand and gravel aquifers. Although ground waters such as aquifers are usually of good quality, some may contain high levels of iron and manganese and some may be contaminated with nitrates and other chemicals from farming and agricultural activities.

Boreholes extract the water from far deeper than any other source of private water supply. Boreholes, often 100mm to 150mm in diameter, can be drilled as deep as 50m.

Collecting water from surface water sources

There are several water sources that can be classified as surface water sources.

Rivers

These offer greater, more reliable yields than boreholes but can be susceptible to pollution and often show a variable quality of water. Pollution often depends on the catchment area and the activities in the general vicinity. Water that comes from ground where there is little peat and agricultural activities is usually of good quality. Peaty ground tends to yield acidic water because of the concentration of CO_2 and this can lead to high concentrations of dissolved metals such as lead.

Microbiological contamination

Contamination by microscopic organisms, such as bacteria, viruses, or fungus.

Lowland water is most likely of poor quality and may show seasonal variations in terms of quality and colour, with late autumn being a peak time for colour changes. **Microbiological contamination** may also be high during periods of heavy rainfall.

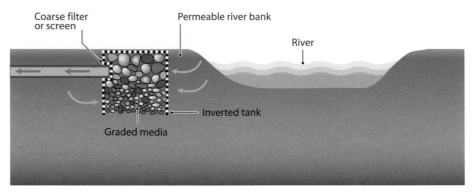

Coarse filter or screen

Permeable river bank

River

Inverted tank

Graded media

River and stream water collection

Because of the potential problems associated with rivers, a surface water source should only be considered as a potential drinking water supply where no ground water source exists. The water will require a minimum of filtration and disinfection treatments. This must be designed for a worst-case scenario when the water is raw, especially after snow-melt and heavy rainfall. Installing a small tank or reservoir will help the settlement process and will do much to reduce the variations in water quality. The tank will require regular inspections and cleaning because of the solid matter within the water.

Streams

Small streams often show a variable quality to the water because of animal and human activity within their catchment area, and may also show colour changes due to the levels of humic and fulvic acids that are used in agriculture as soil supplements.

Pollution and natural variations in the quality of the water are the most common problems that can occur with both stream and river water sources, and these need to be considered carefully when siting the water supply intake point. Water can be pumped directly from the stream or collected from the ground in the immediate vicinity of the stream or riverbank. This is desirable in certain situations where the geology allows a natural filtration process – the water is therefore cleaner than if taken directly from the river (see the diagram above). The intake should not be positioned in an area where water turbulence may be created, especially during periods of heavy rainfall, ie on the bend of a river or where there are sudden changes in water level. Water intakes must be protected by a strainer to prevent the ingress of fish, vermin and debris, and the inlet pipe must feed a settlement tank that allows the particulate matter within the water to settle.

The outlet of the tank should be situated above the floor of the tank and must be fitted with a strainer to prevent sediment contamination. The tank itself must be constructed from a material that will not contaminate or impair the quality of the water, and must be designed in such a way as to prevent the ingress of vermin and debris.

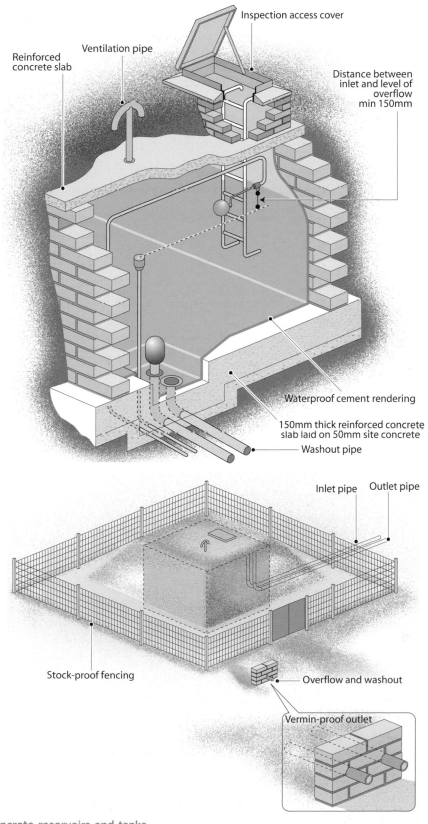

Reinforced concrete slab

Ventilation pipe

Inspection access cover

Distance between inlet and level of overflow min 150mm

Waterproof cement rendering

150mm thick reinforced concrete slab laid on 50mm site concrete

Washout pipe

Inlet pipe

Outlet pipe

Stock-proof fencing

Overflow and washout

Vermin-proof outlet

Concrete reservoirs and tanks

Springs

Where the water table and the surface coincide, a spring forms. The presence of fissures in the Earth's surface usually dictates where natural springs occur. The most reliable springs are from deep aquifers, while aquifers nearer the surface are susceptible to drying up after a short period without rain, especially if the water is flowing from fissured limestone or granite.

Spring water is usually of good microbiological and chemical quality, although shallow aquifers may suffer from variations in water quality due to surface contamination. The probability of agricultural contamination must be carefully considered, especially when the aquifer evacuates at the surface. It must also be remembered that some shallow 'springs' may actually be land surface drains. Here, the water quality is likely to be unacceptable.

The treatment of spring water is usually much simpler than treatment of river or stream water, because there is much less suspended solid matter.

Spring water must be protected from surface contamination once it has reached ground level. It is necessary to consider leakage from septic tanks and agricultural activity. A small chamber built over the spring will protect the source from surface contamination and also serve as a collection source and header tank. It should be constructed so the water enters from the base or the side. The chamber top must be above ground level and should be fitted with a lockable, watertight access cover. An overflow must be provided and sized to accommodate the maximum flow of the spring. The outlet should be fitted with a strainer and positioned above the floor of the chamber.

The chamber must be constructed of a material that will not contaminate or impair the quality of the water and must be designed in such a way as to prevent the ingress of vermin and debris.

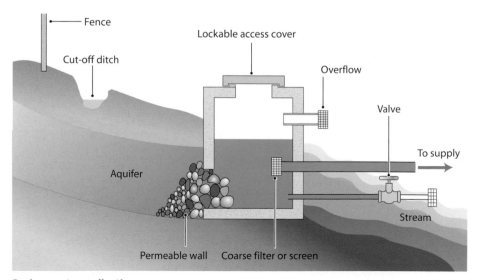

Spring water collection

The land in the immediate vicinity of the chamber must also be fenced off and a small ditch dug upslope to intercept and divert surface water run-off.

Remember that all water sources should be protected from the ingress of vermin and surface water contamination.

Storing treated water

Plumbing systems in domestic dwellings that use water from private water supplies need to include a method of storing the treated water to provide a water reserve in the event of planned or emergency maintenance, or problems with the water source or the water treatment. Storage will also cater for fluctuations in water demand.

Storage may take several forms:

- a small covered reservoir, providing sufficient head to serve the property
- an externally sited break cistern, with the water being pumped into the property by a booster set
- a suitably positioned storage tank or cistern in the roof space of a property from which water flows under gravity to the taps and outlets.

The tank, reservoir or cistern should have sufficient volume to accommodate the maximum demand, and any period when the water supply may be interrupted. The storage cistern/tank/reservoir may become contaminated during construction and must therefore be disinfected before use. This is usually achieved by filling the system with a solution of water/chlorine at 20mg/l of chlorine and leaving it to stand for several hours, preferably overnight. After this time, the chlorine should be drained and the system thoroughly flushed using treated water.

External break cisterns and cisterns in the roof space must be protected against contamination and insulated against freezing in cold weather and undue warming in warm weather. Cisterns and tanks should also be fitted with a lockable, well-fitting but not airtight lid to prevent the ingress of insects and vermin, and overflows and warning pipes must be protected by a mesh screen.

Storage tanks must be inspected every six months and cleaned if necessary to prevent the build-up of silt and debris. This should be followed by disinfection.

Methods of treating water for use in buildings

Larger water supplies served from a private water source, ie those that serve many properties or commercial/industrial establishments, are often treated by 'point of entry' treatment methods. These are very similar to those used by the local water undertaker, and were discussed in detail at Level 2. Remember that:

- the design of the treatment process should be based on a full investigation of site conditions
- the chemical and microbiological content of the water must be established and tests performed to determine the effectiveness of any treatment process and the chemical dosing requirements.

For small supplies, to a single dwelling for instance, the treatment is often precautionary and should include disinfection. The disinfection stage should only be skipped if it can be shown without reasonable doubt (by risk assessment and frequent testing) that the water supply is likely to be consistently pathogen free.

This section of the unit looks at the different methods of water treatment that are often used with private water supplies. Table 1 shows the methods of water treatment available, and their effectiveness.

Table 1: Water treatment methods

	Bacteria	Viruses	Algae	Coarse particles	Turbidity	Colour	Aluminium	Ammonia	Arsenic	Iron manganese	Nitrate	Pesticides	Solvents	Taste and colour
Coagulation and flocculation*	+	+	+	++	++	++	++		+	++				
Sedimentation				++	+		+			+				
Gravel filter/ screen			+	++	+		+			+				
Rapid sand filtration	+	+	+	++	+		+			+				
Slow sand filtration	++	++	++	++	++		+			+				
Chlorination	++	++	+			+		++						
Ozonation	++	++	++			+						++		++
Ultraviolet (UV)	++	++	+											

	Bacteria	Viruses	Algae	Coarse particles	Turbidity	Colour	Aluminium	Ammonia	Arsenic	Iron manganese	Nitrate	Pesticides	Solvents	Taste and colour
Activated carbon						+						+	+	++
Activated alumina									++					
Ceramic filter	++		++	++	++									
Ion exchange								+	+		++			
Membranes	++	++	++	++	++	++	++		+	++	++	++		++

+   Partly effective.

++ Preferred technique/effective.

*   Pre-oxidation may be required for effective removal of aluminium, arsenic, iron and manganese.

Without exception, water from all sources will need treatment before it is acceptable for human consumption. The health risks presented by poor quality water may be due to microbiological or chemical contaminates. Microbiological contamination is the most important issue, as this can lead to infectious diseases such as Legionnaires' disease and cholera. Chemical contamination often leads to more long-term health risks. Substances that affect the appearance, odour or taste often make the water unpalatable to the consumers. Particulates in the water may also present a health risk, as these could also be contaminated with microbiological organisms. In these circumstances, disinfection becomes more difficult. Final disinfection must always be included in any treatment system to effectively kill off any microorganisms that remain. Disinfection solutions containing chloride provide a residual effect that will preserve the quality of the water during storage and distribution (in larger systems).

Treatment of the water is based on the physical removal of contaminants by:

- filtration
- settling, often assisted by the addition of chemicals
- the biological removal of microorganisms.

Water treatment usually consists of a number of key stages:

- initial pre-treatment by settling
- pre-filtration through coarse media
- sand filtration
- disinfection or chlorination.

This process is known more commonly as the multiple barrier principle, and is designed to provide effective water treatment by not relying on a single, less effective, process or being susceptible to failure of one stage in the process. For example, if a system consists of coagulation/flocculation, sedimentation, sand filtration and finally chlorination, failure of the rapid sand filter for example does not mean that untreated water will be supplied to the property. Other processes will remove the majority of the suspended particles and, therefore, many of the microbiological contaminants, and disinfection will remove any that remain. Provided the sand filter is repaired quickly, there will be little in the way of deterioration in water quality.

Table 2: Private Water Supply Regulations 1991: water quality parameters

Parameters	Units of measurement	Concentration or value maximum unless otherwise stated
Colour	Mg/l Pt/Co scale	20
Turbidity	FTU	4
Odour (inc. hydrogen sulphide)	Dilution no.	3 at 25°C
Taste	Dilution no.	3 at 25°C
Temperature	°C	25
Hydrogen ion	pH value	9.5 5.5 (min)
Sulphate	mg SO_4/l	250
Magnesium	mg Mg/l	50
Sodium	mg Na/l	150
Potassium	mg K/l	12
Nitrite	mg NO_2/l	0.1
Nitrate	mg NO_2/l	50
Ammonia	mg NH_4/l	0.5
Silver	µg Ag/l	10
Fluoride	µg F/l	1500
Aluminium	µg Al/l	200
Iron	µg Fe/l	200

Parameters	Units of measurement	Concentration or value maximum unless otherwise stated
Copper	µg Cu/l	3,000
Manganese	µg Mn/l	50
Zinc	µg Zn/l	500
Phosphorus	µg P/l	2,200
Arsenic	ug As/l	50
Cadmium	µg Cd/l	5
Cyanide	µg CN/l	50
Chromium	µg Cr/l	50
Mercury	µg Hg/l	1
Nickel	µg Nl/l	50
Lead	µg Pb/l	50
Pesticides	µg/l	0.1
Conductivity	uS/cm	1,500 at 20°C
Chloride	mg/Cl/l	400
Calcium	mg Ca/l	250
Total hardness	mg Ca/l	min 60
Alkalinity	mg HCO_3/l	min 30
Total coliforms	number/100ml	0
Faecal coliforms	number/100ml	0
Faecal streptococci	number/100ml	0

Coagulation and flocculation

This combined process is used to remove colour, **turbidity**, algae and other microorganisms from surface water. It involves the addition of a chemical coagulant to encourage the formation of a precipitate, or floc, which entraps the impurities. In certain conditions, iron and aluminium can also be removed in this way. The floc is then removed from the water by sedimentation and filtration.

Turbidity

Turbidity refers to how clear or cloudy water is as a result of the amount of total suspended solids it contains – the greater the amount of total suspended solids (TSS) in the water, the cloudier it will appear. Cloudy water can therefore be said to be turbid.

Sedimentation

Sedimentation tanks are designed to slow down water velocity to allow the solids that it contains to sink to the bottom and settle under gravity. Simple sedimentation may also be used to reduce turbidity.

Sedimentation tanks are usually rectangular in shape with a length-to-width ratio of 2:1, and are usually 1.5–2m deep. The inlet and outlet must be on opposite sides of the tank, with the inlet designed to distribute the incoming flow as evenly as possible across the tank. The outlet should be designed to collect the cleared water across the entire width of the tank. The tank will also need to be covered to prevent external contamination.

Sedimentation tanks require cleaning when their performance begins to deteriorate. A 12-month period between cleaning operations is normally sufficient.

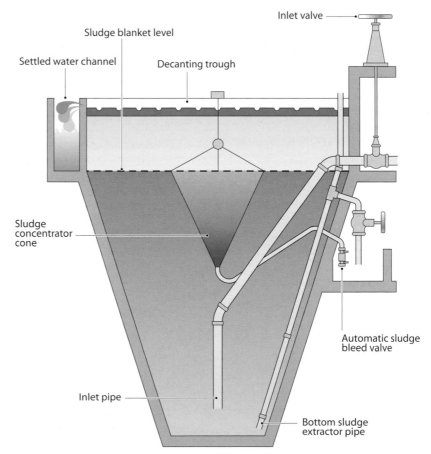

A sedimentation tank

Filtration

Filtration is an important process that removes turbidity and algae from raw, untreated water. There are many different types of filtration system, including the following:

- slow sand filters
- rapid (gravity) sand filters

- pressure filters
- absorption filters
- reverse osmosis.

The difference between the types is not just a matter of the speed of the filtration process – the underlying method also varies. Slow sand filtration, for example, is a biological process while rapid sand filtration is physical treatment process.

Slow sand filters

Slow sand filters are often preceded by micro-straining or coarse filtration. These filters are used primarily to remove microorganisms, algae and turbidity, and make for a slow but very reliable method of water treatment. This method is often suited to small supplies providing that there is sufficient area to properly construct the filtration tanks.

Slow sand filters consist of tanks containing sand with a size range of 0.15mm to 0.30mm and to a depth of around 0.5m to 1.5m. For single dwellings circular modular units, usually used in tandem, are available. These have a diameter of around 1.25m. As the raw water flows downwards through the sand, microorganisms and turbidity are removed by a simple filtration process in the top few centimetres of sand. Eventually a biological layer of sludge develops, which is extremely effective at removing microorganisms in the water. This layer of sludge is known as the schmutzdecke. The treated water is then collected in underdrains and pipework at the bottom of the tank. The schmutzdecke will require removing at periods of between two and ten weeks as the filtration process slows. The use of tandem filters means that one filter can remain in service while the other is cleaned, and allow time for the schmutzdecke to re-establish.

Slow sand filters should be sized to deliver between 0.1 and 0.3m³ of water for every 1m² of filter per hour.

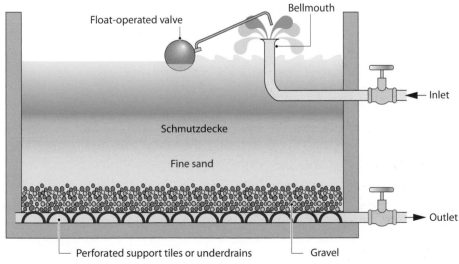

A slow sand filter

Rapid (gravity) sand filters

Rapid sand filters are mainly used to remove the floc from coagulated water, but they can also be successfully used to remove algae, iron, manganese and water turbidity from raw water. Activated carbon in granular form is used to remove any organic compounds. Some filters also incorporate an alkaline medium to increase the pH value of acidic water.

Rapid sand filters are usually constructed from rectangular tanks containing coarse silica sand with a size range of 0.5–1mm laid to a depth of between 0.6 and 1m. As the water flows downwards through the filter, the solids remain in the upper part of the sand bed where they become concentrated. The treated water collects at the bottom of the filter and flows through nozzles in the floor. The accumulated solids are removed either manually every 24 hours or automatically when the head loss reaches a predetermined level. This is achieved by backwashing.

A variety of proprietary units are available containing filtering media of different types and sizes. In some filters the water flows upwards, improving the efficiency.

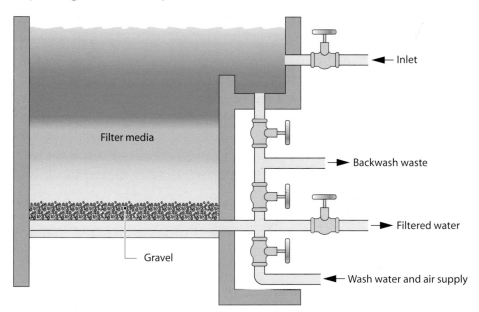

Filter media

Gravel

Inlet

Backwash waste

Filtered water

Wash water and air supply

A rapid (gravity) sand filter

Pressure filters

These are sometimes used where it is important to maintain a head of pressure to avoid the need to pump the water into the supply. The filter bed is enclosed in a cylindrical pressure vessel. Some small pressure filters are capable of delivering as much as $15m^3/h$. The cylinder is typically made of specially coated steel, and smaller units can be manufactured from GRP. They operate in a similar way to rapid sand filters.

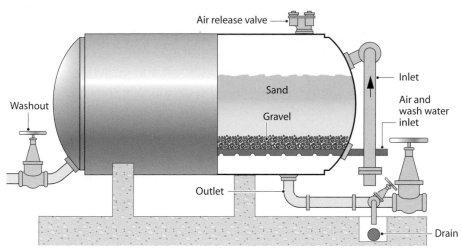

Air release valve

Inlet

Sand

Air and wash water inlet

Washout

Gravel

Outlet

Drain

A pressure filter

Absorption filters

These fall into two distinct categories:

- **Activated carbon** – these filters remove contaminants by the process of physical absorption. Granular activated carbon (GAC) is the most commonly used medium, although block carbon and powdered carbon can also be used. Most filters use replaceable cartridges that are easily changed when the old one is exhausted.

  Activated carbon filters will remove suspended solids, chlorine and some organic contaminants including pesticides. They will also remove some humic acids, which are responsible for giving water derived from peat its brown appearance. Unfortunately, activated carbon provides a good medium for the development of microorganisms and there is some concern that this can cause health problems should the bacteria be reintroduced into the water. Water inhaled in the form of an aerosol during activities such as washing is also a concern. Because activated carbon removes chlorine, bacterial growth has also been found on filters treating chlorinated water. Some manufacturers state that activated carbon should not be used if the water is of unknown quality or contains microbiological organisms.

An activated carbon filter

- **Activated alumina** – these filters can be used where the water contains contaminants such as arsenic, or other chemicals such as fluoride. They are manufactured using aluminium hydroxide.

Reverse osmosis (RO)

Large RO units have been used for many years for producing good drinking (potable) water from low-quality water. They can also be used to produce drinking water from saline (salt) water where the supplies of fresh water are inadequate. RO will successfully remove a wide range of organic and chemical contaminants such as sodium, calcium, fluoride, nitrates, pesticides and solvents.

Semi-permeable

Allowing passage of certain small particles, but acting as a barrier to others.

Reverse osmosis units work by forcing the water under pressure through a **semi-permeable** membrane, The membrane is usually manufactured from polyamide. This material is preferred, as some membranes such as cellulose actively support the growth of bacteria. The filters do not need replacing but the membrane may require periodic chemical cleaning and de-scaling.

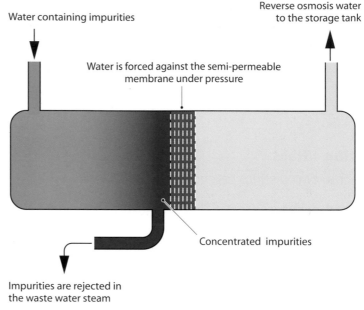

Water containing impurities

Reverse osmosis water to the storage tank

Water is forced against the semi-permeable membrane under pressure

Concentrated impurities

Impurities are rejected in the waste water steam

The principle of reverse osmosis

The flow rate through the units tends to be low in domestic RO units, with the treated water being collected in a storage tank to cushion supply and demand. A level sensor within the storage tank itself usually controls the RO unit. One disadvantage with reverse osmosis is that it produces a lot of wasted water. For every litre of fresh drinking water, 3 litres of wasted water are produced. An alternative would be to use the wasted water for non-potable supplies such as WC flushing.

The water treated by reverse osmosis tends to be soft. For this reason, some units contain a re-hardener to increase both hardness and alkalinity, thus making it less aggressive on metals such as lead.

Disinfection

The greatest danger to water for human consumption (potable water) is from contamination by human and animal faeces. These contain the organisms of many communicable diseases and so the use of disinfection to kill or inactivate these **pathogenic** organisms is vital.

Pathogenic

Causing disease.

There are several disinfection methods available for treating water from a private source. Disinfection by the use of chlorine is the most widely used for large water users, but is much less common for smaller, single-dwelling supplies. Smaller water supplies commonly use ultraviolet (UV) irradiation and ozonation.

This section of the unit looks at three very different methods of disinfection.

Chlorination

There are several methods of chlorination that can be used depending on the size of the supply being disinfected. Chlorination is commonly achieved by the use of liquefied chlorine gas, sodium hypochlorite solution or calcium hypochlorite granules. Chlorine gas is supplied in pressurised cylinders and is extremely dangerous, so you must take great care when storing or handling it. The gas is taken from the cylinder by a chlorinator, which measures and controls the flow of gas.

Sodium hypochlorite solution is delivered in drums, and only a one-month supply should be stored at any one time. Exposure to sunlight can result in a decrease in its disinfection capabilities and also results in the loss of available chlorine and an increase in the concentration of chlorate, relative to chlorine. There are many different methods of dosing the water with sodium hypochlorite. A simple gravity-fed system where the solution is drip-fed at a constant rate has proven very successful provided that the rate of flow and the water quality and composition remain constant.

Calcium hypochlorite can be supplied in powder form, tablets or granules. Calcium hypochlorite is very stable provided it is kept dry, and several months' supply may be stored. You should take care, however, to avoid contact with moisture, as this reacts with it to form chlorine gas. Dosing with a calcium hypochlorite doser is very simple and involves dissolving a measured amount of calcium hypochlorite in a measured volume of water, which is then introduced into the supply. Tablet form is preferred, as the rate of dissolve is very predictable.

With all systems, the resultant free residual chlorine after disinfection should remain within the range of 0.2–0.5mg/l and a contact time of 30 minutes is recommended. The design of the disinfection system is most important, and it must evenly distribute the chorine within the water supply. It must not allow chlorine concentrations to build up in dead zones.

Ozonation

This system uses ozone gas to disinfect the water supply. Ozone is a powerful disinfectant and oxidant that completely kills bacteria and viruses. It cannot, however, be relied on as the only means of disinfection where the water contains cryptosporidium. Ozone may also help to reduce the levels of taste, colour and odour in the water.

Ozone is a gas produced by discharging alternating current through dry air. Small proprietary units use 230V single-phase AC current, but larger units run on 400V three-phase supplies. The ozone-containing

KEY POINT

Cryptosporidium is a gastrointestinal illness that affects humans and cattle and presents itself as severe diarrhoea. It usually affects children between the ages of 1 and 5, but it can affect anyone and the symptoms can be very severe in people with weak immune systems.

air is then mixed with water in a contact column. To be completely effective, the column should give at least four minutes' contact time, giving a residual ozone content of 0.4mg/l. Unless the water is used quickly, the ozone will decompose rapidly and so it is recommended that the disinfectant process is reinforced using a small amount of chlorine.

Small-scale ozone units are available for single-dwelling private water supplies, but they are not widely used. This is generally because of the high power usage and complexity of the equipment.

A domestic ozone generator

UV irradiation

This is the preferred method of disinfection of private water supplies for small, single, domestic dwellings. They use low-frequency UV light to change the cellular structure of microbiological organisms, effectively destroying them.

UV disinfection is affected by the water quality and its flow rate, and the water must be of good quality with low turbidity and colour. Pre-filtration is necessary with this method.

Special low-mercury (Hg) lamps are used to generate the UV radiation in an enclosed chamber usually manufactured from stainless steel. The lamps look very similar to fluorescent tubes but are made from UV-transparent quartz instead of phosphor-coated glass. The lamps generate UV radiation at a wavelength of 254 nanometres (nm). The optimum germicidal wavelength is between 250 and 265nm. The temperature of the lamps is around 40°C, and they are separated from the water by sleeves to prevent them from being cooled by the water. The UV effect of the lamps deteriorates with age and after 10–12 months the efficiency is down to 70% of a new lamp. Therefore, lamp replacement is recommended every 12 months.

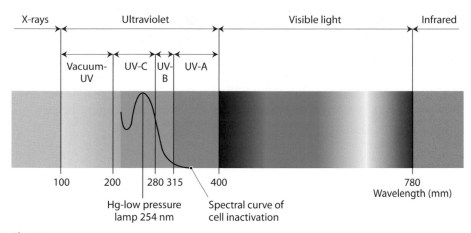

The UV spectrum

As with all systems, disinfection will only be effective provided a sufficient dose of UV is applied. A dosage of 16–40mWs/cm^2 (milliwatt seconds per cm^2) is recommended, but this depends on factors such as the ability of the microorganisms to withstand UV light.

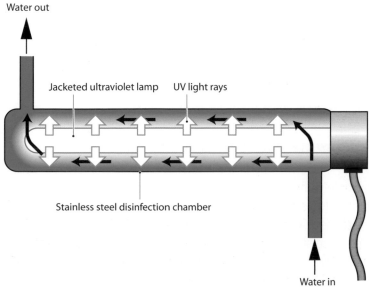

Domestic UV sterilisation units

A domestic UV chamber

The method of operation of different types of boosted cold water supply systems

The UK has over 500,000 people whose only source of potable drinking water is from a private supply. These, as we have already seen, fall into four main categories:

- wells
- springs
- streams
- boreholes.

Occasionally, properties may have access to more than one supply, and in some cases a license may be required to extract the water from a given source.

The delivery of feed waters into a domestic property can be by one of two methods:

- a pumped supply direct from the borehole or well
- a gravity supply from a catchment tank in a spring or stream.

There are no hard and fast rules as to which method is the best to use in any given situation. If the water is being delivered from a borehole then, obviously, a pumped method of supply will be used. A gravity supply may be used where the water source is higher than the property, with the water flowing by gravity from the catchment tank in the water source (ie a spring) to either an external storage/break cistern or tank, before being pumped into the property or direct to a storage cistern located within the property. The water would then be distributed to all outlets from the storage cistern by gravity supply.

Pumped supplies

There are two methods of pumped supply from a well or a borehole:

- pumped supply with pressure control
- pumped supply with level control.

Pumped supply with pressure control

This type of system provides directly drawn water at the point of use. Pressure is maintained within the system by the use of an accumulator (often called a pressure vessel) and a pump. The accumulator is a vessel that contains air under pressure, and water. The water is contained within a neoprene rubber bag inside the accumulator, which expands when water is pumped into it under pressure. The air is then compressed and the pressure rises. As the water within the accumulator is used, the pressure will drop. At a predetermined pressure, the pump will start and the accumulator is refilled, raising the pressure to its operating level. These systems generally operate at 1.5–3 bar. This system is preferred when water treatment is being considered.

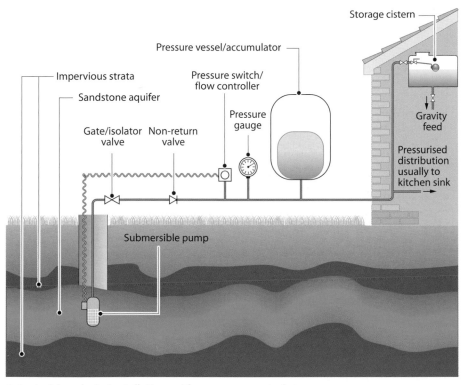

A typical borehole installation with pressure control

Control of the system is automatic. The system contains a submersible or surface-mounted pump to bring the water to the surface, filtration and sterilisation equipment (usually UV), a pressure transducer to sense pressure drop across the installation, a pressure gauge and an accumulator. The kitchen sink is usually installed with water under pressure directly from the accumulator. All other outlets are supplied from a low-pressure supply from a storage cistern situated in the roof space. A non-return or check valve must be fitted upstream of the accumulator.

Pumped supply with level control

This system uses a float switch to monitor the level of the water in a storage cistern. The storage cistern is normally situated in the roof space of a dwelling. The float switch operates a surface-mounted pump, which fills the tank until the level of the float switch is reached. All water for the dwelling passes through the storage cistern, and this supplies all outlets with a low-pressure supply. Water fed directly from the borehole to a kitchen sink under pressure is not possible with this type of installation.

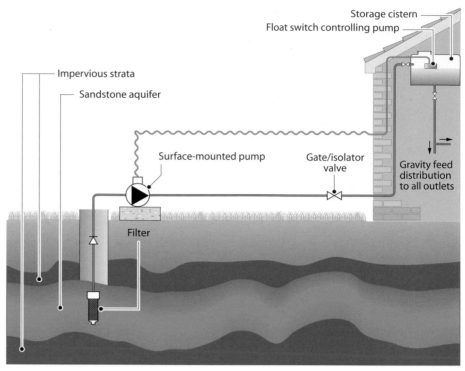

A typical borehole installation with level control

Because all the water for the dwelling is supplied at low pressure, this system can also be used with supplies that are fed from a catchment tank in a stream or spring via an external break/storage cistern. It is also possible to use water directly from a catchment tank without the use of a pump provided that the source of water is higher than the dwelling. Remember, however, that some form of filtration and sterilisation of the water is necessary. A non-return or check valve must be fitted upstream of the pump.

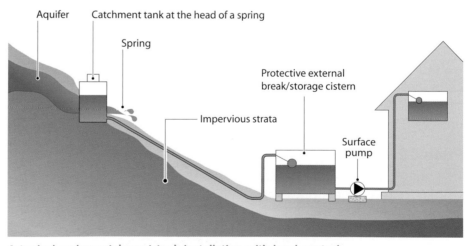

A typical spring catchment tank installation with level control

Components in boosted (pumped) cold water supply systems from private sources for single-occupancy dwellings

This final part of the Outcome deals with private water supplies. You will investigate the components used with private water supplies to single domestic dwellings:

SmartScreen Unit 303
Presentation 3

- small booster pump sets which incorporate all controls and components
- a boosted system with separate controls and components
- accumulators used to increase system flow rate (see pages 165–166).

Vertical, horizontal and submersible pumps

As described earlier in the unit, there are two different types of pump that can be used with private water supplies and, more specifically, boreholes and springs:

- submersible pumps
- surface pumps:
  - horizontal single-stage types
  - vertical multi-stage types.

Surface pumps for private water supplies are available either as single components or as packaged units containing all the necessary equipment pre-fitted. The latter are the easiest to install and only require the final plumbing and electrical connections.

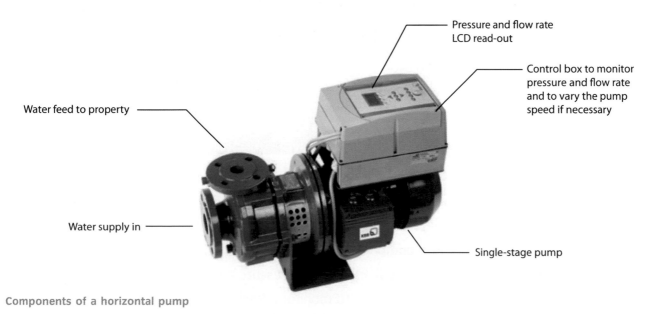

Pressure and flow rate LCD read-out

Control box to monitor pressure and flow rate and to vary the pump speed if necessary

Water feed to property

Water supply in

Single-stage pump

Components of a horizontal pump

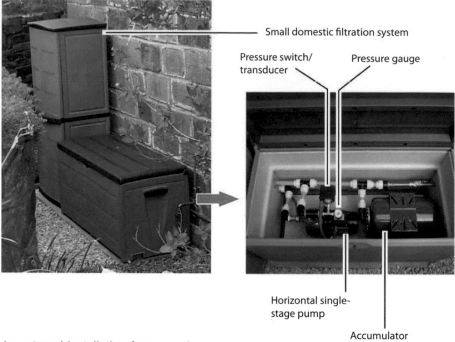

Small domestic filtration system

Pressure switch/
transducer

Pressure gauge

Horizontal single-
stage pump

Accumulator

An external installation from a spring water source

Submersible pumps may be purchased as separate components or in pack form, where all the separately matched equipment is supplied ready to assemble.

A typical submersible pump kit

A typical pump package would normally consist of the following components:

- the pump
- a transducer to sense pressure and flow
- a control box to monitor pressure differentials and flow rate
- an accumulator to assist with providing sufficient system pressure for the installation
- a float switch to prevent the pumps running dry.

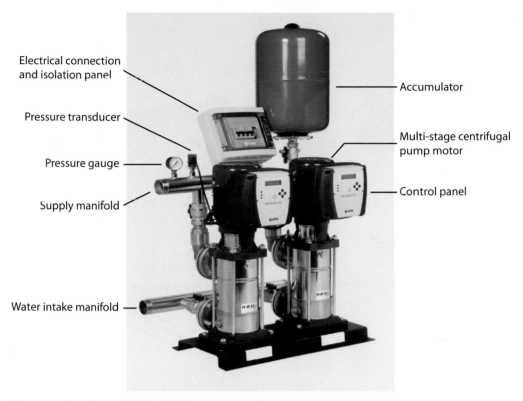

Electrical connection and isolation panel

Pressure transducer

Pressure gauge

Supply manifold

Water intake manifold

Accumulator

Multi-stage centrifugal pump motor

Control panel

Components of a vertical multi-stage pump set

Situations where rainwater harvesting may be appropriate

The UK water supplies arc under stress. Increasing personal consumption, changes to the rainfall patterns caused by climate change and population growth are just some of the contributing factors to an increasingly bleak water supply outlook. Government policy recognises the problem and this is reflected in various pieces of legislation such as the Building Regulations, the Code for Sustainable Homes and the Building Research Establishment Environmental Assessment Method (BREEAM) for sustainable buildings. These call for a reduction in use of the UK mains water supply, wherever possible, by using economising methods and using non-wholesome water for activities such as WC flushing, clothes washing and garden watering.

Because of this, the water industry has reinvented the ancient practice of rainwater harvesting (RWH) in a modern, reliable system that automatically cleans and stores water that, although not wholesome, can be used for most non-drinking water usage. This is typically done by collecting water that falls from the roofs of buildings.

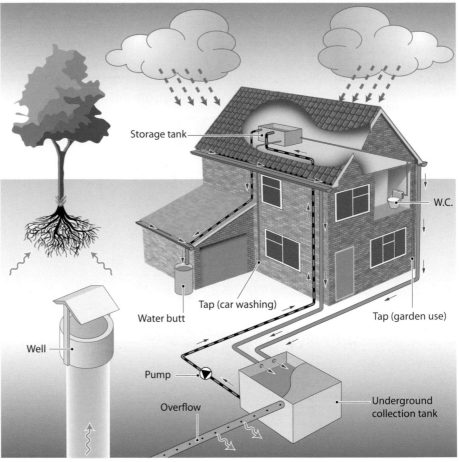

Rainwater harvesting

Know the requirements for backflow protection in plumbing services (LO3)

SmartScreen Unit 303

PowerPoint 3 and Handout 9

There are seven assessment criteria for this Outcome:

1 Determine the fluid risk levels as laid down in water legislation.

2 Compare point-of-use protection with whole-site and zone protection.

3 Propose where non-mechanical backflow prevention devices (air gaps) may be used.

4 Propose where mechanical backflow prevention devices may be used.

5 Explain the regulations for RPZ or RPZD valves.

6 Determine methods of preventing cross-connection in systems that contain non-wholesome water sources.

7 Analyse over-the-rim and ascending-spray sanitary appliances in relation to the Water Regulations.

Range	
Fluid risk levels	Levels 1–5
Non-mechanical	Types AA, AB, AD, AG, AUK1, AUK2, AUK3; DC pipe interrupter
Mechanical	Types BA, CA, DB, EA/EB, EC/ED, HA, HUK1, HC (4.4)
Methods	System plans, colour coding, labelling
Sanitary appliances	Bidet, toilet with cleansing device

The Water Supply (Water Fittings) Regulations 1999 were implemented to harmonise the working practices of plumbers in England and Wales (Scotland has its own virtually identical Scottish Water Byelaws 2004). Before the regulations became law, each local water undertaker had its own set of water byelaws, which were written around the 101 Model Water Byelaws issued in 1986 by the UK government. These gave some commonality to working practices but failed to address the regional variations that often led to confusion regarding what could and could not be done within a particular area or region. On 1 July 1999, the office of the Deputy Prime Minister issued the first ever Water Regulations to be enforced in the UK, effectively eliminating these variations to enshrine a common practice. These were linked to a British Standard, BS 6700 – Design, installation, testing and maintenance of services supplying water for domestic use within buildings and their curtilage, to provide the definitive guide to hot and cold water systems and installations in England and Wales.

Remember, the Water Supply (Water Fittings) Regulations 1999 were put in place to ensure that the plumbing systems we install and maintain prevent the following:

- contamination of water
- wastage of water
- misuse of water
- undue consumption of water
- erroneous metering of water.

Of these, by far the biggest issues surround contamination. Until 1999, there were three classes of fluid. At that time, water was either considered wholesome, suspicious or dangerous. With the implementation of the Water Regulations, these three fluid categories became five to bring the UK into line with the rest of Europe.

This part of the unit investigates the different fluid categories as defined by the Water Supply (Water Fittings) Regulations 1999 and

KEY POINT

BS 6700 has been superseded by BS EN 806:2012 – Specifications for installations inside buildings conveying water for human consumption. This is divided into five parts: Part 1: General, Part 2: Design, Part 3: Pipe sizing – simplified method, Part 4: Installation, Part 5: Operation and maintenance.

the methods, both physical and mechanical, you can employ to prevent contamination of domestic cold water installations by back pressure and back siphonage.

Fluid risk levels as laid down in water legislation

Any water that is not cold wholesome drinking water supplied by a water undertaker can be classed as a potential hazard. The Water Supply (Water Fittings) Regulations 1999 list five fluid categories, described below.

Fluid category 1

Fluid category 1 is wholesome water supplied by a water undertaker, complying with the Water Quality Regulations made under Section 67 of the Water Industry Act 1991. It must be clean, cold and potable. All water undertakers have a duty to supply water that conforms to these regulations, which ensures wholesome water suitable for domestic use or food production purposes. Whenever practicable, water for drinking should be supplied direct from the water undertaker's mains without any intervening storage.

Fluid category 2

Fluid category 2 is water that would otherwise be classified as fluid category 1, but its aesthetic quality has been impaired because of either:

- a change in temperature
- a change in appearance, taste or odour owing to the presence of substances or organisms.

These changes are aesthetic only and do not constitute a health risk. Typical situations in which this may occur in domestic properties include:

- water heated in a secondary hot water system
- mixed fluid category 1 and 2 water discharged from combination taps or showers
- water that has been softened by a domestic common salt regeneration process.

Fluid category 3

Fluid category 3 is water that constitutes a slight health hazard because of the concentration of low-toxicity substances. Water in this category is not suitable for drinking or any other domestic purpose or application. Contaminants include:

- ethylene glycol (anti-freeze), copper sulphate or similar chemical additives such as heating inhibitors, cleansers and de-scalers
- sodium hypochlorite and other common disinfectants.

Typical fluid category 3 situations include:

- In houses, apartments and other domestic dwellings:
  - water in the primary circuits of heating systems, whether chemicals have been administered or not
  - water in wash basins, baths and shower trays
  - washing and dishwashing machines
  - home dialysis machines
  - hand-held garden hoses with a flow-controlled spray or shut-off valve
  - hand-held fertilisers.
- In premises other than single-occupancy domestic dwellings:
  - domestic fittings and appliances such as wash basins, baths or showers installed in commercial, industrial or other premises may be regarded as fluid category 3 – however, if there is a potential for a higher risk, such as in a hospital, medical centre or other similar establishment, then a higher fluid category risk should be applied in accordance with the regulations
  - house garden or commercial irrigation systems without insecticides.

Fluid category 4

Fluid category 4 is water that constitutes a significant health hazard because of the concentration of toxic substances, which can include:

- chemicals, **carcinogenic** substances or pesticides (including insecticides and herbicides)
- environmental organisms of potential health significance.

Carcinogenic
A substance that causes cancer.

Typical fluid category 4 situations include:

- general:
  - primary circuits of heating systems in properties other than a single-occupancy dwelling
  - fire sprinkler systems that contain anti-freeze chemicals.
- house gardens:
  - mini irrigation systems without fertiliser or insecticides, including pop-up sprinkler systems and permeable hoses
- food processing:
  - food preparation
  - dairies
  - bottle washing plants.

- catering:
  - commercial dishwashers
  - refrigerating equipment
- industrial and commercial installations:
  - dyeing equipment
  - industrial disinfection equipment
  - photographic and printing applications
  - car washing and degreasing plant
  - brewery and distilling processes
  - water treatment plant or softeners that use other methods than salt
  - pressurised firefighting systems.

Fluid category 5

Fluid category 5 represents a serious health risk because of the concentration of pathogenic organisms, radioactive material or very toxic substances. These fluids include water that contains:

- faecal material or any other human waste
- butchery or any other animal waste
- pathogens from any source.

Typical fluid category 5 situations include:

- general:
  - industrial cisterns and tanks
  - hose union bib taps in a non-domestic installation
  - sinks, WC pans, urinals and bidets
  - permeable pipes in any non-domestic garden whether laid at or below ground level
  - greywater recycling systems.
- medical:
  - laboratories
  - any medical or dental equipment with submerged inlets
  - bedpan washers and slophoppers
  - mortuary and embalming equipment
  - hospital dialysis machines
  - commercial clothes washing equipment in care homes and similar premises
  - baths, wash basins, kitchen sinks and other appliances that are in non-domestic installations.
- food processing:
  - butchery and meat trade establishments

- slaughterhouse equipment
- vegetable washing.
- catering:
  - dishwashing machines in healthcare premises and similar establishments
  - vegetable washing.
- industrial/Commercial:
  - industrial and chemical plants
  - laboratories
  - any mobile tanker or gully cleaning vehicles.
- sewerage treatment works and sewer cleaning:
  - drain cleaning plant
  - water storage for agricultural applications
  - water storage for firefighting systems.
- commercial agricultural:
  - commercial irrigation outlets below or at ground level and/or permeable pipes, with or without chemical additives
  - insecticide or fertiliser applications
  - commercial hydroponic systems.

Note: The lists of examples for each fluid category are not exhaustive.

The distinction between fluid category 4 and fluid category 5 is often difficult to interpret. In general we can assume that fluid category 4 is such that the risk to health, because of the level of toxicity or the concentration of substances, is such that harm will occur over a prolonged period of days to weeks to months, whereas the risk from fluid category 5, because of the high concentration of substances or the level of toxicity, is such that serious harm could occur after a very short exposure of minutes to hours to days – or even a single exposure.

Remember that fluid category 1 is clean, cold, wholesome water direct from the water undertaker's main, and no other fluid category must come into contact with it or contamination may occur.

Compare point-of-use protection with whole-site and zone protection

There are many commercial and industrial processes in which the whole or part of a plumbing system can present a high risk of backflow to other parts of the installation or even the water undertaker's mains supply, despite the fact that the installation reaches the required standards. In these circumstances whole-site or zone protection must be installed on those parts that are deemed to be high risk.

Whole-site protection

The term whole-site protection simply means that the water undertaker's main is protected at all times from backflow or back siphonage from any fluid category that is not fluid category 1 by a suitable backflow device. Protection should be at the point of entry of the cold water supply.

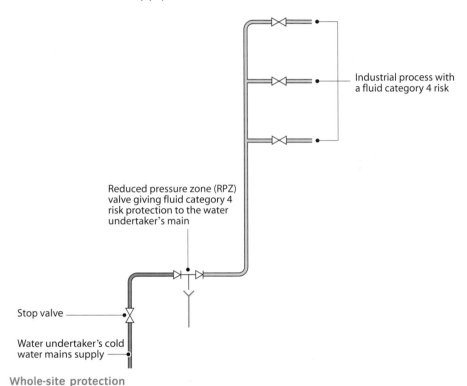

Industrial process with a fluid category 4 risk

Reduced pressure zone (RPZ) valve giving fluid category 4 risk protection to the water undertaker's main

Stop valve

Water undertaker's cold water mains supply

Whole-site protection

If whole-site protection is required, it is important that the water undertaker is informed at the application/notification for water supply stage. They will assess the application for a water supply and advise on what fluid category of backflow protection device must be installed to comply with the Water Supply (Water Fittings) Regulations 1999. The backflow protection device must be installed before the system is commissioned.

Zoned protection

Zoned backflow protection simply means that where different fluid categories exist within the same building, premises or complex, these have their own backflow protection devices to protect any part of the system that is fluid category 1. Zoned protection is also required where any water supply pipe is supplying more than one separately occupied premises.

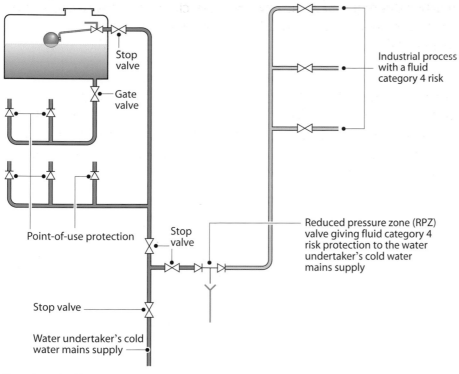

Industrial process with a fluid category 4 risk

Stop valve

Gate valve

Point-of-use protection

Stop valve

Reduced pressure zone (RPZ) valve giving fluid category 4 risk protection to the water undertaker's cold water mains supply

Stop valve

Water undertaker's cold water mains supply

Zoned protection

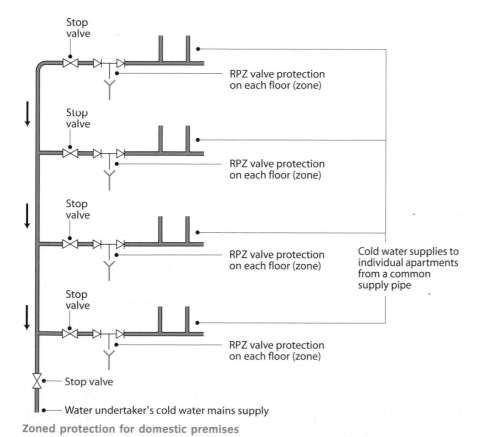

Stop valve

RPZ valve protection on each floor (zone)

Stop valve

RPZ valve protection on each floor (zone)

Stop valve

RPZ valve protection on each floor (zone)

Cold water supplies to individual apartments from a common supply pipe

Stop valve

RPZ valve protection on each floor (zone)

Stop valve

Water undertaker's cold water mains supply

Zoned protection for domestic premises

Point-of-use protection

This is the simplest form of backflow protection. Point-of-use backflow protection devices are used to protect an individual fitting or outlet against backflow and are usually located close to the fitting

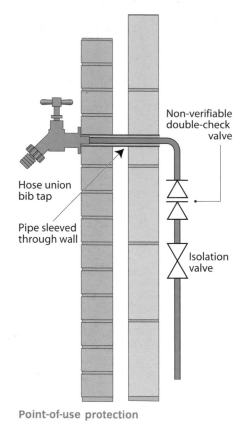

Non-verifiable double-check valve

Hose union bib tap

Pipe sleeved through wall

Isolation valve

Point-of-use protection

SmartScreen Unit 303
Handout 10

that it protects, such as a single-check valve on a mixer tap to protect against fluid category 2 or a double-check valve on a domestic hose union bib tap as protection against fluid category 3.

Eliminating the risk of contamination of wholesome water

The Water Regulations and, more specifically, the Water Regulations Guide, can help us to choose the right course of action based on the risk. The manufacturers, too, help in this regard by designing and manufacturing their appliances, taps and valves to conform to the Water Regulations. For example, most kitchen and bidet taps are designed and made with fluid category 5 risk in mind and most bath and wash basin taps are designed and made with fluid category 3 in mind.

In most cases, where baths, wash basins, bidets and kitchen sinks are concerned, a simple air gap will protect the mains cold water supply. The size of the air gap, however, is dependent on the size of the tap, the appliance type and its likely contents.

Where to use non-mechanical backflow prevention devices (air gaps)

An air gap is simply a physical unrestricted open space between the wholesome water and the possible contamination. The greater the air gap, the greater the level of protection offered. It does not require the use of a mechanical backflow prevention device. Table 3 shows the most important air gaps and how we can apply them.

Table 3: Schedule of non-mechanical backflow prevention arrangements and their respective fluid category protection

Type		Description of backflow prevention arrangements and devices	Fluid category protected against	
			Back pressure	Back siphonage
a	AA	Air gap with unrestricted discharge above spill-over level	5	5
b	AB	Air gap with weir overflow	5	5
c	AD	Air gap with injector	5	5
d	AG	Air gap with minimum-size circular overflow determined by measure or vacuum test	3	3
e	AUK1	Air gap with interposed cistern (eg a WC suite)	3	5

Type		Description of backflow prevention arrangements and devices		Fluid category protected against	
				Back pressure	Back siphonage
f	AUK2	Air gaps for taps and combination fittings (tap gaps) discharging over domestic sanitary appliances, such as a wash basin, bidet, bath or shower tray shall not be less than the following:		X	3
		Size of tap or combination fitting	Vertical distance of bottom of tap outlet above spill-over level of receiving appliance		
		Not exceeding G½	20mm		
		Exceeding G½ but not exceeding G¾	25mm		
		Exceeding G¾	70mm		
g	AUK3	Air gaps for taps or combination fittings (tap gaps) discharging over any higher-risk domestic sanitary appliances where a fluid category 4 or 5 is present, such as: • any domestic or non-domestic sink or other appliance • any appliances in premises where a higher level of protection is required, such as some appliances in hospitals or other healthcare premises shall be not less than 20mm, or twice the diameter of the inlet pipe to the fitting, whichever is the greater.		X	5
h	DC	Pipe interrupter with permanent atmospheric vent		X	5

Notes:

1. X indicates that the backflow prevention arrangement or device is not applicable or not acceptable for protection against back pressure for any fluid category within water installations in the UK.

2. Arrangements incorporating type DC devices shall have no control valves on the outlet of the device; they shall be fitted not less than 300mm above the spill-over level of a WC pan, or 150mm above the sparge pipe outlet of a urinal, and discharge vertically downwards.

3. Overflows and warning pipes shall discharge through, or terminate with, an air gap, the dimension of which should satisfy a type AA air gap.

Each of the air gaps here will have two fluid categories attached to it – one for back pressure and one for back siphonage. The difference between the two is simple:

- **Back pressure** – this is caused when downstream pressure is greater than the upstream or supply pressure in the water undertaker's main or the consumer's potable water supply. Back pressure can be caused by:
  - a sudden loss of upstream pressure, eg due to a burst pipe on a water undertaker's mains supply
  - an increase in downstream pressure caused by pumps or expansion of hot water
  - a combination of the above.

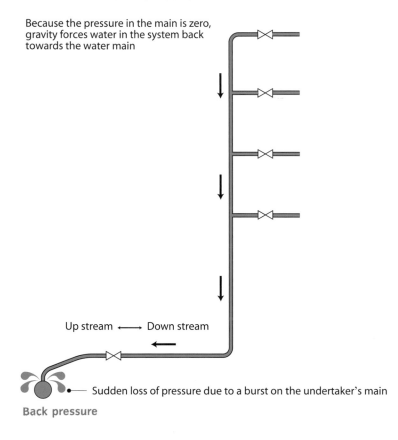

Because the pressure in the main is zero, gravity forces water in the system back towards the water main

Up stream ⟷ Down stream

Sudden loss of pressure due to a burst on the undertaker's main

Back pressure

- **Back siphonage** – this is backflow caused by a negative pressure creating a vacuum or partial vacuum in the water undertaker's mains cold water supply. It is similar to drinking through a straw. If a sudden loss of pressure on the mains supply were to occur while a submerged outlet was flowing, water would backflow upwards through the submerged outlet and down into the water undertaker's main.

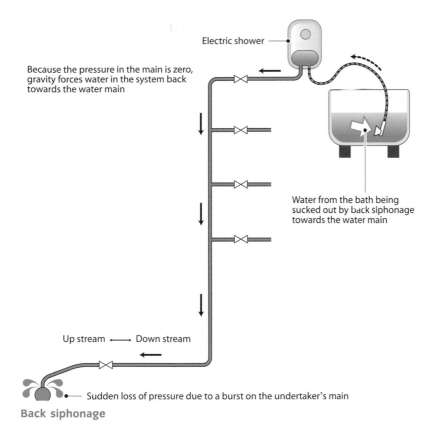

Electric shower

Because the pressure in the main is zero, gravity forces water in the system back towards the water main

Water from the bath being sucked out by back siphonage towards the water main

Up stream ←→ Down stream

Sudden loss of pressure due to a burst on the undertaker's main

Back siphonage

Type AA air gap

This gives protection against fluid category 5 and is a non-mechanical backflow prevention arrangement of water fittings where water is discharged through an air gap into a cistern, which has, at all times, an unrestricted spill-over to the atmosphere. The air gap is measured vertically downwards from the lowest point of the inlet discharge orifice to the spill-over level. Remember the following.

- The type AA air gap is suitable for all fluid categories.
- The size of the air gap is subject to the size of the inlet (see Table 4).
- The flow from the inlet into the cistern must not be more than 15° from the vertical.

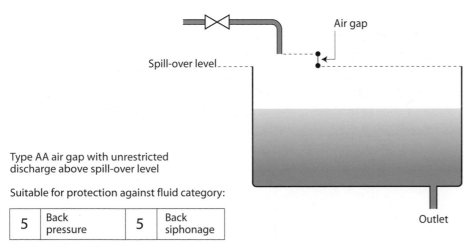

Air gap

Spill-over level

Type AA air gap with unrestricted discharge above spill-over level

Suitable for protection against fluid category:

5	Back pressure	5	Back siphonage

Outlet

A type AA air gap

Table 4: Air gaps at taps, valves, fittings and cisterns

Situation	Nominal size of inlet, tap, valve or fitting	Vertical distance between the tap or valve outlet and the spill-over level of the receiving appliance or cistern
Domestic situation with fluid categories 2 and 3 (AUK2)	Up to and including G½	20mm
	Over G½ and up to G¾	25mm
	Over G¾	70mm
Non-domestic situation with fluid categories 4 and 5 (AUK3)	Any size of inlet pipe	Minimum diameter of 20mm or twice the diameter of the inlet pipe, whichever is the greater of the two

A good example for the use of a type AA air gap would be in an animal drinking trough, where the discharge of water into the trough is in a raised housing on the edge of the trough. The housing is covered to prevent the animals having access to the water supply.

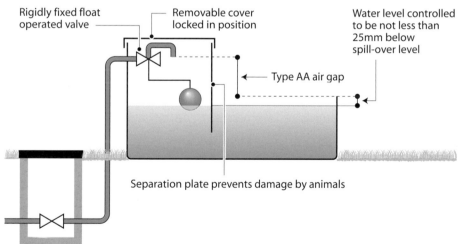

An animal trough

Animal trough schematic

Type AB air gap

A type AB air gap with weir overflow gives protection against fluid category 5. It is a non-mechanical backflow prevention arrangement of water fittings compliant with type AA, except that the air gap is the vertical distance from the lowest point of the discharge orifice which discharges into the receptacle, to the critical level of the rectangular weir overflow.

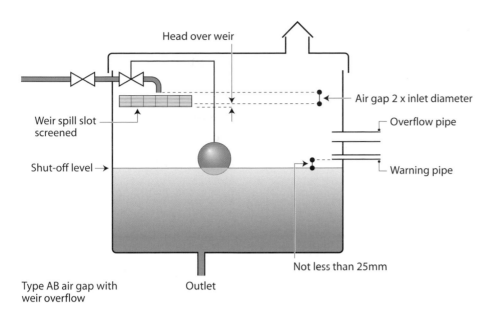

Head over weir

Air gap 2 x inlet diameter

Weir spill slot
screened

Overflow pipe

Shut-off level →

Warning pipe

Not less than 25mm

Type AB air gap with
weir overflow

Outlet

Suitable for protection against fluid category:

5	Back pressure	5	Back siphonage

A type AB air gap with weir overflow

The type AB air gap is suitable for high-risk fluid category 5 situations and is particularly suited to installations where the contents of the cistern need to be protected from contaminants such as insects, vermin and dust. A good example of this is feed and expansion cisterns in industrial/commercial installations, or where high-quality water is required, such as in dental surgeries.

The size of the weir needs to be calculated based on the inlet size. This is usually completed using a weir overflow calculator. An example of a weir calculator can be seen at http://www.airgapcalculator.co.uk/inletcalc/index.html.

A type AB air gap with weir overflow on a cistern

Type AD air gap

A type AD air gap with injector is defined as a non-mechanical backflow prevention arrangement of water fittings with a horizontal injector and a physical air gap of 20mm or twice the inlet diameter, whichever is the greater. It gives protection against back pressure and back siphonage up to fluid category 5. This device is commonly known as a jump jet.

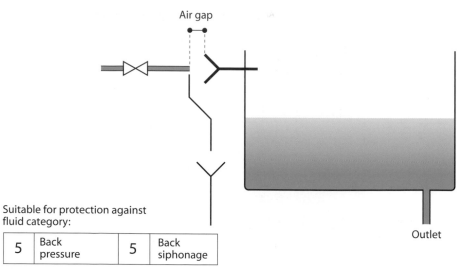

Air gap

Outlet

Suitable for protection against fluid category:

| 5 | Back pressure | 5 | Back siphonage |

A type AD air gap with injector

The principal uses of this type of air gap arrangement are in commercial clothes washing and dishwashing machines. It also has the potential to be used in catering equipment such as steaming ovens.

Type AG air gap

A type AG air gap arrangement with minimum-size circular overflow means a non-mechanical backflow prevention arrangement of water fittings with an air gap – together with an overflow, the size of which is determined by measure or a vacuum test. This arrangement gives protection against fluid category 3.

The type AG air gap fulfils the requirements of BS 6281-2 1982 – Devices without moving parts for the prevention of contamination of water by backflow. Specification for type B air gaps. In a cistern that is open to the atmosphere, the vertical distance between the lowest point of discharge and the critical water level should comply with one of the following requirements.

- It should be sufficient to prevent back siphonage.
- It should not be less than the distances specified in Table 4, depending on cistern type.

Note the following points about type AG air gaps.

- The air gap is related to the size of the inlet supply and is the minimum vertical distance between the critical water level and the lowest part of the discharge outlet of the FOV, as specified in Table 4.
- The critical water level is the level that is reached when the FOV has failed completely and the water is running freely at maximum full-bore flow rate and pressure.
- AG air gaps must comply with the requirements of BS 6281.

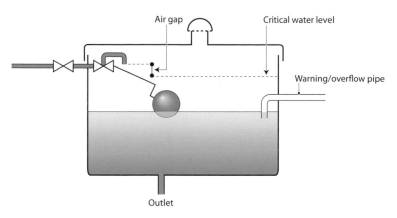

Suitable for protection against fluid category:

3	Back pressure	3	Back siphonage

A type AG gap with minimum size circular overflow determined by measure or vacuum test

Where storage cisterns are installed, it is likely that the critical water level would differ from installation to installation because of varying flow rates and pressures of the incoming supply and the differing lengths and gradients of the overflow pipe. With this type of installation, the type AG air gap is not practical because the critical water level cannot be accurately calculated. It is the critical water level that would determine the position on the cistern of the FOV and the distance between the FOV and the overflow.

AUK1 air gap

A type AUK1 air gap with interposed cistern is a non-mechanical backflow prevention arrangement consisting of a cistern incorporating a type AG overflow and an air gap. The spill-over level of the receiving vessel is located not less than 300mm below the overflow pipe and not less than 15mm below the lowest level of the interposed cistern. It is suitable for protection against fluid category 5 for back siphonage and fluid category 3 for back pressure.

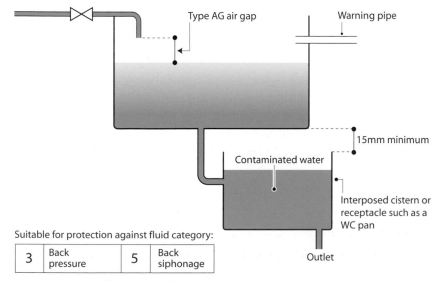

Suitable for protection against fluid category:

3	Back pressure	5	Back siphonage

An AUK1 air gap with interposed cistern

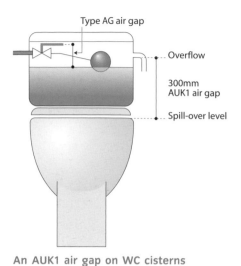

An AUK1 air gap on WC cisterns

This arrangement is most commonly found on WC installations, with the WC pan being the receiving vessel containing fluid category 5 water. A conventional domestic WC suite consists of a 6l/4l dual flushing cistern, a part 2, 3 or 4 FOV with an AG air gap and an overflow arrangement. This creates an AUK1 interposed cistern or, in other words, a cistern that can be supplied from a mains supply or another protected cistern without the need for additional backflow protection.

AUK2 air gap

A type AUK2 air gap for taps and combination fittings (tap gaps) discharging over domestic sanitary appliances means the height of air gap between the lowest part of the outlet of a tap, combination fitting, shower head or other fitting discharging over a domestic sanitary appliance or other receptacle, and the spill-over level of that appliance, where a fluid category 2 or 3 risk is present downstream. An AUK2 air gap is only suitable for back siphonage up to fluid category 3 and must comply with the distances stated in Table 4.

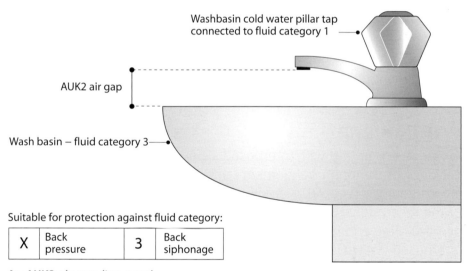

Suitable for protection against fluid category:

X	Back pressure	3	Back siphonage

An AUK2 air gap (tap gaps)

AUK3 air gap

A type AUK3 higher-risk tap gap means the height of an air gap between the lowest part of the outlet of a tap, combination fitting, shower head or other fitting discharging over any appliance or other receptacle, and the spill-over level of that appliance, where a fluid category 4 or 5 risk is present downstream.

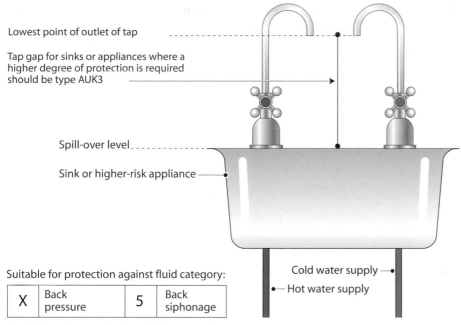

Lowest point of outlet of tap

Tap gap for sinks or appliances where a higher degree of protection is required should be type AUK3

Spill-over level

Sink or higher-risk appliance

Cold water supply

Hot water supply

Suitable for protection against fluid category:

X	Back pressure	5	Back siphonage

An AUK3 air gap (higher-risk tap gaps)

In a domestic dwelling, AUK3 air gaps are most common at the kitchen sink in the form of high-necked pillar taps, sink mixer taps or sink monobloc taps. Sink mixers and monoblocs have a swivel spout. If a cleaner's sink, Belfast sink or London sink is being installed, it is important that any bib taps installed are positioned so as to maintain an AUK3 air gap.

A typical kitchen sink showing an AUK3 air gap

Taps and combination fittings discharging on non-domestic appliances and any appliances in premises where a higher level of protection is required, such as appliances in hospitals or other healthcare premises, require a type AUK3 tap gap.

DC pipe interrupter with a permanent atmospheric vent

A type DC pipe interrupter with permanent atmospheric vent means a non-mechanical backflow prevention device with a permanent unrestricted air inlet, with the device being installed so that the flow of water is in a vertical downward direction. This kind of device is used where there is a threat of back siphonage from a fluid category 5.

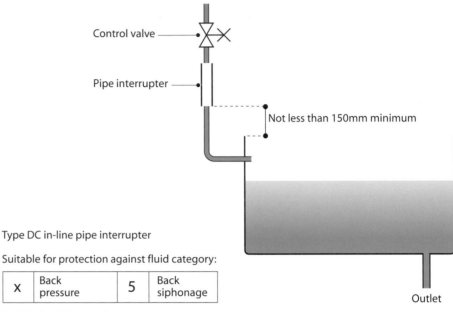

Control valve

Pipe interrupter

Not less than 150mm minimum

Outlet

Type DC in-line pipe interrupter

Suitable for protection against fluid category:

X	Back pressure	5	Back siphonage

A DC pipe interrupter

The idea behind the DC pipe interrupter is to create an air inlet should a back siphonage situation occur. When water begins to backflow upwards, the DC pipe interrupter allows air into the system to break the siphonic action, thus preventing contamination.

A typical DC pipe interrupter

Type DC pipe interrupter: this device must be fitted with the lowest point of the air aperture not less than 150mm above the free discharge or spill-over level of an appliance and have no valve, flow restrictor or tap on its outlet.

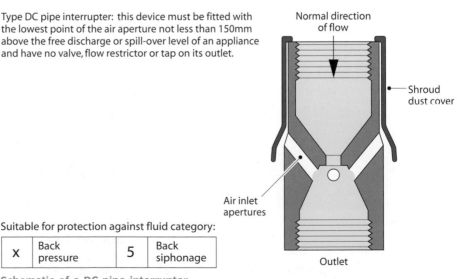

Normal direction of flow

Shroud dust cover

Air inlet apertures

Outlet

Suitable for protection against fluid category:

X	Back pressure	5	Back siphonage

Schematic of a DC pipe interrupter

The DC pipe interrupter is a non-mechanical fitting. It does not contain any moving parts, and the parts are manufactured from corrosion-resistant brass. Typical uses include WCs and urinal installations. Bear in mind the following points:

- the valve should be fitted in the vertical position, discharging downwards
- it must be installed at least 300mm above the overflowing level, or 150mm if fitted above a urinal
- no tap or valve should be installed downstream of the interrupter
- pipe size reductions downstream of the interrupter are not allowed
- the length of the pipe downstream after the interrupter should be as short as possible
- the interrupter should be accessible for replacement and repair
- DC pipe interrupters must comply with BS 6281-3.

Where to use mechanical backflow prevention devices

An air gap is the most effective method of preventing contamination of the water supply, and most installers will try to achieve this within their installations and designs, but there are many cases where air gaps are not practical as a method of protection. In these instances, installers may opt to install a mechanical backflow prevention device. These provide a physical barrier to backflow. However, do remember that mechanical backflow prevention devices have limitations and can be subject to failure.

SmartScreen Unit 303

Handout 11

This section of the unit looks at some of the more common mechanical backflow prevention devices and where we can install them.

Table 5: Schedule of mechanical backflow prevention arrangements and fittings and their respective fluid category protection

Type		Description of backflow prevention arrangements and devices	Suitable for protection against fluid category	
			Back pressure	Back siphonage
a	BA	Verifiable backflow preventer with reduced pressure zone	4	4
b	CA	Non-verifiable disconnector with difference between pressure zones not greater than 10%	3	3
c	DB	Pipe interrupter with atmospheric vent and moving element	X	3
d	DC	Pipe interrupter with permanent atmospheric vent	X	5

Type		Description of backflow prevention arrangements and devices	Suitable for protection against fluid category	
			Back pressure	Back siphonage
e	EA/EB	Verifiable and non-verifiable single-check valves	2	2
f	EC/ED	Verifiable and non-verifiable double-check valves	3	3
g	HA	Hose union backflow preventer. Only permitted for use on existing hose union bib tap in house installations	2	3
h	HUK1	Hose union bib tap incorporating a double-check valve arrangement. Only permitted as a replacement for existing bib taps in house installations	3	3
i	HC	Diverter with automatic return (normally integral with some domestic appliance applications only)	X	3

Notes:

'X' indicates that the backflow prevention device is not acceptable for protection against back pressure for any fluid category.

Arrangements incorporating a type DB device shall have no control valves on the outlet of the device. The device shall not be fitted less than 300mm above the spill-over level of an appliance and must discharge vertically downwards.

Relief ports from BA and CA devices should terminate with an air gap, the dimensions of which should satisfy a type AA air gap.

Type BA verifiable backflow preventer with reduced pressure zone

RPZ valve

A reduced pressure zone valve is a backflow protection device used to protect a category 1 fluid from fluid category 4 contamination.

RPZ valve cutaway

Better known as an **RPZ (reduced pressure zone) valve**, this is a mechanical, verifiable, backflow prevention device, offering protection to water supplies up to and including fluid category 4. Verifiable simply means that the valve can be checked (verified) via test points to see if it is working correctly.

Most RPZ valves consist of three separate elements:

- two check valves
- a differential relief valve
- three test points.

The first check valve is spring loaded to generate a specific pressure drop across this part of the valve. This creates a reduced pressure zone downstream in the middle chamber of the valve and on the downstream side of the differential relief valve. The incoming mains supply maintains pressure on the upstream side of the differential valve and, as long as the mains pressure is higher, the differential relief valve will remain closed.

If, under static conditions, the mains pressure reduces until it is just 0.14 bar above the pressure in the reduced pressure zone, the differential relief valve will open and release the contents of the middle chamber to drain. Should backflow occur past the first check valve element, the pressure on both sides of the differential valve will equalise and the differential relief valve will open to discharge the water.

If complete mains failure occurs, the contents of the middle chamber are discharged to drain providing that both check valve elements are functioning correctly. However, should the upstream check valve become faulty, the pressure in the middle chamber will equalise to that of mains pressure and the differential relief valve will open and continuously discharge water at a steady rate. If the downstream check valve fails under zero mains pressure conditions, the differential relief valve will open and water will discharge from the downstream side of the system until the pressure there also becomes zero.

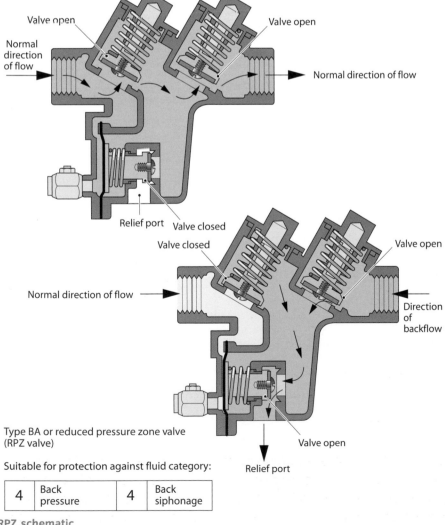

Type BA or reduced pressure zone valve
(RPZ valve)

Suitable for protection against fluid category:

4	Back pressure	4	Back siphonage

RPZ schematic

Testing, commissioning, maintenance and annual inspection can only be carried out by a trained and approved installer. Anyone who tests RPZ valves must be certificated. Specialist training is available from various test centres across the UK. Further recommended reading is the Water Regulations Advisory Scheme Information and Guidance Note 9-03-02.

Type CA non-verifiable disconnector with difference between pressure zones not greater than 10%

These devices are very similar to BA devices (RPZ valves) in that they provide a positive disconnection chamber between the downstream water and the upstream water. The disconnection area between the two main check valves is open to the atmosphere under fault conditions, thereby maintaining an air gap should a loss of upstream pressure occur. Like the RPZ valve, any water discharged would run to drain via a tundish. These devices are suitable for fluid category 3.

A typical use of a type CA disconnector is a permanent connection between a sealed central heating system and the water undertaker's cold water supply.

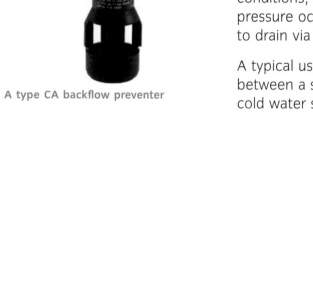

A type CA backflow preventer

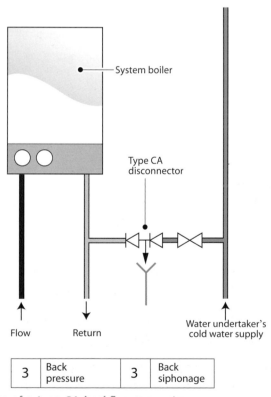

Use of a type CA backflow preventer

3	Back pressure	3	Back siphonage

Type DB pipe interrupter with atmospheric vent and moving element

The type DB pipe interrupter is a backflow prevention device specifically designed for fluid category 4 applications. The concept of the DB interrupter is very simple. Water enters a tube which has one end blanked off. Around the tube is a series of small holes over which a flexible rubber membrane is stretched. As the water flows into the tube, it is forced through the holes and this flexes the rubber membrane to allow water to flow. If the supply pressure suddenly stops, the membrane contracts against the holes to effectively prevent backflow. Any backflowing water is then released to the atmosphere through another series of holes in the outer casing of the device. They are approved for use as protection against back siphonage but not back pressure.

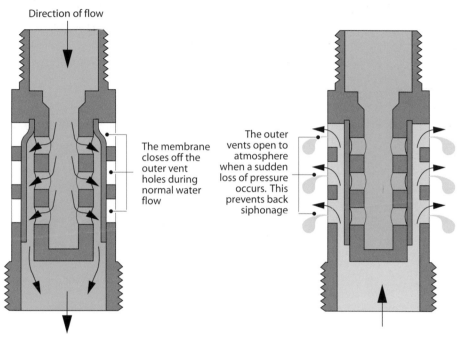

Direction of flow

The membrane closes off the outer vent holes during normal water flow

The outer vents open to atmosphere when a sudden loss of pressure occurs. This prevents back siphonage

A type DB pipe interrupter schematic

DB pipe interrupters are generally used externally as attachments to hose union bib taps and must not be used on appliances that have a control valve restriction, such as a clothes washing machine. They are resistant to frost damage. They must be fitted vertically and have no valves fitted downstream of the device.

Some DB interrupters are manufactured with bayonet-type attachments for domestic garden perforated hose irrigation systems.

A type DB pipe interrupter

Type EA and type EB verifiable and non-verifiable single-check valves

These two valves are the most simple of all mechanical backflow prevention devices and can be used to protect against fluid category 2 for both back pressure and back siphonage. Generally regarded as point-of-use protection, they consist of a spring-loaded one-way valve that will allow water to flow from upstream to downstream only. If back siphonage or back pressure occurs, the valve will shut to prevent a reverse water flow. When no water is flowing, the valve remains in the closed position. Both types are almost identical in appearance. The difference between them is that the type EA device has a test nipple situated on the upstream side of the valve so that it can be tested while in position to verify that it is working correctly. The type EB non-verifiable single-check valve does not have a test point but can be used in the same way as the type EA single-check valve.

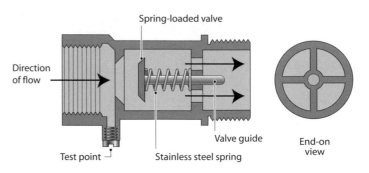

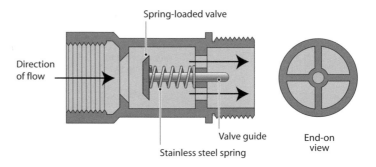

Suitable for protection against fluid category:

Type EA/EB single-check valve			
2	Back pressure	2	Back siphonage

A type EA verifiable single-check valve and a type EB non-verifiable single-check valve

Both valves are manufactured from DZR (de-zincification-resistant) brass and have either type A compression fittings or female BSP (British Standard Pipe) threads for connection to the pipework. The valves should conform to BS 6282-1 for use in hot or cold water systems up to 90°C.

In domestic premises the risk from fluid category 2 generally occurs where the hot and cold supplies are taken to a single terminal fitting such as mixer taps or shower valves. This is known as a cross-connection. However, care must be taken when installing single-check valves to hot water supplies as the expansion of the water can cause excessive pressure on the check valve, causing it to fail. Other uses include the cold water connections to drinks machines.

Verifiable and non-verifiable single-check valves

Type EC and type ED verifiable and non-verifiable double-check valves

Type EC and type ED double-check valves are mechanical backflow prevention devices consisting of two single-check valves in series, which will permit water to flow from upstream to downstream but not in the reverse direction. They are used primarily to protect against fluid category 3 for both back pressure and back siphonage.

The type EC verifiable double-check valve has two test nipples – one on the upstream side of the first check valve and another in the chamber between the first and second check valves. These are used to verify that the valve is working correctly. The type ED non-verifiable double-check valve does not have a test point but can be used in the same way as the type EA single-check valve.

Suitable for protection against fluid category:

Type EC/ED single-check valve			
3	Back pressure	3	Back siphonage

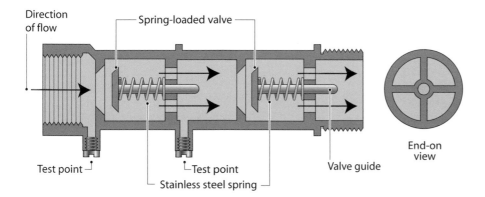

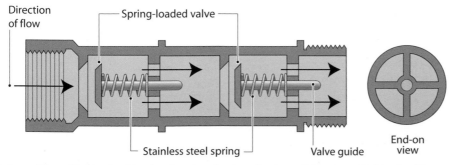

A type EC verifiable double-check valve (top) and a type ED non-verifiable double-check valve (bottom)

Typical uses in domestic installations include garden hose union bib taps and sealed heating systems fitted in conjunction with a temporary filling loop. When used with sealed heating systems, the double-check valve must be fitted to the cold water supply connection to the filling loop and not to the sealed heating connection.

Type HA hose union backflow preventer

As the name suggests, this mechanical backflow prevention device screws on to the outlet thread of a hose union bib tap. It is only permitted for use on existing hose union bib taps in house installations that do not have any form of backflow protection. It is used to protect against back pressure at fluid category 2 and back siphonage at fluid category 3.

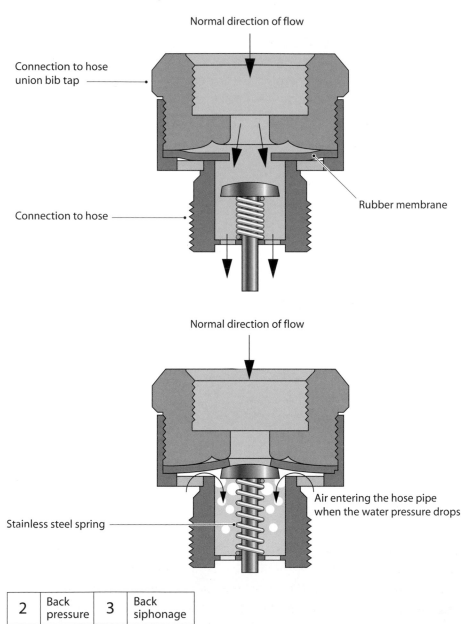

2	Back pressure	3	Back siphonage

A type HA hose union backflow preventer

Type HUK1 hose union bib tap incorporating a double-check valve arrangement

This hose union bib tap incorporates two single-check valves, one situated at the inlet to the tap and one at the outlet. A screw-type test point is also included on the tap body. These devices are fitted in the same way as a normal hose union bib tap. However, they are not suitable for new installations and can only be used as replacements where a hose union bib tap already exists. This is simply because the Water Supply (Water Fittings) Regulations 1999 state that any mechanical backflow prevention device should be fitted *within* the envelope of the building to prevent damage by freezing. They are suitable as protection against fluid category 3 for both back pressure and back siphonage.

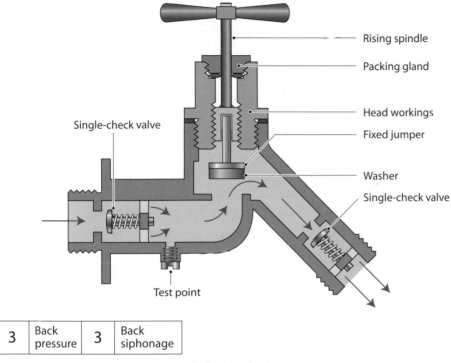

3	Back pressure	3	Back siphonage

A type HUK1 hose union bib tap with double-check valve arrangement

Type HC diverter with automatic return

A type HC diverter with automatic return is a mechanical backflow prevention device used in bath/shower combination tap assemblies which automatically returns the bath outlet open to atmosphere if a vacuum occurs at the inlet to the device.

The type HC diverter with automatic return is usually incorporated into the design of an appliance or fitting. It is not a stand-alone fitting that can be added to the installation. A good example of a type HC diverter would be a bath/shower mixing valve with a diverter valve to operate the shower. While pressure is maintained, the diverter valve remains open and the water is fed to the shower hose. Should loss of pressure occur, the diverter valve closes and any excess water in the shower hose returns to the bath through the open tap, thus preventing the water from backflowing down the cold supply pipe. They are suitable for fluid category 3 to prevent back siphonage only.

A bath/shower mixer tap with a type HC diverter

Section 6.4 of the Water Regulations: guidance clauses relating to backflow prevention

Note: Where tables and figures are mentioned, these refer to those shown in the Water Regulations.

General

G15.1 Except where expanded water from hot water systems or instantaneous water heaters is permitted to flow back into a supply or distributing pipe, every water fitting through which water is supplied for domestic purposes should be installed in such a manner that no backflow of fluid from any appliance, fitting or process can take place.

G15.2 Avoidance of backflow should be achieved by good system design and the provision of suitable backflow prevention arrangements and devices, the type of which depends on the fluid category to which the wholesome water is discharged. A description of fluid risk categories is shown in Schedule 1 of the regulations and some suggested examples relating to the fluid categories are shown in Tables 4a to e.

G15.3 The type of backflow protection for a given situation is related to the fluid risk categories downstream of the backflow prevention device.

G15.4 Schedules of backflow prevention arrangements and backflow prevention devices, and the maximum permissible fluid risk category for which they are acceptable, are shown in Table 4 and Table 6.

G15.5 Wherever practicable, systems should be protected against backflow without the necessity to rely on mechanical backflow protection devices; this can often be achieved by point-of-use protection such as a 'tap gap' above the spill-over level of an appliance. Minimum air gaps for different sizes of taps and applications are shown in Table 4.

G15.6 In cistern-fed systems secondary backflow prevention can often be achieved for appliances by the use of permanently vented distributing pipes.

G15.7 Mechanical backflow protection devices which, depending on the type of device, may be suitable for protection against back pressure or back siphonage, or both, should be installed so that:

■ they are readily accessible for inspection, operational maintenance and renewal

■ except for types HA and HUK1, backflow prevention devices for protection against fluid categories 2 and 3, they should not be located outside premises

■ they are not buried in the ground

■ vented or verifiable devices, or devices with relief outlets, are not installed in chambers below ground level or where liable to flooding

■ line strainers are provided immediately upstream of all backflow prevention devices required for fluid category 4. Where strainers are provided, servicing valves are to be fitted upstream of the line strainer and downstream of the backflow prevention device

■ the lowest point of the relief outlet from any reduced pressure zone valve assembly or similar device should terminate with a type AA air gap located not less than 300mm above the ground or floor level

Note: For information on the installation and maintenance of reduced pressure zone devices (RPZ valve assemblies) see Information and Guidance Note 9-03-02 published by the Water Regulations Advisory Scheme.

Appliances incorporating or supplied with water through pumps

G15.8 Where pumped showers, or other appliances supplied through or incorporating pumps, are installed care should be taken in positioning branches from distributing pipes.

Bidets (including WCs adapted as bidets) with flexible hose and spray handset fittings and with submerged water inlets

G15.9 Bidets with flexible hose and spray handset fittings and/or water inlets below the spill-over level of the appliance, are a fluid category 5 risk and should not be supplied with water directly from a supply pipe.

G15.10 Bidets of this type may:

- be supplied with cold and/or hot water through type AA, AB or AD backflow prevention arrangements serving the bidet only

- be supplied with cold water from an independent distributing pipe serving the bidet only or a common distributing pipe serving the bidet and which may also serve a WC or urinal flushing cistern only

- be supplied with hot water from a water heater, which is supplied from an independent distributing pipe, that serves the bidet only (see Figure 6.1a)

- where the bidet is at a lower elevation than any other outlets or appliances, be supplied with water from a common cold and/or hot water vented distributing pipe providing that:

  - the elevation of the spill-over level of the bidet, if there is no flexible hose, or

  - the elevation of the spray outlet, with the hose extended vertically above the spill-over level of the bidet, whichever is the highest

is not less than 300mm below the point of connection of the branch pipe serving the bidet to the main distributing pipe serving other appliances.

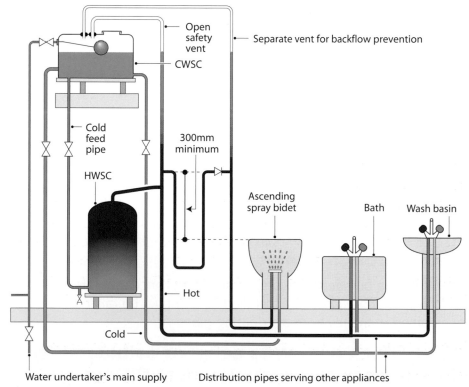

Connections to an ascending spray bidet

Bidets with water inlets above spill-over level only

G15.11 Bidets in domestic locations with taps or mixers located above the spill-over level of the appliance, and not incorporating an ascending spray inlet below spill-over level or spray and flexible hose, may be served from either a supply pipe or a distributing pipe provided that the water outlets discharge with a type AUK2 air gap above the spill-over level of the appliance. See Table 4.

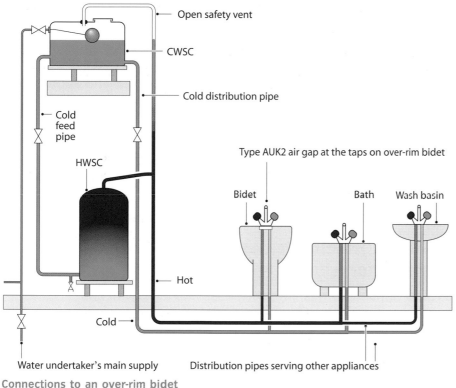

Connections to an over-rim bidet

WCs and urinals

G15.12 The water supply to a manually operated WC or urinal flushing valve may be derived either from a supply pipe or a distributing pipe. The flushing valve should be located above the WC pan or urinal and must incorporate, or discharge through, a pipe interrupter with a permanent atmospheric vent – see type DC in Table 4 and Table 5. The lowest part of the vent opening of the pipe interrupter should be located not less than 300mm above the spill-over level of the WC pan or not less than 150mm above the sparge outlet of a urinal.

Note: Flushing valves cannot be used in domestic WCs or urinals.

Shower heads or tap inlets to baths, wash basins, sinks and bidets

G15.13 Except where suitable additional backflow protection is provided, all single tap outlets, combination tap assembly outlets, or fixed shower heads terminating over wash basins, baths or bidets in

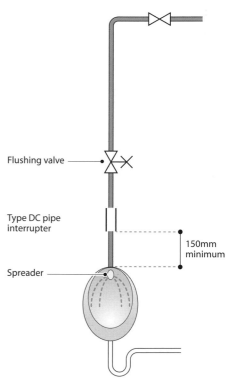

Use of flushing valves and DC pipe interrupters

domestic situations should discharge above the spill-over level of the appliance with a tap gap (type AUK2) as scheduled in Table 4. For a sink in a domestic or non-domestic location, and for any appliances in premises where a higher level of protection is required, such as some appliances in hospitals or other healthcare premises, a tap gap (type AUK3) is required, see Table 4.

Submerged inlets to baths and wash basins

G15.14 Submerged inlets to baths or wash basins in any house or domestic situation are considered to be a fluid category 3 risk and should be supplied with water from a supply or distributing pipe through a double-check valve. Submerged inlets to baths or wash basins in other than a house or domestic situation, and sinks in any location, are considered to be a fluid category 5 risk and appropriate backflow protection will be required.

Drinking water fountains

G15.15 Drinking water fountains should be designed so that the outlet of the water delivery jet nozzle is at least 25mm above the spill-over level of the bowl. The nozzle should be provided with a screen or hood to protect it from contamination.

Washing machines, washer-dryers and dishwashers

G15.16 Household washing machines, washer-dryers and dishwashers are manufactured to satisfy a fluid category 3 risk. Where they are likely to be used in a non-domestic situation, appropriate backflow protection for a higher fluid risk category should be provided.

Hose pipes for house garden and other applications

G15.17 Hand-held hoses should be fitted with a self-closing mechanism at the outlet of the hose.

A typical self-closing hand-held garden spray

Commercial and other installations excluding house gardens

G15.18 Any taps and fittings used for supplying water for non-domestic applications, such as commercial, horticultural, agricultural or industrial purposes should be provided with:

- backflow protection devices appropriate to the downstream fluid category

- where appropriate, a zone protection system.

G15.19 Soil watering systems installed in close proximity to the soil surface (that is, where the watered surface is less than 150mm below the water outlet discharge point), for example, irrigation systems, permeable hoses etc, are considered to be a fluid category 5 risk and should be supplied with water only through a type AA, AB, AD or AUK1 air gap arrangement.

House garden installations

G15.20 Taps to which hoses are, or may be, connected and located in house gardens are to be protected against backflow by means of a double-check valve. The double-check valve should be located inside a building and protected from freezing.

G15.21 Where, in existing house installations, a hose pipe is to be used from an existing hose union tap located outside a house and which is not provided with backflow protection, either:

- the existing hose union tap should be provided with a double-check valve located inside the building, or

- the tap should be replaced with a hose union bib tap that incorporates a double-check valve (type HUK1), or

- a hose union backflow preventer (type HA) or a double-check valve should be continuously fitted to the outlet of the tap.

G15.22 Where fixed or hand-held devices are used with hose pipes for the application of fertilisers or domestic detergents, the minimum backflow protection provided should be suitable for protection against a fluid category 3 risk. Backflow protection against a fluid category 5 risk should be provided where these devices are used for the application of insecticides.

G15.23 Where mini-irrigation systems, such as porous hoses, are installed in house garden situations only, a hose union tap with backflow protection in accordance with clauses G15.20 or G15.21 combined with a pipe interrupter with atmospheric vent and moving element device (type DB) at the connection of the hose to the hose union tap – or not less than 300mm above the highest point of the delivery point of the spray outlet or the perforated surface of the porous hose, whichever is the highest – is acceptable.

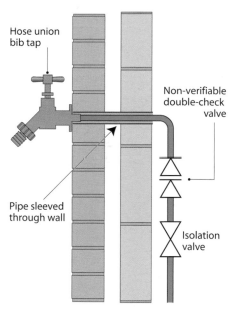

Hose union bib tap

Non-verifiable double-check valve

Pipe sleeved through wall

Isolation valve

A domestic hose union bib tap installation

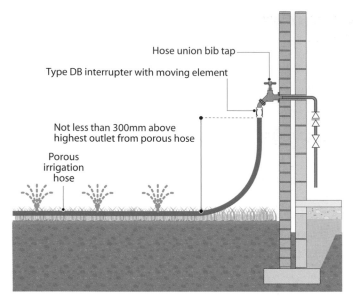

A porous hose installation

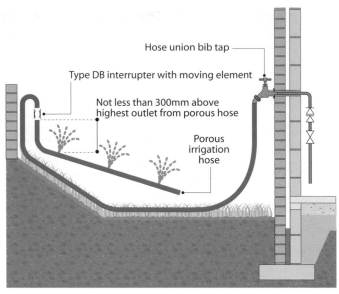

A porous hose installation on rising ground

Whole-site and zone protection

G15.24 A whole-site or zone backflow prevention device should be provided on the supply or distributing pipe. This may be a single-check valve or double-check valve, or another no less effective backflow prevention device, according to the level of risk as judged by the water undertaker, where:

- a supply or distributing pipe conveys water to two or more separately occupied premises (whether or not they are separately chargeable by the water undertaker for a supply of water), or

- a supply pipe conveys water to premises, which under any enactment are required to provide a storage cistern capable of holding sufficient water for not less than 24 hours of ordinary use.

G15.25 The provision of zone or whole-site backflow protection should be in addition to individual requirements at points of use and within the system.

G15.26 Zone protection may be required in other than domestic premises where particular industrial, chemical or medical processes are undertaken.

Fire protection systems

G15.27 Wet sprinkler systems (without additives), first-aid fire hose reels and hydrant landing valves are considered a fluid category 2 backflow risk. Wet sprinkler systems with additives to prevent freezing are considered a fluid category 4 risk.

Hydro-pneumatic

A pressure intensifier that enables generation of great force.

G15.28 Fluids contained within large cylindrical **hydro-pneumatic** pressurised vessels are considered to be a fluid category 4 risk.

G15.29 Where fire protection systems and drinking water systems are served from a common domestic supply pipe, the connection to the fire systems should be taken from the supply pipe immediately on entry to the building, and appropriate backflow protection devices should be installed.

The regulations for RPZ valves or RPZDs

Reduced pressure zone (RPZ) valves and reduced pressure zone devices (RPZDs) must be installed in accordance with the Water Supply (Water Fittings) Regulations 1999. This is detailed in the WRAS Information and Guidance Note 9-03-02, from which the following notes are taken. Remember that commissioning and testing of an RPZ valve must only be carried out by an accredited tester approved by the water undertaker as being competent to test.

General requirements

The installation and use of RPZ valves and RPZDs must be notified in advance to the water undertaker and may be subject to additional installation terms and conditions. The end user must be made aware of these terms and conditions regarding installation and maintenance. Some supplies may not be suitable for RPZ installation due to an initial lack of pressure, as RPZ valves operate on the principle of a pressure differential. Therefore, the water pressure at the intended location must be known prior to installation.

Testing methods and maintenance regimes must be in accordance with the water undertaker's requirements. Failure to comply with such regimes increases the risk of contamination by backflow and may result in the water supply to the building being temporarily suspended. This may also result in the water undertaker requesting removal of the RPZ valve and the installation of an alternative arrangement.

Water fittings must satisfy the requirements of the regulations. Reputable manufacturers have their valves and fittings tested to show WRAS approval and compliance with the requirements of the regulations.

Approved contractors

- The regulations allow water undertakers to designate approved contractors and these are allowed to certify that the installation (excluding commissioning and testing) or alteration they have carried out complies with the regulations. On completion, the approved contractor must provide a copy of the compliance certificate to the end user or person who requested that the installation be carried out. A copy must also be sent to the water undertaker.

- Under the regulations, consent to alter or extend plumbing systems is not required if the work is to be carried out by an approved contractor, and this includes backflow protection devices up to fluid category 4 risks. However, it is a requirement under the Approved Installation Method (AIM) for RPZ valves and devices that prior notice shall be given for the installation of all RPZ valves and devices to ensure that the backflow risk assessment is correct.

Areas of acceptable use

- An RPZD can be installed at any point of use where there is a fluid category 4 risk. Fluid categories were defined earlier in this unit (see pages 196–199) and can be found in Schedule 1 of the Water Supply (Water Fittings) Regulations 1999.

- Where necessary, an RPZ valve can provide protection against backflow from the whole premises (whole-site protection) or from a part of it (zone protection). This does not replace any need for adequate point-of-use protection.

- RPZ valves intended for any other fluid than cold water must be suitably approved.

- In exceptional circumstances and at the discretion of the water undertaker, an RPZ valve may be used for protection against risks that are greater than fluid category 4 for a limited period.

- In all circumstances, the responsibility for adequate protection downstream of the RPZ valve to ensure the water quality rests with the end user.

Table 6 gives examples of fluid category 4 risks where an RPZ valve may be used. This list is not exhaustive and further advice should be sought from the local water undertaker.

Table 6: Examples of fluid category 4 risks where RPZ valves may be used

General	• Fire sprinkler systems using anti-freeze solutions • Primary circuits and central heating systems, in other than a house, with design heat output greater than 45kWh (150,000Btu/h)
Domestic or residential gardens	• Mini-irrigation systems without fertiliser • Insecticide application, eg via pop-up sprinklers or porous hoses
Food processing	• Bottle washing apparatus • Dairies • Food preparation

Catering	• Bottle washing apparatus • Dishwashing machines (not for healthcare patients) • Potato peeling machines (pre-washed produce) • Refrigerating equipment
Industrial and commercial installations	• Brewery and distillation plant • Car washing and degreasing plant • Commercial clothes washing plant, excluding use for medical or healthcare items • Dyeing equipment • Pressurised firefighting systems • Printing and photographic equipment • Water treatment plant using other than salt

Installation

■ RPZ valves shall not be installed in a place or position which is:

- liable to flooding
- above electrical equipment
- liable to mechanical or other damage
- exposed to freezing, unless measures are taken to prevent the assembly from freezing
- concealed.

■ Take care with regard to the design of installations where pressure fluctuations are likely. The installation of a single-check valve upstream of the RPZ device can prevent fluctuations that may lead to frequent discharges from the relief valve. Seek advice from the RPZ valve manufacturer.

■ Any discharge from the relief valve must be visible. There must also be an air gap at the exit port of the relief valve. Any tundish installed must be in accordance with BS EN 1717.

■ RPZ valves must be installed horizontally, with the relief port/ discharge pointing downwards, unless the valve is designed to be fitted vertically. An inline strainer should be fitted before the valve to prevent any debris fouling the valve mechanism.

■ RPZ valves must be installed above floor level at a height that allows effective inspection, testing and maintenance. The minimum height from the ground should not be less than 300mm with a maximum height of 1500mm.

■ Every RPZ valve assembly must be fitted with isolation valves at both inlet and outlet.

■ After installation, the assembly must be flushed and disinfected (if required) prior to being taken into service. This should be in accordance with BS EN 806 (BS 6700).

Site inspection and testing

■ Each RPZ device must have a unique reference number for identification purposes. This should either be permanently attached or engraved onto the device.

■ Site testing of RPZ valves must only be carried out by an approved tester. This should be done at least annually or at more frequent intervals as specified by the water undertaker.

■ Subject to agreement with the water undertaker a device may be tested prior to the expiry of a current test period. The acceptable timing should be:

- up to 30 days prior to the expiry of any test period of 6 months or more

- up to 14 days prior to the expiry of any test period of less than 6 months. When testing takes place within these timescales, the new certificate shall be dated from the expiry date of the one it is replacing.

■ RPZ valves that are used on hot water supplies must be tested under normal operating conditions with due regard being given to health and safety during operating and testing.

■ Assemblies shall be inspected to establish:

- accessibility

- the measurements of air gaps at drain points

- the satisfactory function of the strainer (debris to be removed if necessary).

■ The function of the RPZ valve and associated fittings shall be tested and recorded by the accredited tester to establish:

- the tightness of the isolating valves, dependent on the field test method used

- the function of the relief valve (opening and closing):
  - a) the relief valve shall be watertight at both the commencement and conclusion of the test
  - b) the relief valve should start to open at a differential pressure between zone 1 and zone 2 of not less than 0.14 bar.

- watertightness of both check valves in a no-flow situation:
  - a) No.1 (upstream) check valve shall maintain a minimum direction of flow pressure differential of not less than 0.2 bar in the direction of flow, greater than the opening point of the relief valve (the buffer).

b) No. 2 (downstream) check valve shall be watertight against a back pressure and shall maintain a minimum direction of flow pressure differential of not less than 0.07 bar. However, if the No. 2 check valve is a type EB check valve conforming to the relevant parts of BS EN 13959, the pressure differential shall be not less than 0.005 bar.

- any supplementary information required by the water undertaker.

■ In the event of a RPZ valve failing a test, it should be repaired or replaced and satisfactorily re-tested. Where this cannot be done within 72 hours of the initial test failure, the water undertaker must be informed immediately. The water undertaker will assess the nature of the test failure and whether the risk of backflow requires the water supply to the RPZ valve to be shut off or other measures taken to minimise the risk.

Commissioning and test data

■ On completion of a test, a test report certificate must be completed by the testing engineer. Copies of the certificate must be supplied to the person responsible for the valve and the water undertaker within 10 working days.

■ A test record can be left on/adjacent to the RPZ valve.

Record of installation and test data

■ Records should be kept for all RPZ installations and assemblies.

■ A copy of the current test certificate must be available on site.

■ Installation, commissioning and test data must be forwarded to the water undertaker and copies retained by the tester and the water undertaker's customer for at least five years.

■ Records of RPZ valve testing should contain the following information:

- the results of tests performed
- details of calibration dates of test equipment, which will be at least annually and following damage or repair
- comments relevant to installation, maintenance and operation
- information to identify the valve, the tester and the tester's accreditation number
- the date and time of commencement and of completion of inspection and test(s)
- the date when the next test is due.

Preventing cross-connection in systems that contain non-wholesome water sources

A cross-connection is a direct, physical connection between wholesome, potable water and water that is considered non-potable, such as recycled water or harvested rainwater. In extreme circumstances, this can result in serious illness and even death. Cross-connections occur during correct plumbing design and installation, such as the hot and cold connections to a shower valve or a mixer tap (cross-connection between fluid category 1 and fluid category 2) and these, for the most part, are protected by the correct use of mechanical backflow prevention devices. However, some modern plumbing systems require much more thought and planning, rather than simply the installation of a check valve. The Water Supply (Water Fittings) Regulations 1999 demand that cross-connections from a water undertaker's mains to recycled water and rainwater harvesting systems and even connections to private water supplies are eliminated completely in order to safeguard the wholesome water supply. There are several ways in which we can do this:

- correct design of systems, taking into account the requirements of the regulations in place
- careful planning and routing of pipework and fittings
- careful use of mechanical backflow prevention devices and air gaps
- using the correct methods of marking and colour coding pipework and systems.

Of these, identification of pipework is most important, especially when additions to the system are required or during routine and emergency maintenance operations.

Colour coding pipework

All pipes, cisterns and control valves that are used for conveying water that is not considered to be wholesome must be readily identifiable from pipes or fittings used with a potable water supply. There are two ways in which this can be achieved:

- By the use of labels or colour banding pipes in accordance with BS 1710 – Identification of pipelines and services. Above-ground pipes and fittings should be labelled at junctions and either side of valves, service appliances and bulkheads.
- By the use of pigmented materials and pipes. British Standards recommend that contrasting types or colours of pipework are used to make identification easier.

Pipeline colour codes to BS 1710		
Wholesome drinking water		Green – Blue – Green
Hot water supply		Green – White – Crimson – White – Green
Reclaimed water		Green – Black – Green
Effluent		Black
Chemical dosing		Violet
Fire fighting		Green – Red – Green

Pipeline colour codes to BS 1710

Note: Blue medium-density polyethylene (MDPE) water supply pipe must not be used under any circumstances to convey anything other than wholesome drinking water, nor must it be used to form ducts for conveying pipes with any other fluids or cables.

Analyse over-the-rim and ascending-spray sanitary appliances in relation to the Water Regulations

Note: Where tables and figures are mentioned, these refer to those shown in the Water Regulations.

Bidets (including WCs adapted as bidets) with flexible hose and spray handset fittings and with submerged water inlets

G15.9 Bidets with flexible hose and spray handset fittings and/or water inlets below the spill-over level of the appliance are a fluid category 5 risk and should not be supplied with water directly from a supply pipe.

G15.10 Bidets of this type may:

- be supplied with cold and/or hot water through type AA, AB, or AD backflow prevention arrangements serving the bidet only, or

- be supplied with cold water from an independent distributing pipe serving the bidet only or a common distributing pipe serving the bidet and which may also serve a WC or urinal flushing cistern only, or

- be supplied with hot water from a water heater, which is supplied from an independent distributing pipe, that serves the bidet only (see Figure 6.1a), or
- where the bidet is at a lower elevation than any other outlets or appliances, be supplied with water from a common cold and/or hot water vented distributing pipe, providing that:
  - the elevation of the spill-over level of the bidet, if there is no flexible hose, or,
  - the elevation of the spray outlet, with the hose extended vertically above the spill-over level of the bidet, whichever is the highest

is not less than 300mm below the point of connection of the branch pipe serving the bidet to the main distributing pipe serving other appliances.

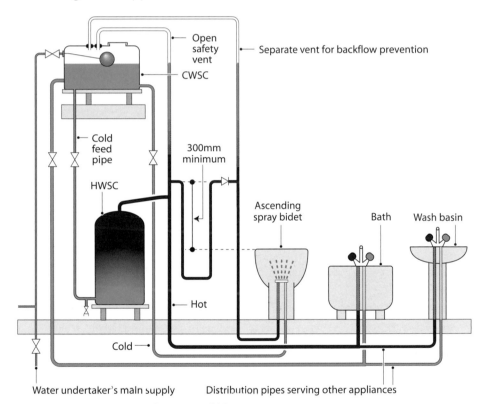

Connections to an ascending spray bidet

Bidets with water inlets above spill-over level only

G15.11 Bidets in domestic locations with taps or mixers located above the spill-over level of the appliance, and not incorporating an ascending spray inlet below spill-over level or spray and flexible hose, may be served from either a supply pipe or a distributing pipe provided that the water outlets discharge with a type AUK2 air gap above the spill-over level of the appliance. See Table 4.

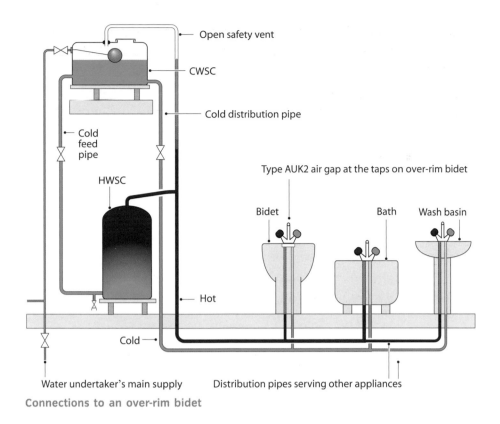

Connections to an over-rim bidet

Know the uses of specialist components in cold water systems (LO4)

There are three assessment criteria for this Outcome.

1 Explain the working principles of cold water system specialist components.

2 Describe factors to consider when selecting specialist cold water components.

3 Identify the maintenance requirements for specialist cold water components.

SmartScreen Unit 303
Presentation 5

Range	
Specialist components	Infrared operated taps, concussive taps, combination bath tap and shower head, flow-limiting valves, spray taps, urinal-water conservation controls, shower pumps – single and twin impellor, pressure-reducing valves, shock arrestors/mini expansion vessels

Range	
Factors	Design requirements, customer preference, water conservation, time of use, anti-vandalism, accessibility, hygiene
Maintenance requirements	Regular servicing, manufacturers' guidelines, inspection, testing

Your work as a plumber will cover a multitude of installations, systems and components. Sometimes you may be asked to install specialist components that you may only come into contact with on a limited number of occasions. Even so, it is important that you become familiar with these 'specialist' components to ensure that you can position and install them correctly, according to the manufacturer's instructions and in line with any regulations or recommendations.

This part of the unit looks at a selection of components that may be unfamiliar to you. You will investigate how they operate and the best ways of installing them in accordance with the recommendations in place.

Each of the specialist components will be dealt with individually, along with any installation and maintenance requirements it has.

Infrared-operated taps

Low-voltage (6V DC current) infrared-operated outlets are becoming popular for use in public conveniences, hotels and public buildings. They use infrared sensors to operate solenoid valves. The solenoid valves open for a defined length of time to allow a certain quantity of water to flow through them. They are frequently used to flush WCs and urinals and to operate taps and shower fitments. Infrared-operated outlets have several advantages over standard taps and outlets:

- They are easy to operate.
- They stop the spread of germs and bacteria.
- They can help with water conservation.
- They can prevent scalding injury.

How do they work?

The way infrared-operated outlets work is quite simple.

1 The sensor eye emits an infrared beam that is approximately 200–260mm wide. When an object, eg a hand, is within range of the infrared sensing zone, the beam is interrupted and a signal wire transfers an electronic signal to a solenoid valve.

2   The solenoid valve acts as a latching mechanism that allows a restricted flow of water to flow through it. As soon as the valve receives the electronic signal, it snaps open, allowing the water to flow to the outlet.

3   When the object leaves the sensing zone, the infrared beam returns to normal, the electronic signal ceases and the valve closes.

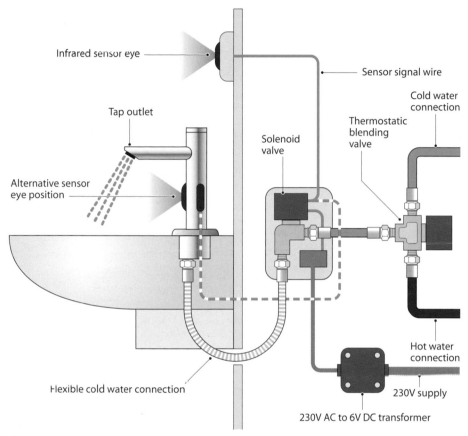

The operation of an infrared tap

Non-concussive (self-closing) taps

Non-concussive taps are self-closing taps that are usually used in public washrooms, hotels and places of work where there is a risk that the tap may either be left open or there is a high risk of vandalism. They are operated by pressing the tap head downwards, which opens the tap via a spring-loaded plunger. After a period of time, the spring then lifts the plunger to close the tap. The time for which the tap is open can be adjusted up to a maximum of about 20 seconds of water flow.

Most non-concussive taps use an internal cartridge system, so repairs and maintenance are fairly simple. When the tap requires maintenance, the cartridge can be replaced easily by removing the tap head and withdrawing the cartridge.

Non-concussive taps are available as single taps or as thermostatic monobloc mixer taps.

A non-concussive-type thermostatic mixer tap

Table 7: The advantages and disadvantages of non-concussive taps

Advantages	Disadvantages
• They are self-closing, so water is not wasted due to the tap being left open • Most models are vandal proof • They can help save money on water costs if a water meter is fitted	• The rapid closing of the tap can cause water hammer and pipework reverberation • They require regular maintenance • They should not be used where there is a risk of fouling by grease or dirt • The tap may block with scale deposits in hard-water areas • Because of the restricted amount of water released when the tap is used, waste pipes may not reach a self-cleansing velocity and may block with residue

Combination bath tap and shower head

This type of tap is more commonly known as a bath/shower mixer tap. There are many different styles, including:

- pillar type – a traditional style predominantly used in period bathrooms
- deck mixer type – there are many types of deck bath/shower mixer to suit all types of modern bathroom styles
- thermostatic type – this bath/shower mixer has the benefit of thermostatic control by the inclusion of a temperature-sensitive wax cartridge
- wall mounted – specifically designed for baths without tap holes.

Pillar-type bath/shower mixer tap

Deck-type bath/shower mixer tap

Thermostatic-type bath/shower mixer tap

Wall-mounted bath/shower mixer tap

Bath/shower mixer taps are designed for use where the pressures of the hot and cold water are equal. They should not be fitted where the cold water is direct from the mains supply and the hot water is fed from a vented, low-pressure hot water storage cylinder. This type of installation creates an imbalance in the water supply and correct mixing of both hot and cold water cannot take place. It can also cause the hot water to be pushed back into the cylinder by the high pressure of the cold water supply.

Where there is a risk that the bath water can be siphoned back into the water undertaker's mains cold water supply through the shower hose of the mixer tap, the hot and cold water connections to the bath/shower mixer should be fitted with double-check valves or, as an alternative, the shower hose should be fixed by a retaining ring so that the head cannot be placed below the overspill level of the bath.

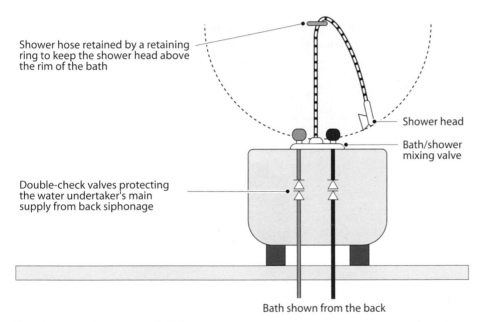

Shower hose retained by a retaining ring to keep the shower head above the rim of the bath

Shower head

Bath/shower mixing valve

Double-check valves protecting the water undertaker's main supply from back siphonage

Bath shown from the back

Bath/shower mixer tap installation

Most modern bath/shower mixer taps have a type HC diverter with automatic return backflow protection device built into them as part of the design of the tap (see page 221).

Flow-limiting valves

Flow-limiting valves are designed to limit the flow of water to appliances irrespective of the pressure upstream. Most are designed with an integral service valve and interchangeable cartridges that allow the same valve body to be used for differing flow rates depending on the appliance that it is serving. The flow rate can be changed by using a different cartridge insert. They can be used on both hot and cold water supplies and are ideal for limiting the flow to:

- wash basins
- baths and sinks
- bidets
- WCs (when used with a strainer cartridge as required by Regulation R25.6).

The flow-limiting cartridges are capable of delivering a flow rate of between 0.07 and 0.43l/s. By limiting the maximum flow rate, the guesswork is removed from designing variables in the system such as pipe sizing, cistern capacities and pump size. The use of flow-limiting valves also helps to:

- balance the flow rates between hot and cold supplies
- assist with balancing within a building, thereby preventing some appliances – eg showers – from consuming all available water while other appliances at a higher level or further downstream are starved of water
- save money on water and energy costs.

A flow-limiting valve with integral service valve

Spray taps

Spray taps are designed to reduce the flow of water from the outlet of a tap to a fine spray. This helps to conserve water, often reducing the water consumption of the tap by as much as 20%. There are many different types of spray taps for use in both bathrooms and kitchens. Some taps use a special insert to make the water appear bubbly. This is known as an aerator and simply introduces air into the water flow as the tap is running. Kitchen spray taps often have a pull-out nozzle for rinsing plates and dishes.

A typical wash basin spray tap

Urinal-water conservation controls

Many urinal installations do not have any form of water control and so they flush continuously even during periods when the building is unoccupied. Quite often, the flow rate is higher than that specified by the Water Supply (Water Fittings) Regulations 1999. Under the regulations, a urinal should use no more than 7.5 litres per bowl/hour and 10 litres/hour for a single bowl. The urinal should also have some form of limiting device to prevent unnecessary flushing during periods when the building is not used. In practice, flow rates from urinals are not measured and are often deliberately increased in an attempt to control unwanted odours.

There are many designs of flush controller available. These can use:

- a timer to match the hours of occupancy
- infrared detectors to detect the presence of people to open a solenoid valve for a short period of time

Infrared urinal flushing system

- a mechanical means to detect reduced hydraulic water pressure caused by taps being opened to open a valve controlling water to a urinal cistern.

Some controls even allow a cistern to fill only very slowly if no movement is detected for a preset period of time.

Where a large number of urinals are installed, separate controllers may be necessary to prevent all the urinals flushing simultaneously when only one person enters the room.

Whichever system of water control is used, it must be correctly set up following both the manufacturer's instructions and the requirements of the regulations.

Shower pumps – single and twin impellor

Shower booster pumps are used to give a high flow-rate shower, in conjunction with shower mixing valves. There are two types of shower booster pump.

Twin impeller pump on the inlet to the mixer valve

This kind of pump increases the pressure of the hot and cold water supplies to the mixer valve independently. The water is then mixed to the correct temperature in the valve before flowing to the shower head.

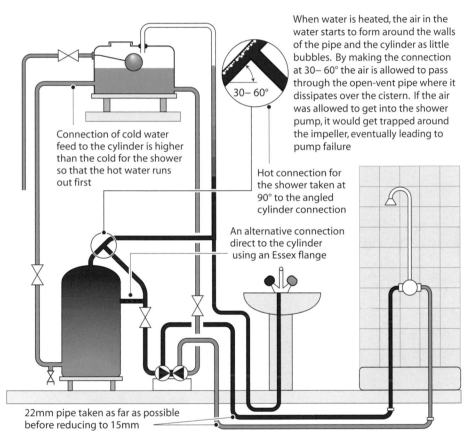

When water is heated, the air in the water starts to form around the walls of the pipe and the cylinder as little bubbles. By making the connection at 30– 60° the air is allowed to pass through the open-vent pipe where it dissipates over the cistern. If the air was allowed to get into the shower pump, it would get trapped around the impeller, eventually leading to pump failure

30– 60°

Connection of cold water feed to the cylinder is higher than the cold for the shower so that the hot water runs out first

Hot connection for the shower taken at 90° to the angled cylinder connection

An alternative connection direct to the cylinder using an Essex flange

22mm pipe taken as far as possible before reducing to 15mm

A pump-assisted shower installation with twin impeller, inlet shower booster pump

A twin impeller, inlet shower booster pump

You need to take care when making the hot connection to the cylinder. There are two ways in which this can be done. The first method involves installing the hot water draw-off from the cylinder at an angle of between 30 and 60°, with the hot shower pump connection being made at an angle of 90° with a tee piece (see diagram). This allows any air in the system to filter up to the vent and away from the hot shower pump inlet.

The second method involves making a direct connection to the cylinder using a special fitting called an Essex flange. With this method, the hot water is taken directly from the hot water storage vessel, avoiding any air problems that may occur.

An Essex flange

Single impeller pump off the outlet from the mixer valve

These boost the water after it has left the mixer valve. They are usually used with concealed shower valves and fixed deluge-type, large water-volume shower heads.

A large water volume 'deluge' shower head

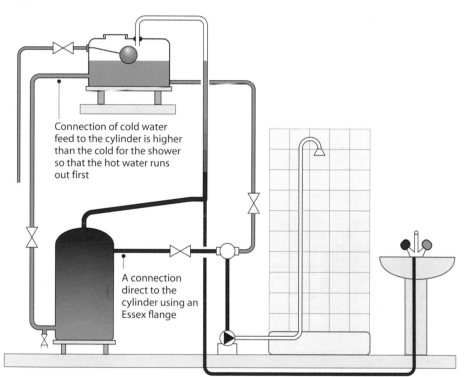

Connection of cold water feed to the cylinder is higher than the cold for the shower so that the hot water runs out first

A connection direct to the cylinder using an Essex flange

Pump-assisted shower installation with single impeller, outlet shower booster pump

A single impeller shower booster pump

In both of these installations, the pump increases the pressure of the water, which means that the minimum 1m head is not necessary. However, a minimum head of 150mm is required to lift the flow switches, as these switch the pump on.

Using shower pumps in negative head situations

With some installations it is possible to install the pump where a negative head exists. A negative head is where the cistern is lower than the pump. In this instance a means of starting the pump must be in place.

A negative head shower installation

Negative head shower pumps have the inclusion of a small accumulator or pressure vessel that piggybacks the pump. They work by sensing a sudden drop in pressure. When the shower valve is opened, the pressure within the system, caused by the pumping power of the shower pump, suddenly drops. The accumulator then immediately forces a small amount of water out via a small plastic tube, to activate the flow switch located inside the pump casing. Once the flow switch is activated, the shower pump will run at negative head. When the shower is turned off, the system remains charged at the shower pump pressure until the shower valve is opened again and the sudden pressure drop occurs, starting the pump once more.

Some shower pumps also use a small pressure transducer (pressure switch), which activates the shower pump directly, rather than using a flow switch.

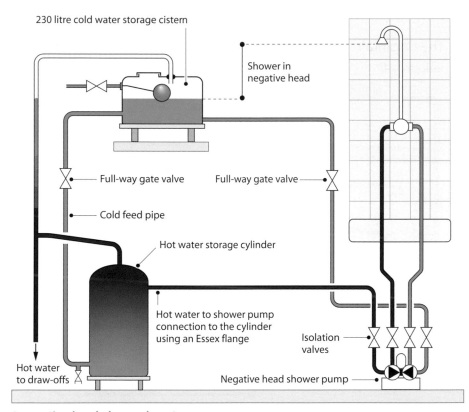

A negative head shower booster pump

Pressure-reducing valves

The water pressure in England and Wales varies considerably. Older properties that have not had their mains cold water service updated for many years could have less than 1 bar pressure. This is insufficient to run even the most basic of modern appliances such as combination boilers and even some electric showers. At the other extreme, some areas where a new water main has been laid could receive up to 10 bar pressure. High water pressures (anything above 4 bar pressure), although very good for firefighting, can be damaging to domestic plumbing systems. It can cause erosion corrosion where the flow of the water wears away pipes and fittings, especially where changes of direction occur. It can also cause leaks to water heaters, increased noise within the system, dripping taps and water hammer. Therefore, any pressures above that needed to provide sufficient flow to water fittings and appliances becomes damaging and wasteful, and reduces considerably the life expectancy of the system as a whole. This adds to the cost of water due to water wastage, and increases energy usage.

A pressure-reducing valve (PRV) is used to reduce a high upstream pressure to a lower downstream pressure. It acts as a buffer between the supply pressure and the system or appliance it is fitted to during both flow and non-flow conditions. Pressure-reducing valves perform two functions.

- They reduce high supply pressures upstream to a lower, more functional pressure for distribution.
- They maintain a set pressure, ensuring that the pipework and appliances are not subjected to excessive stress and can operate at a more moderate, acceptable pressure.

How do PRVs work?

PRVs take high-pressure water and reduce it to a lower pressure under flow and no-flow conditions, which means that they effectively stop the water pressure from creeping up when there is no water flow. This type of control is known as 'drop tight'. Most PRVs use a balanced spring and a diaphragm to control downstream water pressure. They work by sensing water pressures on either side of the diaphragm.

The diaphragm separates all of the parts in contact with the water from the control spring and the valve's mechanism. The valve body is internally protected from debris by a stainless steel mesh strainer.

Under no-flow conditions, the downstream pressure applies back pressure on the valve seat and the diaphragm. This overcomes the pressure of the spring and forces the seat against the diaphragm to prevent the downstream pressure from increasing.

Under flow conditions, the back pressure is reduced and the spring forces the valve open to allow water flow.

> **KEY POINT**
>
> A loading unit (LU) is a number or a factor which is allocated to an appliance. It relates to the flow rate at the terminal fitting, the length of time in use and the frequency of use.

A pressure-reducing valve

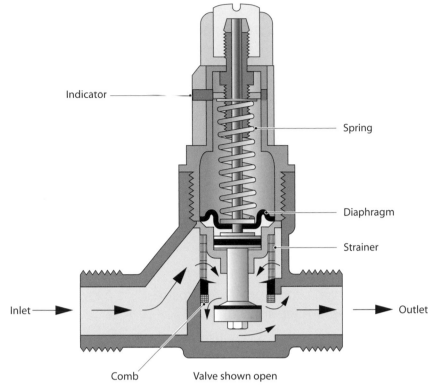

Pressure-reducing valve cutaway

Shock arrestors/mini expansion vessels

Before looking at how shock arrestors (or water hammer arrestors as they are more commonly known) work, it may help to understand why some systems require them.

Water hammer

Water hammer (or fluid hammer) is a pressure surge or wave caused when a fluid (either liquid or gas) is suddenly forced to stop or change direction. This can occur in plumbing systems when a valve is closed very quickly at the end of a pipeline. When the valve is closed, a pressure wave propagates within the pipework. This is known as hydraulic shock. The shockwave travels in the opposite direction to the water flow, often with disastrous consequences. The shockwave can cause major problems with repeated water hammer, vibration and noise, which can lead to joint failure and pipework damage. To put this into perspective, a system that is normally operating at 3 bar can suffer shockwaves equalling twice this pressure. If the pipework is installed poorly or has not been clipped correctly, the noise, vibration and subsequent damage can often be much worse. A shock arrestor can help prevent water hammer by cushioning the effects of the shockwave.

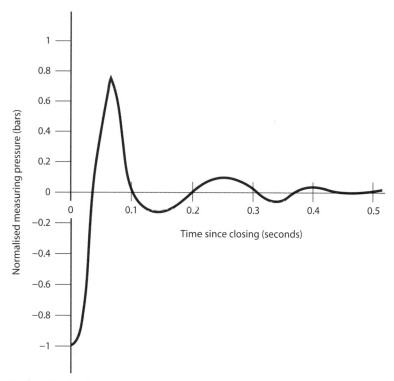

Hydraulic shockwave

Shock arrestors

Shock arrestors are often manufactured from corrosion-resistant brass, copper or stainless steel. They contain a piston or a diaphragm, which is cushioned by a calculated amount of inert gas or air. When the shockwave hits the arrestor, the piston (or diaphragm) moves with the shockwave (or hydraulic impulse) to dampen its effects by absorbing the kinetic energy. This allows the shockwave to dissipate safely without damaging the pipework and fittings.

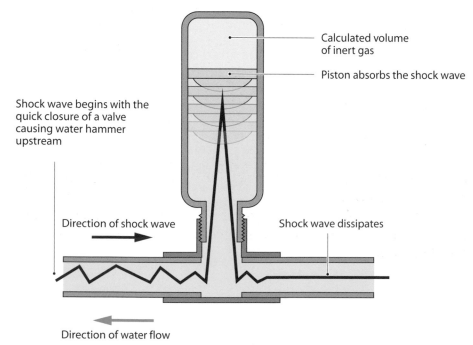

How the shock arrestor works

A shock arrestor/mini expansion vessel

Using the shock arrestor as a mini expansion vessel

Some types of shock arrestor can be used successfully as mini expansion vessels. These are usually required where there is a small amount of water expansion associated with hot water pipework on mains fed hot water supplies (unvented hot water storage systems, combination boilers or multi-point instantaneous hot water heaters). The expanded water increases the internal pressure within the pipework, and this can cause damage to terminal fittings such as shower mixing valves and backflow protection devices. A mini expansion vessel/shock arrestor allows the expansion to take place within the vessel, thereby protecting systems and equipment from internal damage. They are a requirement of most major shower manufacturers.

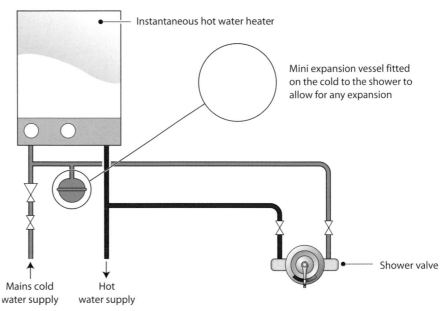

Instantaneous hot water heater

Mini expansion vessel fitted on the cold to the shower to allow for any expansion

Shower valve

Mains cold water supply

Hot water supply

Use of a mini expansion vessel

Mini expansion vessels can also be fitted on cold water installations to counteract the problems of increased system pressure due to water expansion. This occurs when the mains cold water, usually at a temperature of around 4°C in the winter and 16°C in the summer, is left to stand in the pipework for a period of time. The ambient air temperature in the building can be considerably higher, which causes the water to expand slightly. This expansion can damage backflow protection devices such as single-check valves, and terminal fittings such as ceramic disc-type taps, especially where the run of pipework is extensive.

Know the fault-diagnosis and rectification procedures for cold water systems and components (LO5)

There are four assessment criteria to this Outcome:

1 Interpret documents to identify diagnostic requirements of cold water system components.

2 Describe diagnostic checks on cold water system components.

3 Explain reasons why a treatment device may not be operating as required.

4 Explain blue water corrosion.

Range	
Documents	Manufacturers' instructions, industry standards
Diagnostic checks	Pressures, flow rates, levels, correct operation
Reasons	Limescale build-up, salt crystals require replenishment, failure to backwash

Whenever mechanical or electrical components are installed on a cold water system, the risk of breakdown and failure is ever present. Periodic preventative maintenance can alleviate some of these problems, but occasionally components simply wear out and must be replaced.

When identifying faults that have occurred on cold water systems, the customer can prove an invaluable source of information as they can often describe when and how the fault first manifested itself and any characteristics it has shown. Discussion with the customer often results in a successful repair without the need for extensive diagnostic tests. The customer should be asked about:

- the immediate history of the fault:
  - when it first occurred
  - how they noticed it
  - what characteristics it showed
- if they noticed any unusual noises. This may well indicate the type of component failure that has taken place
- if they attempted any repairs themselves. If so, what did they do? This is important because if repairs have been attempted, they may well have to be undone to successfully diagnose the problem

■ what the result of the fault was. Again, this is an important aspect because it can often indicate where the fault lies. For instance, if the customer has noticed a drop in flow rate or pressure, this might indicate a blockage, a blocked filter, scale growth or a mechanical fault such as a loss of charge from an accumulator or a pump fault.

This part of the unit looks at some of the methods that help us identify system faults.

Identify diagnostic requirements of cold water system components

When attempting to identify faults with cold water systems, the most important documents to consult are the manufacturers' instructions. In most cases these will contain a section on fault-finding that will prove to be an invaluable source of information. Fault-finding using a manufacturer's instructions usually takes three forms:

■ finding known problems that can occur and the symptoms associated with them

■ using methods by which to identify the problem in the form of a flow chart. These usually follow a logical, step-by-step approach, especially if the equipment has many parts that could malfunction, such as a cold water boosting set

■ learning the techniques required to replace the malfunctioning component.

A replacement parts list will also be present for those components that can be replaced. When ordering parts, it is advisable to use the model number of the equipment and the parts number from the replacement parts list. This will ensure that you purchase the correct part.

You can also consult BS 6700, as it contains important information about minimum flow rates required by certain appliances. Again, this should be used in conjunction with the manufacturer's instructions. Remember! Manufacturers' instructions always take precedence over the British Standards and regulations.

Describe diagnostic checks on cold water system components

Routine checks on components and systems can help to identify any potential problems that may be developing within the system. They will help to keep the system operating at its maximum performance level and within the system design specification. Checks that can be performed may include the following.

- Checking components for correct operating pressures and flow rates, using a pressure and flow rate meter or a weir jug. However, take care when using a weir jug as the flow rate of some systems can exceed the flow rate capacity of the weir jug, making an accurate reading impossible.

- Cleaning system components (including dismantling and reassembly). This is a very important part of the diagnostic process, especially with components such as filters and valves. Sediment and scale can quickly build up in certain water areas, and it is important that these components are cleaned on a regular basis to prevent loss of flow rate and pressure. The use of manufacturers' literature is recommended, especially for complex components such as pumps.

- Checking for correct component operation:

A pressure gauge

  - Pumps – performance should be checked against the manufacturer's literature. A slight fall in performance is to be expected with age. Check for signs of damage or wear and tear on the pump, leakage, that the pump switches on and off at the correct pressure and that there are no unusual noises or vibrations when the pump is operating.

  - Pressure switches (transducers) – transducers are often checked using a calibrated pressure gauge on the system pipework to ensure that the system is operating within acceptable limits. A transducer that is malfunctioning will often give a reading of over 12 bar pressure when tested, irrespective of the actual pressure within the system.

A weir jug

  - Float switches – when fitted to a cold water storage cistern (CWSC), these can usually be operated by hand, by moving the float arm, to ensure that the pump starts when the float switch is moved to the 'on' position.

  - Accumulators (pressure vessels) – these are checked at the Schrader valve with a pressure gauge to ensure that the vessel contains the correct pressure.

  - Gauges and controls – gauges should correspond to the known pressures within the system. Any deficiency could indicate that the gauge is faulty.

  - Correct operation of water treatment devices – this must be done by following the recommendations in the manufacturer's instructions.

  - Water filters – these must be regularly cleaned and cleared of all scale and sediment build-up to ensure that flow rates within the system are maintained.

  - Water softeners – again, these should be checked using the manufacturer's instructions, especially where the softener uses external additives. Water softeners can be checked for correct operation by the use of a pH indicator.

- CWSC and break cisterns – when used to supply wholesome water supply, they should be checked and cleaned at regular intervals and at least every six months for large cisterns supplying wholesome water to many properties.

Methods of repairing faults in cold water system components

Repairing system components should be undertaken using the manufacturer's servicing and maintenance instructions, as these will contain the order in which the components should be dismantled and reassembled. With all components, there will be occasions when they cannot be repaired and replacement is the only option. Some of the components that may be repaired and/or replaced are listed in the table below.

Table 8: Cold water components

Component	Known faults	Symptoms	Repairs
Pumps	Worn/broken impeller	• Motor working but water not being pumped • No water at the outlets	• No repair possible • Replace the pump
	Burnt-out motor	• Voltage detected at the pump terminals but pump not working	• No repair possible • Replace the pump
	Cracked casing	• Water leaking from the pump body	• No repair possible • Replace the pump
	Faulty capacitor	• Slow starting pump	• Replace the capacitor if possible • Check manufacturer's instructions

Component	Known faults	Symptoms	Repairs
Accumulators and pressure vessels	Pressure loss due to faulty Schrader valve	• No pressure in the accumulator	• Pump air into the accumulator using a foot pump and check the Schrader valve with leak detector fluid • Check for bubbles • Replace Schrader valve as necessary
	Ruptured bladder/diaphragm	• Water discharging from the Schrader valve • A delay in water flowing at the outlets	• It is possible to replace the bladder/diaphragm of some accumulators • Check the manufacturer's instructions
Transducers (pressure switches)	Ruptured sensor diaphragm	• Pumps fail to operate when a tap is opened	• No repair possible • Replace the transducer
	Incorrect pressure reading due to calibration	• Incorrect pressure reading at the pressure gauge	• Re-calibration of the transducer required • Check with the manufacturer
Float switches	Faulty micro-switch	• Pumps not operating when the float is in the 'on' position	• Replace the micro-switch
	Ruptured float	• Float is not responding to the water level	• Drain down the cistern and replace the float arm
Pressure (bourdon) gauges	Sticking pressure indicator needle	• Gauge not reading the correct pressure and does not move when the pressure is raised or lowered	• No repair possible • Replace the gauge

Component	Known faults	Symptoms	Repairs
RPZ valves	Perforated diaphragm	• Water dripping from the relief port	• It may be possible to replace the diaphragm • Consult the manufacturer • Repairs must be undertaken by a registered operative
Gate valves	Broken spindle	• Head turns freely but does not open and close the valve	• Replace the valve
	Scale on the gate	• Water passing by the valve	• Isolate the water supply, split the valve and clean the gate
FOVs	Worn washers	• Water discharging from the warning/overflow pipe	• Isolate the water supply and replace the washer (two washers if the valve is an equilibrium type)
Stop taps	Vibration from the tap at low flow rate	• Loose jumper plate	• Isolate the water supply and replace the jumper plate
	Stop tap does not isolate the supply	• Worn tap washer and seating	• Isolate the supply, split the tap, replace the washer and re-grind the seating
Check valves	Check valve stuck in the open position	• Water back circulating through the standby pump of a boosting set	• Isolate the valve, drain down and replace

Note: When replacing or repairing valves and controls, it is important to ensure that the water supply is isolated and the section of pipework is completely drained before beginning to repair or replace the valve.

Reasons why a water treatment device may not be operating as required

Base exchange (ion exchange) water softeners operate by passing water containing calcium and magnesium (hard water) through a sealed vessel that contains resin beads. The beads attract and absorb the calcium and magnesium, replacing the hardness with sodium, effectively softening the water. After time, the beads become saturated with the hardness salts, reducing the effectiveness of the water softener. When this happens, the beads need to be regenerated using a brine or salt solution. Backwashing removes the hardness from the beads before the sodium is replenished by the salt in the brine solution. Most modern softeners complete this procedure automatically either by a timed backwashing programme or when the beads become saturated with the hardness salts.

Base exchange water softeners often begin to lose their effectiveness when:

- limescale builds up within the softener
- the salt crystals require replenishment
- the water softener fails to backwash.

It is therefore important that water softeners are maintained at regular intervals to ensure that the softening process remains effective.

Blue water corrosion

Blue water corrosion occurs in copper plumbing systems and is identified by a blue-green coloration to the water when it is first drawn from the tap. It is often (but not always) associated with **pitting corrosion**. The colour of the water is the result of very fine copper corrosion products suspended in the water and is more common in copper pipes that are used infrequently. Occasionally, the signs of blue water corrosion are apparent even after a very short period of stagnation.

The colour of the water can vary greatly from pale blue to dark blue/green. The particles become very visible if the water is left to stand in a container overnight, as they will sink to the bottom of the container leaving a blue/green copper corrosion residue. This can sometimes be evident as blue/green staining on sanitary ware.

Blue water corrosion is more likely to occur in soft water areas where the pH value of the water is high, or areas where the residual chlorine content of the water is low. However, such is the nature of blue water corrosion that copper pipe can develop the symptoms even in other low pH water areas.

Pitting corrosion

This is the localised corrosion of a metal surface. It is confined to a point or small area and takes the form of cavities and pits. Pitting is one of the most damaging forms of corrosion in plumbing, especially in central heating radiators, as it is not easily detected or prevented. Copper tube, although not a ferrous metal, is relatively soft and can suffer from pitting corrosion if flux residue is allowed to remain on the tube after soldering.

Evidence of blue water corrosion on sanitary ware

Treating and minimising blue water corrosion

There is some evidence to show that blue water corrosion could be linked to microbial activity interacting with the copper pipe and that, because the signs of corrosion appear in a non-uniform manner, the internal surface of the pipe may also play a part in its formation. It has been shown that in nearly all cases of blue water corrosion, flushing the system with water in excess of 70°C followed by disinfection with water containing a chlorine-based solution will prevent its appearance. However, if the residual chlorine content is not maintained, it is likely that the corrosion will reappear.

Know the commissioning requirements of cold water systems and components (LO6)

SmartScreen Unit 303
Presentation 6

There are four assessment criteria to this Outcome:

1 Explain the commissioning checks required on boosted cold water systems.
2 State the requirements for disinfection.
3 State the actions to be taken when commissioning procedures identify faults.
4 Describe the information required on a commissioning record for a cold water system.

Range	
Checks	Soundness, flushing, filling, float level, low level switch, pressure vessel, pressure switch, flow rate, pressure
Requirements	Competent person, domestic, non-domestic, refer to BS 6700
Actions	Investigate, rectify, re-check
Information	Type of appliance, location, pressures, flow rates, temperature, installation information (who, when), maintenance requirements, components

Testing and commissioning is probably the most important aspect of any cold water installation. It is at this stage that you see whether the system you have installed is leak free and performs to the specification.

Correct commissioning procedures are laid down by industry standards and manufacturers' instructions, but experience will also play a vital part in the testing and commissioning process.

This Outcome will help you know how to assess the correct commissioning and testing processes in order to bring to life the systems and appliances that you have so carefully and professionally installed.

The commissioning checks required on boosted cold water systems

Inadequate commissioning, system flushing and maintenance operations can affect the quality of drinking water, irrespective of the materials that have been used in the system installation. Building debris and swarf (pipe filings) can easily block pipes and these can also promote bacteriological growth. In addition, excess **flux** used during the installation can cause corrosion and may lead to the amount of copper that the water contains exceeding the permitted amount for drinking water. This could have serious health implications and, in severe cases, may cause blue water corrosion (see pages 257–258).

It is obvious, then, that correct commissioning procedures must be adopted if these problems are to be avoided. There are four documents that must be consulted.

- The Water Supply (Water Fittings) Regulations 1999
- BS 6700 and BS EN 806 (in conjunction with BS 8558)
- Building Regulations Approved Documents G1 and G2
- manufacturers' instructions for any equipment and appliances.

Corrosion of copper pipes

Flux
A paste used to clean oxides from the surface of the copper and to help with the flow of solder into the fitting.

The Water Supply (Water Fittings) Regulations 1999

These are the national requirements for the design, installation, testing and maintenance of cold and hot water systems in England and Wales (Scotland has its own, almost identical, Scottish Water Byelaws 2004). Their purpose is to prevent contamination, wastage, misuse, undue consumption and erroneous metering of the water supply used for domestic purposes. Schedule 2 of the regulations states the following.

- The whole installation should be appropriately pressure tested, details of which can be found in the Water Regulations Guide (Section 4: Guidance clauses G12.1 to G12.3). This requires a pressure test of 1.5 times the maximum operating pressure for the installation or any relevant part.

- Every new water service, cistern, distribution pipe, hot water cylinder or other appliance and any extension or modification shall be thoroughly flushed with drinking water prior to being taken into service.

■ Under certain circumstances, the whole system should be disinfected before being put into use. This will be discussed later in this Outcome.

This document should be read in conjunction with the British Standards.

The British Standards (BS EN 806-4, BS 8558 and BS 6700)

The main British Standard for commissioning, testing, flushing and disinfection of systems is BS EN 806-4:2012 – Specifications for installations inside buildings conveying water for human consumption, in conjunction with guidance document BS 8558:2011 – Guide to the design, installation, testing and maintenance of services supplying water for domestic use within buildings and their curtilages. In particular, the following sections are of relevance.

■ Section 6 Commissioning:
  ● 6.1 Filling and hydrostatic pressure testing of the installations inside buildings conveying water for human consumption
  ● 6.1.1 General
  ● 6.1.2 Steel pipes, stainless steel pipes and copper pipes (linear elastic material)
  ● 6.1.3 Plastic pipes (elastic or visco-elastic material)
■ Section 6.2 Flushing the pipework:
  ● 6.2.1 General procedure
  ● 6.2.2 Flushing with water
  ● 6.2.3 Flushing procedure with water/air mixture
■ Section 6.3 Disinfection:
  ● 6.3.1 General
  ● 6.3.2 Selection of disinfectants
  ● 6.3.3 Methods for using disinfectants
  ● 6.3.4 Disinfection of storage cisterns and distribuling pipes.

In reality, the information has changed very little from the previous British Standard, BS 6700:2006+A1:2009 – Design, installation, testing and maintenance of services supplying water for domestic use within buildings and their curtilages, and this document should be referenced where alternative information is not available.

The Building Regulations

The Building Regulations make reference to cold water services and systems. These are mentioned briefly in Approved Document G1 – Cold water supply, and Approved Document G2 – Water efficiency. Additional recommendations can be found in Annex 1 – Wholesome water, and Annex 2 – Competent persons self-certification schemes.

Manufacturers' instructions

Where appliances and equipment are installed on a system, the manufacturer's instructions are a key document when undertaking testing and commissioning procedures, and it is important that these are used correctly both at installation and when commissioning operations. Only the manufacturer will know the correct procedures that should be used to safely put equipment into operation so that it performs to its maximum specification. Remember the following.

- Always read the instructions before operations begin.

- Always follow the procedures in the correct order.

- Always hand the instructions over to the customer upon completion.

- Failure to follow the instructions may invalidate the manufacturer's warranty.

Visual inspections of boosted cold water systems

Before soundness testing a cold water system, you should carry out visual inspections of the installation. This should include the following.

SmartScreen Unit 303
Worksheet 1

- Walk around the installation. Check that you are happy that the installation is correct and meets installation standards.

- Check that all open ends are capped off and all valves are isolated.

- Check that all capillary joints are soldered and that all compression joints are fully tightened.

- Check that enough pipe clips, supports and brackets are installed and that all pipework is secure.

- Check that the equipment, ie boosting pumps, float switches, accumulators, etc, is installed correctly and that all joints and unions on and around the equipment are tight.

- Check that cisterns and tanks are supported correctly and that FOVs are provisionally set to the correct water level.

- Check that all appliances' isolation valves and taps are off. These can be turned on and tested when the system is filled with water.

- Where underground services have been installed, check that any pipework is at the minimum depth required by the Water Supply (Water Fittings) Regulations 1999.

The water undertaker must be given the opportunity to view and inspect the installation, preferably before it is tested and commissioned to ensure that the Water Supply (Water Fittings) Regulations 1999 have been complied with. Any remedial work pointed out by the water inspector can then be completed without the need to drain the system.

The initial system fill

The initial system fill is always conducted at the normal operating pressure of the system. The system must be filled with fluid category 1 water direct from the water undertaker's mains cold water supply. It is usual to conduct the fill in stages so that the filling process can be managed comfortably. There are several reasons for this.

- Filling the system in a series of stages allows the operatives time to check for leaks stage by stage. Only when the stage being filled is leak free should the next stage be filled.

- Airlocks from cistern-fed supplies are less likely to occur, as each stage is filled slowly and methodically. Any problems can be assessed and rectified as the filling progresses without the need to isolate the whole system and initiate a full drain-down. Allowing cisterns to fill to capacity and then opening any gate valves is the best way to avoid airlocks. This ensures that the full pressure of the water is available and the pipes are running at full bore. Trickle filling can encourage airlocks to form, causing problems later during the fill stage.

- It is possible that less manpower will be required with staged filling procedures. On very large and multi-storey systems, the use of two-way radios greatly helps the operatives during the filling process, and isolation of a potential problem becomes quicker. Operatives should be stationed at the main isolation points to initiate a rapid turn-off should a problem occur.

When the system has been filled with water, it should be allowed to stabilise and any FOVs should be allowed to shut off. The system will then be deemed to be at normal operating pressure.

Once the filling process is complete, another thorough visual inspection should take place to check for any possible leakage. The system is then ready for pressure testing.

Pressure testing procedures

Pressure testing can commence when the initial fill to test the pipework integrity has been completed. Again, on large systems, this is best done in stages to avoid any possible problems.

The requirements of the Water Regulations

The UK Water Regulations require the following:

'Regulation 12 – the water system should be capable of withstanding an internal water pressure not less than 1.5 times the maximum pressure to which the installation or any relevant part is designed to be subjected in operation.'

In practice, this means that a system that has an operating pressure of 2 bar:

2 x 1.5 = 3 bar

Regulation 12 also states that the Regulation shall be deemed satisfied in the case of water systems that do not contain elastomeric (plastic) pipe where the whole system is subjected to the test pressure by pumping, after which the test should continue for 1 hour without further pumping and without any visible leakage and any pressure loss.

Where the system does contain elastomeric (plastic) pipework, there are two acceptable tests that can be conducted. These are classified as Test type A and Test type B.

Cold water systems testing is detailed in BS 6700 and BS EN 806-4.

- Copper tubes and low carbon steel pipes – systems installed using copper tubes and/or low carbon steel pipes should be tested to 1.5 times normal operating pressure (50% above normal operating pressure). The system should be left for a period of 30 minutes to allow for temperature stabilisation and then left for a period of 1 hour with no visible pressure loss.

- Elastomeric (plastic polybutylene) pressure pipe systems – these are tested rather differently from rigid pipes. There are two tests that can be carried out. These are known as test type A and test type B and are detailed in BS 6700. It is important that the correct tests are used for elastomeric pipes, as these can become stressed very easily at high pressures.

  - Test type A – slowly fill the system with water and raise the pressure to 1 bar (100kPa). Check and re-pump the pressure to 1 bar if the pressure drops during this period, providing there are no leaks. Check for leaks. After 45 minutes, increase the pressure to 1.5 times normal operating pressure and let the system stand for 15 minutes. Now release the pressure in the system to one-third of the previous pressure and let it stand for a further 45 minutes. The test is successful if there are no leaks.

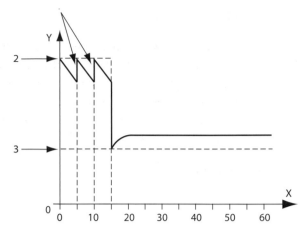

Key
1 Pumping
2 Test pressure 1.5 times maximum working pressure
3 0.5 times maximum working pressure
X Time (minutes)
Y Pressure

Test type A

● Test type B – slowly fill the system with water, pump the system up to the required pressure and maintain the pressure for a period of 30 minutes and note the pressure after this time. The test must continue with no further testing. Check the pressure after a further 30 minutes. If the pressure loss is less than 60kPa (or 0.6 bar), the system has no visible leakage. Visually check for leakage for a further 120 minutes. The test is successful if the pressure loss is less than 20kPa (0.2 bar).

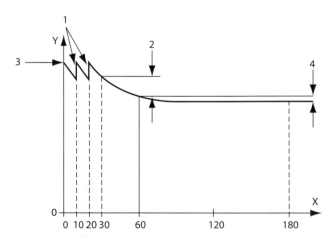

Key
X Time (minutes)
Y Pressure
1 Pumping

2 Pressure drop < 60kPa (0.6bar)
3 Test pressure
4 Pressure drop < 20kPa (0.2bar)

Test type B

Planning the test

Before the test is conducted, a risk assessment should be carried out. Pressure testing involves stored energy, the possibility of blast and the potential hazards of high-velocity missile formation due to pipe fracture and fitting failure. A safe system of work should be adopted and a permit to work sought where necessary. Personal protective equipment (PPE) should also be used.

You should carefully consider the following factors.

- Is the test being used appropriate for the service and the building environment?

- Will it be necessary to divide the vertical pipework into sections to limit the pressures in multi-storey buildings?

- Will the test leave pockets of water that might cause frost damage or corrosion later?

- Can all valves and equipment withstand the test pressure? If not, these will need to be removed and temporary pipework installed.

- Are there enough operatives available to conduct the test safely?

- Can different services be interconnected as a temporary measure to enable simultaneous testing?

- How long will it take to fill the system using the available water supply?

- When should the test be started, taking into account the size of the system? Preparation should also be taken into account, too.

Preparing for the test

- Check that the high points of the system have an air vent or a tap to help with the removal of air from the system during the test. These should be closed to prevent accidental leakage.

- Blank or plug any open ends and isolate any valves at the limit of the test where the test is being conducted in stages.

- Remove any vulnerable equipment and components and install temporary piping.

- Open the valves within the section to be tested.

- Important: check that the test pump is working correctly and that the pressure gauge is calibrated and functioning correctly.

- Attach the test pump to the pipework and install extra pressure gauges if necessary.

- Check that a suitable hose is available for draining-down purposes.

A test pump

The hydraulic test procedure

1 Using the test pump, begin to fill the system. When the pressure shows signs of rising, stop and walk the route of the section under test. Listen for any sounds of escaping air and visually check for any signs of leakage.

2 Release air from the high points of the system or section and completely fill the system with water.

3 When the system is full and free of air, pump the system up to the required test pressure.

4 If the pressure falls, check that any isolated valves are not passing water, and visually check for leaks.

5 Once the test has been proven sound, the test should be witnessed and a signature obtained on the test certificate.

6 When the pressure is released, open any air vents and taps to atmosphere before draining down the system.

7 Refit any vulnerable pieces of equipment, components and appliances.

State the requirements for disinfection

The British Standards require that water piping systems be disinfected in the following circumstances:

- in new installations (except private dwellings occupied by a single family)

- where large extensions or alterations to systems have been undertaken

- where it is suspected that the system has been compromised by contamination, eg fouling by drainage, sewage or animals, or the physical entry by operatives during maintenance operations or repair

- where the system has not been in regular use nor regularly flushed

- where underground pipework has been installed. This does not include localised repairs or where a junction has been installed after the fittings have been disinfected by immersion in a solution of sodium hypochlorite to dilution of 200 parts per million (ppm). 1ppm is equivalent to 1mg/l.

It is considered good practice to clean and disinfect systems on a regular basis as part of routine maintenance operations. It is also a statutory requirement to disinfect systems where large amounts of water are stored and used in a way that could create the risk of *Legionella pneumophila* (the bacterium causing Legionnaires' disease).

The requirements of the disinfection process

Disinfection requires that the system is first thoroughly flushed with water and then refilled and a disinfection agent added as the system refills. The disinfection agent should have an initial concentration of 50ppm, which should be calculated according to the total water capacity of the pipework and any storage vessels. The chlorinated water should then be allowed to stand within the pipework for a contact period of at least 1 hour. The process will have been successfully completed if, after 1 hour, the free residual chlorine content is at least 30ppm at the furthest outlet on the system.

The pipework must not be used during the period of disinfection and notices must be posted stating 'Disinfection in progress – do not use!' to indicate that the system is undergoing disinfection. These should be posted at all outlets or any point where water may be drawn off.

During the disinfection process, the use of household chemicals, bleaches and toilet cleaners must be avoided as these can react with the disinfectant to produce highly toxic fumes. All personnel and residents within the building, including those that are not normally present during working hours, should be informed of the disinfection operation before it commences. All operatives should reference the appropriate **COSHH** data sheets for the disinfectant chemicals beforehand and take appropriate action by wearing the correct PPE during the process, ie face mask, goggles, plastic apron and long arm gauntlets, when handling and mixing the chemicals.

COSHH

COSHH is an acronym that stands for 'Control of Substances Hazardous to Health'. Under the COSHH Regulations 2002, employers have to prevent or reduce their employees' exposure to substances that are hazardous to health.

The disinfectant types

The disinfectants used for water supplies must be approved for use by the Drinking Water Inspectorate and must conform to the following standards:

- EN 900 for calcium hypochlorite, or
- EN 901 for sodium hypochlorite.

Because most disinfection processes use sodium hypochlorite, the rest of the unit looks at this disinfectant.

Sodium hypochlorite solution contains around 5% available chlorine content, which is equivalent to 50,000ppm of chlorine. Obviously, for disinfection of water supply systems this is far too much. To create an initial content of 50ppm for the disinfection process:

$50 \div 1,000,000 \times 100 \div 5 = 0.001$ part of sodium hypochlorite solution to 1 part water. This equates to a ratio of 1 litre of sodium hypochlorite containing 5% chlorine to every 1,000 litres of system water capacity.

For commercial-strength sodium hypochlorite solutions that contain 10% available chlorine, the ratio is much less:

50 ÷ 1,000,000 x 100 ÷ 10 = 0.0005 part of sodium hypochlorite solution to 1 part water. This equates to a ratio of 1 litre of sodium hypochlorite containing 10% chlorine to every 1,000 litres of system water capacity.

Calculating system capacity

To administer the correct concentration of disinfectant, the total system volume in litres must be calculated. This is completed by:

- calculating the capacity in litres up to the water line of any storage cisterns, and
- calculating the capacity in litres of the pipework.

Calculating the capacity of a cistern

Calculation of cistern capacity is a simple process.

- Measure the length of the cistern in metres (m).
- Measure the width of the cistern in metres (m).
- Measure the height to the water line in metres (m).

L x W x H (to water line) = volume (m^3)

Volume x 1,000 = capacity in litres (l).

Example

A cold water storage cistern measures 2m x 1.5m x 1m to the water line. What is the capacity in litres of the cistern?

$$2 \times 1.5 \times 1 = 3m^3$$

$$3 \times 1,000 = 3,000 \text{ litres.}$$

The capacity of the cistern is therefore **3,000 litres**.

Calculating the capacity of the system pipework

Calculating the capacity of the pipework involves measuring the length of the pipework and the pipe size. The bigger the pipe, the more water it will hold. The calculation to determine the capacity of a pipe is as follows:

$$\mathbf{\pi r^2 \times length \times 1,000}$$

Where:

π = 3.142

r = radius of the pipe in metres (m)

Fortunately, the copper tube manufacturers provide the capacities of the common tube sizes per metre within their data sheets, and this is provided in Table 9.

Example

What is the capacity of 70mm diameter pipe measuring 1m in length? The radius of the pipe is half the diameter, therefore:

$$70 \div 2 = 35mm \text{ or } 0.035m$$

So, the calculation looks like this:

$3.142 \times 0.035^2 \times 1 \times 1,000 =$

$3.142 \times (0.035 \times 0.035) \times 1 \times 1,000 =$

$3.142 \times 0.001225 \times 1 \times 1,000 =$

3.85 litres

Table 9: The capacity of copper tube per metre

Standard/spec	Tube dimension (OD)	Carrying capacity (l/m)
BS EN 1057 R250 – Half-hard straight lengths	6	0.018
	8	0.036
	10	0.061
	12	0.092
	15	0.145
	22	0.320
	28	0.539
	35	0.835
	42	1.232
	54	2.091
	66.7	3.247
	76.1	4.197
	108	8.659

The calculation of the capacity of the pipework can be carried out using a table (see Table 10). The section of pipe, its size and length must first be identified and numbered and then added to the table. Consider the diagram on the next page.

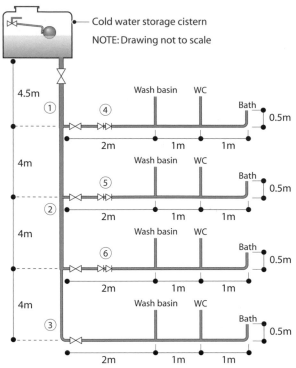

Cold water storage cistern

NOTE: Drawing not to scale

Pipework to the wash basin and WC have been assumed at 1m for each appliance per floor

Pipework sizes have not been calculated and are assumed for the purpose of system capacity calculations

Calculation of pipework capacity

Example

A system of cold water is to be disinfected. The system contains the sizes and lengths of pipe as identified in Table 10.

Table 10: Determining the capacity of the system pipework

Pipe	Size (mm)	Length (m)	Capacity per metre (l)	Capacity of the pipe (l)
1	42	4.5	1.232	5.544
2	35	8	0.835	6.68
3	22	8.5	0.320	2.72
4	22	4.5	0.320	1.44
5	22	4.5	0.320	1.44
6	22	4.5	0.320	1.44
Appliance legs	15	8	0.145	1.16
Total pipework capacity (l)				20.424

Therefore the capacity of the pipework shown in the example drawing is:

20.424 litres

If the capacity of the cistern from the previous example is taken into account, the total system capacity is:

3,000 + 20.424 = **3020.424 litres**.

If we assume that sodium hypochlorite solution with 5% chlorine content is to be used, then to calculate the amount of sodium hypochlorite solution required:

3020.424 x 0.001 = **3.020424 litres**

Therefore, to successfully disinfect the example system, 3,020 litres of sodium hypochlorite solution will need to be added during the system fill.

SUGGESTED ACTIVITY

System capacity calculations: A system of cold water is to be disinfected. The system contains the sizes and lengths of pipe identified in the table below.

Pipe	Size (mm)	Length (m)	Capacity per metre (l)	Capacity of the pipe (l)
1	54	10		
2	42	22.5		
3	35	15.5		
4	28	24.5		
5	22	14.5		
6	15	10		
Appliance legs	15	12		
Total pipework capacity (l)				

Now add to the system capacity total and determine the amount of sodium hypochlorite solution (5% chlorine content) that must be used to disinfect the system.

Disinfection method

1 Thoroughly flush the system with water from the undertaker's main to remove any swarf, flux residue or other contaminants.

2 Calculate the cistern capacity and the amount of disinfectant required to give the required 50ppm.

3 Add this to the cistern as it is filling to give a good mix with the cistern water.

4 Working away from the cistern, open each draw-off until a smell of disinfectant is detected. This should be done at every draw-off to ensure that the system is full of disinfectant.

5   As the water from the cistern is drawn off, it will be necessary to add more disinfectant solution to ensure that the required 50ppm throughout the installation is achieved.

6   Once the system is full and the FOVs on the system have shut, the 1-hour test period can begin.

Testing for residual chlorine

At the end of the 1-hour period, the free residual chlorine content should be 30ppm at the furthest draw-off point. If this is not achieved, then the system must be drained and the test started afresh.

To test for residual chlorine content, a special chemical analysis kit is used, which indicates the strength of the chlorine by the colour it turns when the chemical is added to the water.

The test kit consists of a clear plastic tube with a sample of water from the chlorinated system in it. A tablet is then added to the water and the tube is shaken. As the tablet dissolves, the water will change colour. The colour shows the chlorine strength, as indicated in the following table:

Water colour	Chlorine level (ppm or mg/l)
Clear	None
Faint pink/pink	0.2–1
Pink/red	1–5
Red/purple	5–10
Purple/blue	10–20
Blue/grey-green	20–30
Grey-green/yellow	30–50
Muddy brown	Over 50
Colour develops but then disappears	Excessive

Cuprous chloride corrosion

This occurs because the chloride ions present in sodium hypochlorite solution are very aggressive when in contact with copper and copper alloys, due to the fact that the chloride can form an unstable film on the inside of the pipe. This means that even small amounts of chlorine can cause corrosion problems with copper piping.

Flushing the system after disinfection

Heavily chlorinated water is dangerous to wildlife and fish. Prolonged contact with copper pipe can also cause **cuprous chloride corrosion**, which will continue to attack the tube, even after flushing.

Before draining the system, you should seek advice from the local water undertaker. Alternatively, a neutralising chemical (sodium thiosulphate) can be administered to the system to clean the water before draining takes place. The chemical should be added at the following rate:

System volume x ppm of chlorine x 2 = grams of chemical required

Once the system has been drained, it should be flushed with clean water until the free residual chlorine content is no more than that in the water undertaker's mains cold water supply.

Remember the following:

- Do take care to warn people before disinfection begins. The chemicals are dangerous.

- Do handle chemicals with care, and always use PPE.

- Do calculate the capacity of the system. Using too much chemical disinfectant will not produce better results and may result in the test having to be carried out again.

- Do use chlorinated water when topping up cisterns and pipework during the test. This will help maintain the correct disinfectant level.

- Do check the chlorine level at the end of the test period.

- Do not leave the water in the system for more than 1 hour, and never overnight.

- Do not discharge the test water directly into a water course or drain without first contacting the local water undertaker or the Environment Agency.

- Always complete the disinfection record paperwork correctly.

The flushing procedure for cold water systems and components

The flushing of cold water systems is a requirement of the British Standards. All systems, irrespective of their size, must be thoroughly flushed with clean water direct from the water undertaker's mains supply before being taken into service. This should be completed as soon as possible after the installation has been completed to remove potential contaminants such as flux residue, PTFE, excess jointing compounds and swarf.

Simply filling a system and draining down again does not constitute a thorough flushing. In most cases, this will only move any debris from one point in the system to another. In practice, the system should be filled and the water run at every outlet until the water runs completely clear and free of any discoloration. It is extremely important to ensure that all equipment and appliances and all water fittings are flushed completely. When flushing boosted cold water supplies, water from the undertaker's water main must first be introduced into the break cistern and then boosted, using the boosting pumps, to all points on the system.

It is generally accepted that systems should not be left charged with water once the flushing process has been completed, especially if the system is not going to be used immediately, as there is a very real risk that the water within the system could become stagnant. In practice, it is almost impossible to effect a complete drain-down of a system, particularly large systems, where long horizontal pipe runs may hold water. This in itself is very detrimental as corrosion can often set in and this can also cause problems with water contamination. It is recommended, therefore, that to minimise the risk of corrosion and water quality problems to leave systems completely full and flush through at regular intervals of no less than twice weekly, by opening all terminal fittings until the system has been taken permanently into operation. If this is the case, provision for frost protection must be made.

Taking flow rate and pressure readings

When the system has been commissioned and put into operation, the flow rates and pressures should be checked against the specification and the manufacturer's instructions. This can be completed in several ways.

A bourdon pressure gauge

- Flow rates can be checked using a weir gauge. This is sometimes known as a weir cup or a weir jug, and is simple to use. The gauge has a slot running vertically down the side of the vessel, which is marked with various flow rates. When the gauge is held under running water, the water escapes out of the slot. The height that the water achieves before escaping from the slot determines the flow rate. Although the gauge is accurate, excessive flow rates will cause a false reading because the water will evacuate out of the top of the gauge rather than the side slot.

- System pressures (static) can be checked using a bourdon pressure gauge at each outlet or terminal fitting. Bourdon pressure gauges can also be permanently installed either side of a boosting pump to indicate both inlet and outlet pressures.

- Both pressure (static and running) and flow rate can be checked at outlets and terminal fittings using a combined pressure and flow rate meter.

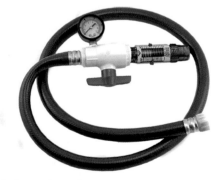

A flow rate and pressure meter

What to do when commissioning procedures identify faults

Commissioning is the part of the installation where the system is filled and run for the first time. It is at this point that you will see whether it works as designed. Occasionally problems will be discovered when the system is fully up and running.

Systems that do not meet correct installation requirements

These can take two forms.

- **Systems that do not meet the design specification** – problems such as incorrect flow rates and pressures are quite difficult to deal with. If the system has been calculated correctly and the correct equipment has been specified and installed following the manufacturer's instructions, problems of this nature should not occur. However, if the pipe sizes are too small in any part of the system, flow rate and pressure problems will develop almost immediately downstream of where the mistake has been made. In this instance, the drawings should be checked and you should confirm with the design engineer that the pipe sizes used are correct before any action is taken. It may also be the case that too many fittings or incorrect valves have been used, causing pipework restrictions.

Another cause of flow rate and pressure deficiency is the incorrect set-up of equipment such as boosting pumps and accumulators. In the first instance, the manufacturer's data should be consulted, and set-up procedures followed in the installation instructions. It is here that mistakes are often made. If problems still continue, the manufacturer's technical support should be contacted for advice. In a very few cases, the equipment specified will be at fault and will not meet the design specification. If this is the case the equipment must be replaced.

- **Poor installation techniques** – installation is the point where the design is transferred from the drawing to the building. Poor installation techniques account for the following:

  - Noise – incorrectly clipped pipework can often be a source of frustration within systems running at high pressures, because of the noise it can generate. Incorrect clipping distances and, often, lack of clips and supports can put strain on the fittings and cause the pipework to reverberate throughout the installation, even causing fitting failure and leakage. To prevent these occurrences, the installation should be checked as it progresses and any deficiencies brought to the attention of the installing engineer. On completion, the system should be visually checked before flushing and commissioning begins.

  - Undue warming of cold water systems – this generally occurs if the cold water pipework has been installed too close to either hot water pipework or heating system pipework. The Health and Safety Executive states that to prevent microbacterial growth due to temperature, the cold water must not exceed 20°C. To prevent this, cold water pipework should be insulated across the entire system including cisterns and storage vessels.

- Leakage – water can cause a huge amount of damage to a building and can even compromise the building structure. Leakage from pipework, if undetected, may cause damp, mould growth and an unhealthy atmosphere. It is therefore important that leakage is detected and cured at a very early stage in the system's life.

  It is almost impossible to ensure that every joint on every system installed is leak free. Manufacturing defects on fittings and equipment and damage sometimes cause leaks. Leakage due to badly jointed fittings and poor installation practice is much more common, especially on large systems where literally thousands of joints have to be made until the system is complete. These can often be avoided by taking care when jointing tubes and fittings, using recognised jointing materials and compounds, and using manufacturers' recommended jointing techniques.

- **Microbiological contamination within a cold water system** – contamination of cold water systems can occur for a variety of reasons:

  - the ingress of insects and vermin into stored water
  - the ingress of debris during installation
  - poor installation practice
  - undue warming of the cold water.

  There are many forms of microbiological contamination, including *Legionella pneumophila*, *E. coli* and *Pseudomonas*, and most are associated with an increase in water temperature, resulting in conditions in which the bacteria can reproduce. In such instances, the system must be disinfected using the techniques outlined on pages 266–272. In all cases of microbiological contamination, advice should be sought from a recognised company dealing with such outbreaks and by consultation with the HSE.

- **Defective components and equipment** – defective components cause frustration and cost valuable installation time. If a component or piece of equipment is found to be defective, do not attempt to repair it, as this may invalidate the manufacturer's warranty. The manufacturer should first be contacted as they may wish to send a representative to inspect the component before replacing it. The supplier should also be contacted to inform them of the faulty component. In some instances where it is proven that the component is defective and was not a result of poor installation, the manufacturer may reimburse the installation company for the time taken to replace the component.

The information required on a commissioning record for a cold water system

Commissioning records for large cold water systems should be kept for reference during maintenance and repair and to ensure that the system meets the design specification. Typical information that should be included on the record is as follows:

- the date, the time and the name(s) of the commissioning engineer(s)
- the location of the installation
- the amount of cold water storage (if any)
- the types and manufacturers of equipment and components installed
- the type of pressure test carried out, and its duration
- disinfection processes, the disinfection chemicals used and the disinfection readings
- the flow rates and pressures at the outlets
- the pressures on both the suction side and the discharge side of any boosting pumps
- the accumulator pressure.

These records should be kept in a file in a secure location.

Be able to carry out commissioning and rectify faults on cold water systems (LO7)

There are four assessment criteria to this Outcome:

1 Install pressure-reducing valves to cold water outlets.
2 Set pressure-reducing valves in accordance with manufacturers' instructions.
3 Install backflow protection devices.
4 Rectify faults on cold water components.

Range	
Components	Concussive taps, flow-limiting valves, pressure-reducing valves

This Outcome is part of the practical assessment that will be conducted within the workshop environment.

There are three practical tasks.

- **Task CW1** – install a pressure-reducing valve and backflow protection devices to cold water outlets (photographic evidence required), at a location to be determined by your assessor. You must:
  - ensure that you have all necessary documentation before beginning the task
  - select the appropriate fittings and confirm the suitability of the cold water supply
  - prepare the work area
  - install the pressure-reducing valve and backflow prevention devices and any relevant pipework required in line with the Water Regulations and manufacturers' instructions
  - hydraulically test the installation with a hydraulic test bucket in line with BS EN 806 (BS 6700)
  - thoroughly flush the system.

- **Task CW2** – commission a pressure-reducing valve in line with the manufacturer's instructions. You will be asked to set the pressure of the valve to a pressure determined by the assessor. You must then check and record the pressure to ensure that it has been set correctly. Finally, you must check and record the flow rate at the outlet.

- **Task CW3** – identify faults on cold water installations that have been previously installed by your assessor. You must:
  - confirm that you have all the documentation required to correctly identify the faults
  - visually inspect the installation to ensure that it meets current standards and regulations
  - identify the fluid category at four outlets determined by the assessor
  - confirm the type of backflow prevention device at each of the four outlets
  - state whether the backflow protection is adequate, and state the backflow protection required where the device fitted is incorrect
  - answer the test your knowledge questions set as part of the task.

Before undertaking the assessment, discuss any concerns you may have with your tutor. Be clear about what is required of you, and ask the appropriate questions. There are time limits on each of the three tasks.

Good luck with your assessment!

Conclusion

As this unit has shown, there is much more you need to learn about cold water systems beyond the simple systems you became accustomed to at Level 2. Cold water systems are often complex in their design and require great skill in their installation and commissioning, and this is reflected in the associated regulations. Add to this the difficulties surrounding private water supplies and it is obvious that the modern plumber must be competent in many different areas of work.

Unit 303 gives an excellent insight into the complexities of modern cold water systems that will build on the knowledge you gained at Level 2 and enable you to make a worthwhile contribution within the work environment.

Test your knowledge questions

1 The Water Industry Act 1991 and the Water Resources Act 1991 have been amalgamated into which single piece of legislation?

2 Which British Standards are linked to the Water Supply (Water Fittings) Regulations 1999?

3 What are the five things that the Water Supply (Water Fittings) Regulations 1999 are intended to prevent?

4 Which set of regulations regulates boreholes and wells?

5 Which backflow prevention device is notifiable under the Water Regulations?

6 Under which document of the Building Regulations are cold water systems notifiable?

7 Which system of cold water uses a pneumatic pressure vessel?

8 What is a solenoid valve?

9 Briefly describe a break cistern.

10 What do float switches provide?

11 What is a transducer?

12 For a single-occupancy dwelling, what is the water extraction rate of a private water supply?

13 Briefly describe a submersible pump.

14 Which sources of private water supplies are described as surface water sources?

15 Treatment of private water supplies involves settling, pre-filtration, sand filtration and disinfection. What is this process better known as?

16 Briefly describe reverse osmosis.

17 What is ultraviolet irradiation used for?

18 Name the two systems of supplying water from a private source.

19 Identify the pump set illustrated.

20 Which type of accumulator is usually installed with private water supplies?

21 Calculate the pre-charge pressure of an accumulator when the normal working pressure is 2.5 bar.

22 Which mechanical backflow prevention device is also known as a type BA device?

23 There are two types of double-check valves. What is the difference between the two types?

24 Water supplied by a water undertaker is classified as which fluid category?

25 Which fluid category constitutes a serious health risk?

26 Backflow can be separated into two categories. What are they?

27 By what term is a self-closing tap also known?

28 Which type of backflow prevention device do modern bath/shower mixer taps incorporate?

29 What is the purpose of a flow-limiting valve?

30 What type of shower booster pump is a twin impeller-type pump?

31 What does a shock arrestor prevent?

32 By what method can the flow rates of the outlets be checked?

33 What is the minimum free chlorine content after a 1-hour disinfection?

34 How should the system be flushed after disinfection?

Assessment checklist

What you now know (Learning Outcome)	What you can now do (Assessment criteria)	Where this is found (Page number)
1. Know the regulations relating to cold water supplied for domestic purposes	1.1 Describe the purpose of the Water Supply (Water Fittings) Regulations 1999.	140–141
	1.2 Describe how the Water Regulations impact on the installation and use of water systems.	141–142
	1.3 Explain the requirements for advance notification of work.	142–145
	1.4 Differentiate between the installer and user responsibilities under the Water Regulations.	145–146
	1.5 Define the legal requirements for drawing water from a water undertaker's water main using a pump or booster.	146–147
2. Know the types of cold water system layouts used in buildings	2.1 Describe the principles of operation of cold water system component layouts used in multi-storey buildings.	148–153
	2.2 Describe the requirements for large-scale cisterns.	153–164
	2.3 Explain the function of components used in boosted cold water systems in multi-storey buildings.	164–167
	2.4 Describe alternative water supplies to buildings.	168–175
	2.5 Propose methods of treating water for use in buildings.	175–187
	2.6 Define the method of operation of different types of boosted cold water supply systems for buildings.	188–193
	2.7 Identify situations where rainwater harvesting may be appropriate.	193–194
3. Know the requirements for backflow protection in plumbing services	3.1 Determine the fluid risk levels as laid down in water legislation.	196–199
	3.2 Compare point-of-use protection with whole-site and zone protection.	199–202
	3.3 Propose where non-mechanical backflow prevention devices (air gaps) may be used.	202–213
	3.4 Propose where mechanical backflow prevention devices may be used.	213–229
	3.5 Explain the regulations for RPZ or RPZD valves.	229–233
	3.6 Determine methods of preventing cross-connection in systems that contain non-wholesome water sources.	234–235
	3.7 Analyse over-the-rim and ascending-spray sanitary appliances in relation to the Water Regulations.	235–237

What you now know (Learning Outcome)	What you can now do (Assessment criteria)	Where this is found (Page number)
4. Know the uses of specialist components in cold water systems	4.1 Explain the working principles of cold water system specialist components.	237–250
	4.2 Describe factors to consider when selecting specialist cold water components.	237–250
	4.3 Identify the maintenance requirements for specialist cold water components.	237–250
5. Know the fault-diagnosis and rectification procedures for cold water systems and components	5.1 Interpret documents to identify diagnostic requirements of cold water system components.	251–252
	5.2 Describe diagnostic checks on cold water system components.	252–256
	5.3 Explain reasons why a treatment device may not be operating as required.	257
	5.4 Explain blue water corrosion.	257–258
6. Know the commissioning requirements of cold water systems and components	6.1 Explain the commissioning checks required on boosted cold water systems.	259–266
	6.2 State the requirements for disinfection.	266–74
	6.3 State the actions to be taken when commissioning procedures identify faults.	274–276
	6.4 Describe the information required on a commissioning record for a cold water system.	277–279
7. Be able to carry out commissioning and rectify faults on cold water systems	7.1 Install pressure-reducing valves to cold water outlets.	251–279
	7.2 Set pressure-reducing valves in accordance with manufacturers' instructions.	
	7.3 Install backflow protection devices.	
	7.4 Rectify faults on cold water components.	

UNIT 304
Complex hot water systems

This combination unit provides the learner with knowledge of hot water systems, components, storage cylinders and temperature control as well as the appropriate commissioning procedures and fault-finding techniques. Learners will consider alternative fuel sources and new technologies.

There are five Learning Outcomes (LOs) for this unit and each LO will be discussed in turn. There are 56 Guided Learning Hours for this unit. The learner will:

1 Understand the types of hot water systems.
2 Understand the operating principles of components found in hot water systems.
3 Know the fault-diagnosis and rectification procedures for hot water systems and components.
4 Know the commissioning requirements of hot water systems and components.
5 Be able to install and inspect hot water systems.

Understand the types of hot water systems (LO1)

SmartScreen Unit 304
Presentation 1

There are seven assessment criteria for this Outcome:

1 Explain the working principles of different types of centralised hot water supply systems used in buildings.
2 Identify types of localised hot water supply systems used in buildings.
3 Compare thermal stores and unvented hot water storage systems.
4 State the recommended design temperatures within hot water systems.
5 Describe the requirements of Part G of the Building Regulations for hot water installations.
6 Evaluate the use of different fuels in domestic hot water systems.
7 Explain the operating principles of hot water digital showers.

Range	
Centralised	Unvented hot water systems, open-vented hot water systems, thermal stores
Localised	Unvented point-of-use heaters, instantaneous heaters
Temperatures	Hot water storage vessel, hot water outflow, secondary return at point of use, maximum hot water temperature to Part G
Requirements	Blending valves, safety thermostat
Fuels	Gas, electric, solar thermal, oil, solid fuel, geothermal, biomass
Operating principles	Pre-set temperatures, remote control activated, digital processor

The working principles of different types of centralised hot water supply systems

Centralised systems are those in which the source of the hot water supply is sited centrally in the property for distribution to all of the hot water outlets. They are usually installed in medium to large domestic dwellings, such as a three-bedroomed house. These systems can be further divided, as follows:

- centralised open-vented hot water systems:
  - direct systems
  - indirect systems:
    - a) double-feed indirect systems
    - b) single-feed, self-venting indirect systems
- centralised unvented hot water systems
- centralised instantaneous hot water systems.

Centralised open-vented hot water systems

In an open-vented storage hot water system, water is heated, generally by a boiler or an immersion heater, and stored in a hot water storage vessel sited in a central location in the property usually in the airing cupboard. Open-vented systems contain a vent pipe which remains open to the atmosphere, ensuring that the hot water cannot exceed 100°C. The vent pipe acts as a safety relief should the system become overheated. It must be sited over the cold-feed cistern in the roof space.

The cylinder is fed with water from the cold-feed cistern. The capacity of the cistern will depend on the capacity of the hot water storage vessel, and BS 6700 recommends that the capacity of the cistern feeding cold water to a hot water storage vessel must be at least equal to that of the hot water storage vessel. There are some important points to note about open-vented hot water systems.

- The open-vent pipe must not be smaller than 22mm, and must terminate over the cold-feed cistern.
- The open-vent pipe must not be taken directly from the top of the hot water storage vessel.
- The hot water draw-off pipe should rise slowly from the top of the cylinder to the open-vent pipe and incorporate at least 450mm of pipe between the storage cylinder and its connection point to the open vent. This is to prevent **parasitic circulation** (also known as one-pipe circulation) from occurring.
- The cold-feed pipe should be sized in accordance with BS 6700. The cold feed is the main path for expansion of water to take place within the cylinder when the water is heated. The heated water from the cylinder expands up the cold-feed pipe, raising the water level in the cold feed cistern.
- The cistern should be placed as high as possible to ensure good supply pressure. The higher the cistern, the greater the pressure at the taps. Poor pressure can be increased by raising the height of the cistern.
- All pipes should be laid with a slight fall (except the hot water draw-off) to prevent airlocks within the system.

Parasitic circulation

When hot water is pumped through one pipe below the radiators, then back to the tank for re-heating.

■ The cold-feed pipe from the storage cistern must only feed the hot water storage cylinder.

■ A drain-off valve should be fitted at the lowest point of the cold-feed pipe.

These systems were investigated at Level 2 but are covered here briefly to refresh your previous learning.

Direct systems

Direct open-vented hot water storage systems use a direct-type hot water storage cylinder. The direct cylinder contains no form of heat exchanger, so is not suitable for use with central heating systems. The connections for the cold feed and draw-off are usually male thread connections, while the primary flow and return connections are female thread connections. They are usually heated by either one or two immersion heaters, depending on the cylinder type, or they

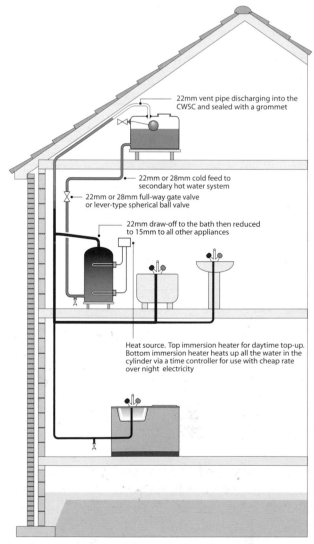

22mm vent pipe discharging into the CWSC and sealed with a grommet

22mm or 28mm cold feed to secondary hot water system

22mm or 28mm full-way gate valve or lever-type spherical ball valve

22mm draw-off to the bath then reduced to 15mm to all other appliances

Heat source. Top immersion heater for daytime top-up. Bottom immersion heater heats up all the water in the cylinder via a time controller for use with cheap rate over night electricity

A direct open-vented hot water storage system

may be heated by a gas-fired hot water circulator. Existing installations may also use a back boiler placed behind a solid fuel fire. Because the water in the boiler comes directly from the hot water storage cylinder, the boiler must be made of a material that does not rust. This is to prevent rusty water being drawn off at the taps. Suitable boiler materials include:

- copper
- bronze
- stainless steel.

Indirect systems

An indirect system uses an indirect-type hot water storage cylinder, which contains some form of heat exchanger to heat the secondary water. There are two distinct types of cylinder:

- double-feed indirect hot water storage cylinder
- single-feed, self-venting indirect hot water storage cylinder.

The heat exchanger contains primary water and so is classified as part of the central heating system to the dwelling.

Indirect (double-feed) hot water storage systems

A system of this type is probably the most common of all hot water delivery systems installed in domestic properties. It uses a double-feed indirect hot water storage cylinder, which contains a heat exchanger, at the heart of the system. The heat exchanger within the cylinder is usually a copper coil but, in older-type cylinders, it can also take the form of a smaller cylinder called an annular. It is called indirect simply because the secondary water in the cylinder is heated indirectly by the primary water via the heat exchanger.

In a double-feed indirect system two cisterns are used – a large cistern for the domestic hot water and a smaller one for the heating. It is now general practice to install indirect cylinders in preference to direct types, even if the indirect flow and return are capped off.

The double-feed indirect hot water storage cylinder allows the use of boilers and central heating systems that contain a variety of metals, such as steel and aluminium, because the water in the cylinder is totally separate from the water in the heat exchanger. This means that there is no risk of dirty or rusty water being drawn off at the taps. The system is designed in such a way that the water in the boiler and primary pipework is hardly ever changed, with the only loss of water being in the feed and expansion cistern through evaporation.

The secondary water is drawn from the hot water storage cylinder to supply the hot taps. It is heated by conduction, as the water in the cylinder is in contact with the heat exchanger.

A feed and expansion cistern feeds the primary part of the system, and this must be large enough to accommodate the expansion of the water in the system when it is heated. The vent pipe from the primary system must terminate over the feed and expansion cistern. An alternative method would be to use a sealed heating system, which is fed with water from the cold water main via a filling loop. Expansion of water is accommodated in an expansion vessel.

Hot water storage cylinders must conform to BS 1566, which specifies the minimum heating surface area of the heat exchanger.

A typical open-vented indirect (double-feed) hot water storage system utilising fully pumped primary circulation is shown below.

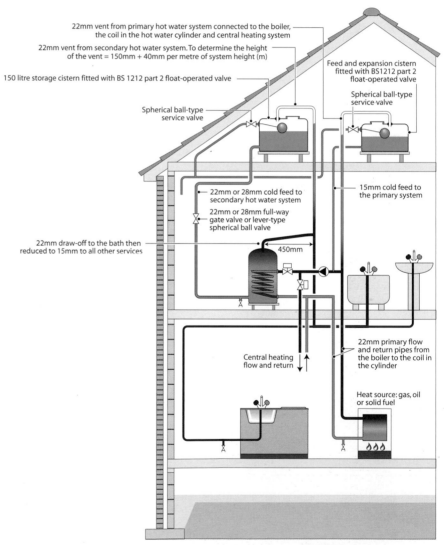

22mm vent from primary hot water system connected to the boiler, the coil in the hot water cylinder and central heating system

22mm vent from secondary hot water system. To determine the height of the vent = 150mm + 40mm per metre of system height (m)

150 litre storage cistern fitted with BS 1212 part 2 float-operated valve

Feed and expansion cistern fitted with BS1212 part 2 float-operated valve

Spherical ball-type service valve

Spherical ball-type service valve

22mm or 28mm cold feed to secondary hot water system

15mm cold feed to the primary system

22mm or 28mm full-way gate valve or lever-type spherical ball valve

22mm draw-off to the bath then reduced to 15mm to all other services

450mm

22mm primary flow and return pipes from the boiler to the coil in the cylinder

Central heating flow and return

Heat source: gas, oil or solid fuel

An indirect open-vented (double-feeds-type) hot water storage system

Indirect (single-feed, self-venting) hot water storage systems

Systems of this type use a single-feed, self-venting indirect cylinder, often referred to by its trade name as a Primatic cylinder. It contains a special heat exchanger, which uses air entrapment to separate the primary water from the secondary water.

It is fitted in the same way as a direct system, with only one cold feed cistern in the roof space but, unlike the direct system, it allows a boiler and central heating to be installed. It does not require a separate feed and expansion cistern. The heat exchanger works in such a way that the primary and secondary waters are separated by a bubble of air that collects in the heat exchanger, preventing them from mixing. According to the Domestic Building Services Compliance Guide, these are no longer allowed to be used for new or replacement cylinders. A double-feed-type cylinder must be used on all replacement installations.

A typical open-vented indirect (single-feed, self-venting) hot water storage system utilising gravity circulation is shown below.

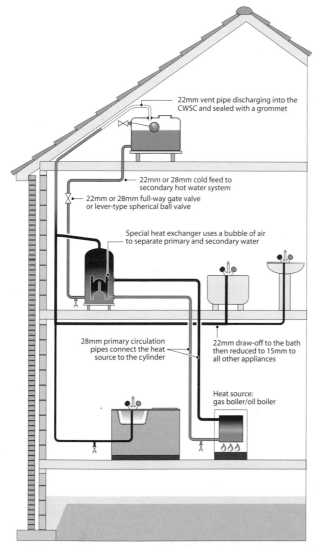

22mm vent pipe discharging into the CWSC and sealed with a grommet

22mm or 28mm cold feed to secondary hot water system

22mm or 28mm full-way gate valve or lever-type spherical ball valve

Special heat exchanger uses a bubble of air to separate primary and secondary water

28mm primary circulation pipes connect the heat source to the cylinder

22mm draw-off to the bath then reduced to 15mm to all other appliances

Heat source: gas boiler/oil boiler

An indirect open-vented single-feed self-venting hot water storage system

Unvented hot water storage systems

An unvented hot water storage system is simply a sealed system of pipework and components that is supplied with water above atmospheric pressure. The system does not require the use of a feed cistern. Instead, it is fed with water directly from a water undertaker's mains supply or with water supplied by a booster pump and a cold water accumulator if the mains pressure is low.

An unvented hot water system differs from open-vented types because there is no vent pipe. Expansion of water due to the water being heated is accommodated either in an external expansion vessel or in an expansion bubble within the storage cylinder. The system also requires other mechanical safety devices for the safe control of the expansion of water and to ensure that the water within the storage cylinder does not exceed 100°C.

There are many factors to consider before this arrangement is installed in a property:

- available pressure and flow rate – this is probably the most important factor, simply because poor pressure and flow rate will affect the operating performance of the installation. Pressure and flow rate readings should be taken at peak times to ensure adequate water supply before recommending this type of system
- the route of the discharge pipework, termination and discharge pipework size
- the type of terminal fittings to be used – this is especially important when **retro-fitting** unvented installations on existing hot water systems, as the existing taps or other fittings may not be suitable
- cost – unvented systems tend to be very expensive.

Retro-fitting

Adding installations to systems that did not have these when manufactured.

Types of unvented hot water storage cylinders

There are two types of unvented hot water storage cylinders, and both are manufactured to BS EN 12897:2006 – Specification for indirectly heated unvented (closed) storage water heaters, and available as direct-fired/heated or indirectly heated vessels:

- unvented hot water storage cylinders using an external expansion vessel
- unvented hot water storage cylinders incorporating an internal expansion air gap.

Most unvented cylinders are manufactured from high-grade duplex stainless steel for strength and corrosion resistance. Some older cylinders may be manufactured from copper or steel with a polyethylene or cementitious lining.

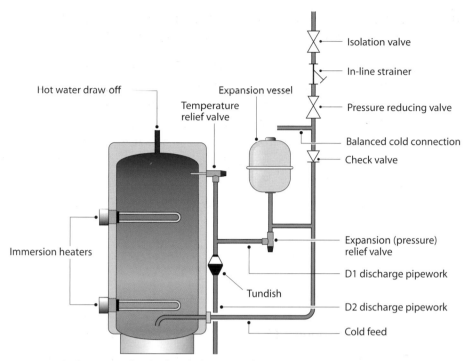

Isolation valve

In-line strainer

Pressure reducing valve

Balanced cold connection

Check valve

Hot water draw off

Temperature relief valve

Expansion vessel

Expansion (pressure) relief valve

D1 discharge pipework

Immersion heaters

Tundish

D2 discharge pipework

Cold feed

A typical unvented cylinder with external expansion vessel

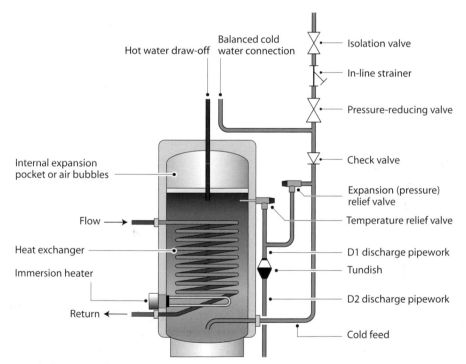

Balanced cold water connection

Isolation valve

In-line strainer

Pressure-reducing valve

Check valve

Hot water draw-off

Internal expansion pocket or air bubbles

Expansion (pressure) relief valve

Temperature relief valve

Flow

Heat exchanger

D1 discharge pipework

Immersion heater

Tundish

D2 discharge pipework

Return

Cold feed

A typical unvented cylinder with internal expansion

Unvented hot water storage cylinders can be purchased as units or packages.

- Units are delivered with all the components already factory fitted, and require less installation time.

- Packages are delivered with all the components separately packaged (except those required for safety, such as the temperature relief valve). These have to be fitted by the installer in line with the manufacturer's instructions.

An unvented hot water unit with components factory fitted

Installation of unvented hot water storage cylinders

The installation of unvented hot water storage systems is subject to the strict requirements of Building Regulations Approved Documents G3 and L, and the Water Supply (Water Fittings) Regulations 1999. Typical pipework layouts are shown below.

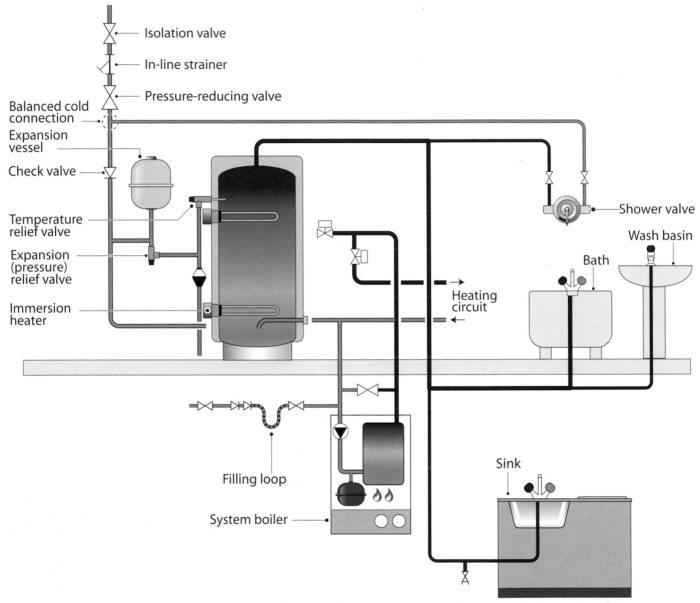

Isolation valve

In-line strainer

Pressure-reducing valve

Balanced cold connection

Expansion vessel

Check valve

Temperature relief valve

Expansion (pressure) relief valve

Immersion heater

Shower valve

Wash basin

Bath

Heating circuit

Sink

Filling loop

System boiler

Installation of an indirectly heated UHWSS with a system boiler

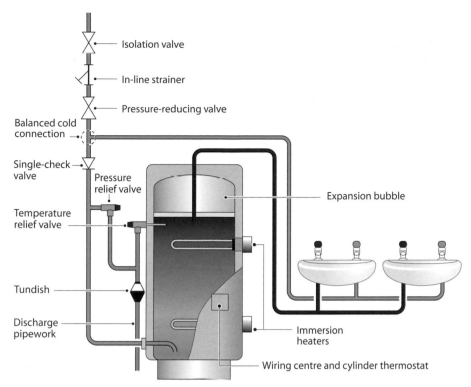

Installation of a directly fired UHWSS with immersion heaters as the primary heat source

The unit or package must be installed in accordance with the manufacturer's instructions that are supplied with the vessel. There may be special instructions from the manufacturer regarding the installation requirements of a particular vessel.

The floor on which the vessel is to be sited must be substantial enough to accommodate the weight of the vessel and its water contents.

The pipework must be fitted in accordance with BS EN 806 (BS 6700) and BS 8558. Unvented hot water storage systems (UHWSS) require at least a 22mm cold water feed supplied by a water undertaker because of the high flow rate and pressure that the vessels operate at. Water can be supplied through a boosting pump and cold water accumulator if necessary (this is discussed later in the unit). A 22mm hot water draw-off is required in all installations but this may be reduced for particular appliances such as wash basins, sinks and bidets. Isolation valves should be fitted at all appliances in line with good practice.

The order that the functional and safety components are installed in is of paramount importance if the system is to operate safely and efficiently. This is shown by the illustration on page 294 for indirectly heated vessels and the illustration on page 295 for directly heated vessels.

Unvented hot water storage systems require the installation of discharge pipework to safely convey any water that may be discharged as the result of a defect or malfunction. Discharge pipework is covered later in this section.

Various types of unvented hot water system

There are different types of unvented hot water system, which are defined by how the water is heated:

- indirect storage systems
- direct storage systems
  - electrically heated
  - gas- or oil-fired
- small point-of-use (under-sink) storage heaters.

Indirect storage systems

Indirect unvented hot water storage systems utilise an indirect unvented hot water storage cylinder at the heart of the system. As with open-vented systems, the cylinder contains a coiled heat exchanger to transfer the heat indirectly from the primary system to the secondary system. This can be done in one of two ways:

- by the use of a gas-fired condensing boiler
- by the use of an oil-fired condensing boiler.

Older, non-condensing boilers may be used if the boiler is an existing appliance, provided that the boiler contains both a control thermostat and a high-energy cut-out (high limit) thermostat to limit the water temperature at the coil should the control thermostat fail. On no account should solid-fuel appliances and boilers be used to provide heat to the coil. The primary hot water system may be either an open-vented or a sealed system.

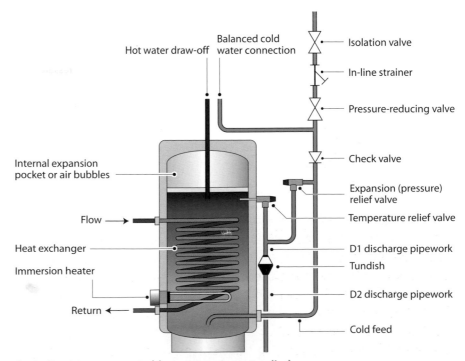

An indirect-type unvented hot water storage cylinder

An immersion heater provides back-up hot water heating for use during the summer or for when the boiler malfunctions.

Direct storage systems

Direct systems use a direct-type unvented hot water cylinder that does not contain any form of heat exchanger. There are two very different types.

- **Electrically heated** – this type of cylinder does not contain a heat exchanger. Instead, the water is heated directly by two immersion heaters controlled by a time switch. One immersion heater is located close to the bottom of the cylinder to heat all of the contents of the cylinder at night, and another is located in the top third to top up the hot water during the day if required via a 1-hour boost button on the time switch. Both immersion heaters are independently controlled and cannot be used simultaneously.

  These immersion heaters are manufactured to BS EN 60335-2-73 and must contain a user thermostat, usually set to 60°C, and a non-resetting thermal cut-out (high limit stat).

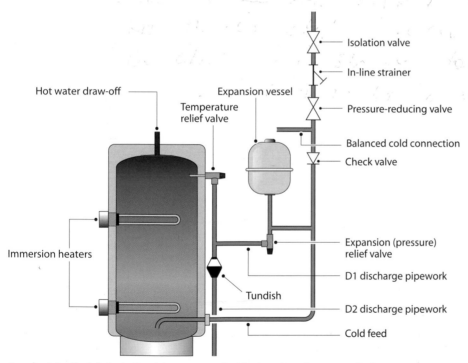

An electrically heated direct-type unvented hot water storage cylinder

- **Gas- or oil-fired** – the design of these water heaters originated in North America. They consist of a hot water storage vessel with a flue pipe that passes through the centre. Expansion of the water is catered for by the use of an external expansion vessel. Below the storage vessel is a burner to heat the water, and this can be fuelled by either gas or oil depending on the type. The burner is controlled by a thermostat and a gas/oil valve. An energy cut-out prevents the water exceeding the maximum of 90°C. The safety and functional controls and components layout is almost identical to other unvented hot water storage systems.

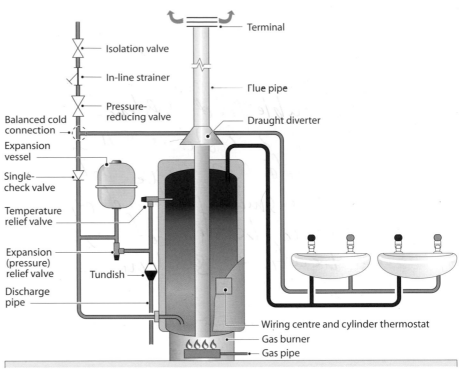

A gas-fired direct-type unvented hot water storage cylinder

Direct unvented under-sink storage heaters

Unvented under-sink hot water storage heaters are connected directly to the mains cold water supply and deliver hot water at near mains cold water pressure. Because they have less than 15 litres of storage, they are not subject to the stringent regulations that govern the installation of larger unvented hot water storage units.

The expansion of water may be taken up within the pipework provided that the pipework is of sufficient size to cope with the water expansion. If not, an external expansion vessel will be required.

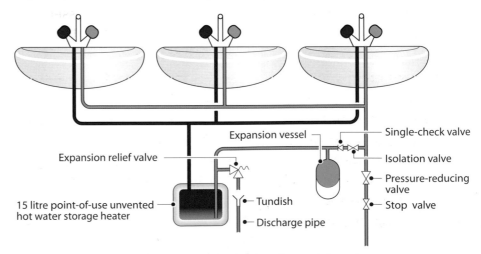

Unvented-type under-sink storage water heater pipework layout

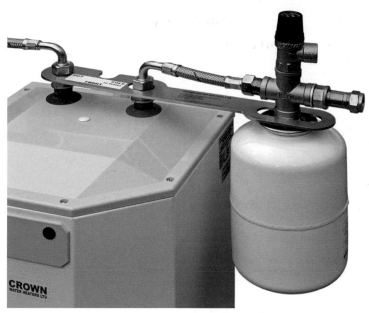

An unvented-type under-sink storage water heater with expansion vessel

Comparisons between open-vented and unvented hot water storage systems

There are important differences between these two types of systems. Table 1 shows comparisons between open-vented and unvented hot water storage systems.

Table 1: Comparisons between open-vented and unvented storage hot water systems

Open-vented systems	
Advantages	**Disadvantages**
• Storage is available to meet the demands at peak times • Low noise levels • Always open to the atmosphere • Water temperature can never exceed 100°C • Reserve of water available if the mains supply is interrupted • Low maintenance • Low installation costs	• Space needed for both the hot water storage vessel and the cold water storage • Risk of freezing • Increased risk of contamination • Low pressure and often poor flow rate • Outlet fittings can be limited because of the low pressure

Unvented systems	
Advantages	**Disadvantages**
• Higher pressure and flow rates at all outlets, giving a larger choice of outlet fittings • Balanced pressures at hot and cold taps • Low risk of contamination • The hot water storage vessel can be sited almost anywhere in the property, making it a suitable choice for houses and flats alike • The risk from frost damage is reduced • Less space required because cold water storage is not needed • Installation is quicker as less pipework is required • Smaller-diameter pipework may be used in some circumstances	• No back-up of water should the water supply be isolated • If the cold water supply suffers from low pressure or flow rate, the system will not operate satisfactorily • There is the need for discharge pipes that will be able to accept very hot water and there will be restrictions on their length • High maintenance • Higher risk of noise in the system pipework • Initial cost of the unvented hot water storage vessel is high

Centralised instantaneous hot water systems

These can be divided into the following types:

- gas-fired instantaneous multi-point hot water heaters – these heat the water instantaneously

- gas- or oil-fired combination boilers – these operate in a similar fashion to instantaneous hot water heaters but also have a central heating capability

- thermal stores – sometimes referred to as water-jacketed tube heaters

- gas- or oil-fired combined primary storage units – these are very similar in operation to thermal stores.

Gas-fired instantaneous multi-point hot water heaters

With this type of hot water heater, cold water is taken from the water undertaker's main and heated in a heat exchanger according to demand, before being distributed to the outlets. As long as taps are running, hot water will be delivered to them, and there is no limit to the amount of hot water that can be delivered. There is also no storage capacity.

Expansion of water as a result of being heated is accommodated by back pressure within the cold water main. However, if this is not adequate or the cold water system contains pressure-reducing valves or check valves, an expansion vessel must be fitted.

The heater works on **Bernoulli's principle** by using a Venturi tube to create a pressure differential across the gas valve when the cold water is flowing into the heater.

Bernoulli's principle

Bernoulli's principle states that when a pipe is suddenly reduced in size, the velocity of the water increases but the pressure decreases. The principle can also work in reverse. If a pipe suddenly increases in size, the velocity will decrease but the pressure will increase slightly.

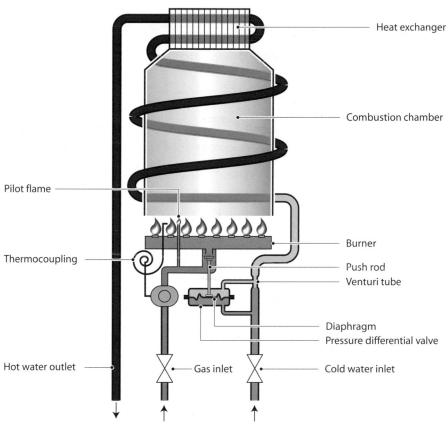

A gas instantaneous hot water heater

Gas- or oil-fired combination boilers

Combination (combi) boilers are dual-function appliances. They provide instantaneous hot water and central heating within the same appliance. In normal working mode, combi boilers are central heating appliances, supplying a proportion of their available heat capacity to heat the central heating water. When a hot tap is opened, a diverter valve diverts the boiler water around a second heat exchanger, which heats cold water from the water undertaker's cold water main to supply instantaneous hot water at the hot taps. In this mode, the entire heat output is used to heat the water. Temperature control is electronic and this automatically adjusts the burner to suit the output required. Typical flow rates are around 9 litres per minute (35°C temperature rise). Some combi boilers incorporate a small amount of storage and this can double the flow rate to around 18 litres per minute.

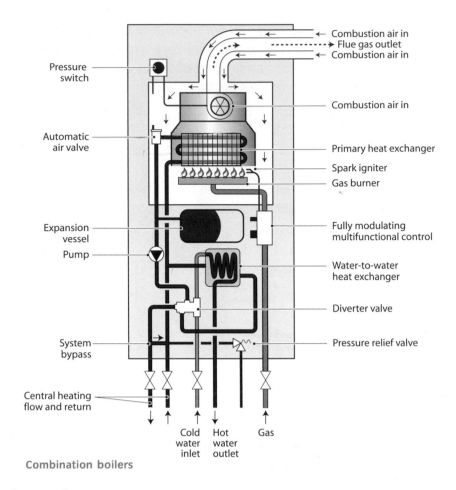

Combination boilers

Thermal stores

Sometimes known as water-jacketed tube heaters, thermal stores work by passing mains cold water through two heat exchangers, which are encased in a large storage vessel of primary hot water fed from a boiler. They are very similar to indirect systems, but they work in reverse.

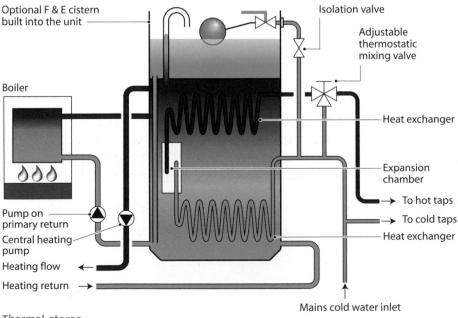

Thermal stores

Inside the unit are two heat exchangers, which the mains cold water passes through, and a small expansion chamber. The expansion chamber allows for the small amount of expansion of the secondary water. The primary water can reach temperatures of up to 82°C, which can, potentially, be transferred into the secondary water. Because of this, an adjustable thermostatic mixing valve blends the secondary hot water with mains cold water so that the water does not exceed 60°C.

Gas- or oil-fired combined primary storage units

These are very similar in design to thermal stores and work in exactly the same way in that cold water from the mains supply is passed through a heat exchanger. The difference is that the unit has its own heat source in the form of a gas burner to heat the primary water, eliminating the need for a separate boiler.

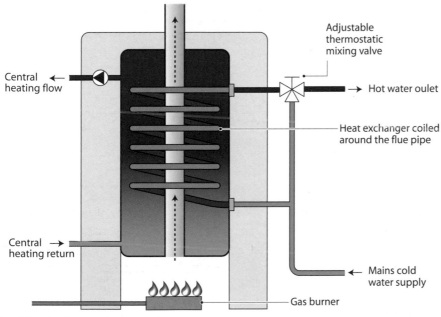

Central heating flow

Central heating return

Adjustable thermostatic mixing valve

Hot water oulet

Heat exchanger coiled around the flue pipe

Mains cold water supply

Gas burner

Combined primary storage units

Types of localised hot water supply systems used in buildings

Localised systems are often called single-point or point-of-use systems. They are designed to serve one outlet at the position where they are needed, and are usually installed where the appliance is some distance away from the centralised hot water supply.

Localised systems can be divided into two categories:

- instantaneous-type localised water heaters
- storage-type localised water heaters.

Localised hot water systems

Systems of hot water supply that are installed at the place where they are needed.

Instantaneous hot water systems

A system of hot water supply that heats cold water directly from the cold water main via a heat exchanger. There is no storage capacity.

Instantaneous-type localised water heaters

These can be fuelled by either gas or electricity, and are generally described as inlet controlled. This simply means that the water supply is controlled at the inlet to the heater. The water is heated as it flows through the heater and will continue to be heated as long as it is flowing. When the control valve is closed, the water flow stops and the heat source shuts down.

This type of heater is generally used to supply small quantities of hot water for appliances such as wash basins and showers. Typical minimum water pressure is 1 bar.

An instantaneous localised water heater

Storage-type localised water heaters

This type of heater is often referred to as a displacement-type heater, as the hot water is displaced from the heater by cold water entering the unit. Typical storage capacities are between 7 and 10 litres (for the over-sink type). Storage-type localised water heaters can be divided into the following categories:

- **Over-sink heaters** – as the name suggests, these are fitted over an appliance such as a sink. The water is delivered from a spout on the heater. A common complaint with these heaters is that they constantly drip water from the spout. This is normal, as the heater must be open to the atmosphere at all times to accommodate the expansion when the water is heated. The dripping water is due to the expansion taking place and will stop once the heater has reached its operating temperature.

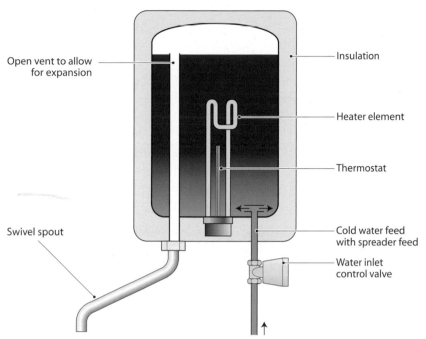

Open vent to allow for expansion

Insulation

Heater element

Thermostat

Swivel spout

Cold water feed with spreader feed

Water inlet control valve

A typical over-sink storage water heater

- **Under-sink heaters** – these heaters work in exactly the same way as over-sink heaters. The main difference is that they usually require a special tap or mixer tap that allows the outlet to be open to the atmosphere at all times to allow for expansion. The inlet for water to the heater is still controlled from the tap. Typical capacities are up to 15 litres.

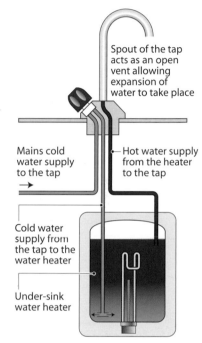

Spout of the tap acts as an open vent allowing expansion of water to take place

Mains cold water supply to the tap

Hot water supply from the heater to the tap

Cold water supply from the tap to the water heater

Under-sink water heater

A typical under-sink storage water heater

Compare thermal stores and unvented hot water storage systems

The first and most important point to remember about thermal stores and unvented hot water storage systems is that they are both mains fed and so they are only as good as the mains cold water supplying them. The controls and components that are installed as part of the system do not enhance either flow rate or pressure – in fact, the opposite is true. Valves and other components often prove restrictive and actually decrease both flow rate and pressure. Therefore, it is true to say that if the flow rate and pressure entering the building are poor, delivery of water at the taps will also be poor, regardless of the type of system installed and the claims of the manufacturer.

Both thermal stores and unvented hot water storage systems use water supplied directly from the mains cold water supply. Systems that deliver mains-pressure hot water supply differ in almost all respects from any other type of hot water system. The next section looks more closely at these two important systems.

Hot water systems

Systems that deliver the highest performance require a cylinder, but the function and purpose of the cylinder differs greatly depending on the system. To the untrained eye, thermal stores and unvented hot water storage systems look very similar. They are both full of hot water and cylindrical in shape, but that's where the similarity ends. It's how the hot water is produced that makes them different.

Unvented hot water storage systems

These systems use a storage cylinder under mains pressure. As you saw earlier in the unit, the hot water storage vessel is filled with water direct from the mains cold water supply, with the water inside the cylinder being heated either directly via an immersion heater or indirectly via a heat exchanger which is heated by an external heat source such as a gas boiler or a solar collector. The heated water is then forced out (or displaced) by cold water entering the cylinder when a hot tap or outlet is opened. By definition, the hot water supplied at the taps is at the same pressure and flow rate as the mains cold water entering the unit.

Unvented hot water cylinders are built to withstand considerable pressure. Because of this most cylinders are manufactured from stainless steel or heavy-grade copper sheet, and for a good reason. If the water inside the cylinder were to become overheated due to a control or component malfunction, there would be a risk that the water would boil at a temperature higher than 100°C. Subsequent vessel rupture would mean that the water inside the cylinder would turn instantly to steam with enough force to demolish the average three-bedroomed house! To combat this risk, unvented cylinders are fitted with various controls and components to vent the cylinder and prevent such a problem occurring. To ensure that unvented systems are fitted correctly and safely, all systems of this type must comply with Approved Document G3 of the Building Regulations, be fitted by a registered installer, and notice must be given to the local Building Control. This applies to any system with over 15 litres of water storage.

Vented mains pressure systems using a thermal store

With these systems, the cylinder itself is vented to the atmosphere so the dangers of overheating and excessive pressure build-up cannot occur. This is the main difference between this type and unvented systems – the vast body of water inside the vented cylinder is not the water that is delivered to the taps. This is merely the means of heating the mains cold water instantaneously as it passes through a specially designed heat exchanger within the cylinder. The cylinder is, in essence, a vented thermal battery that is heated by an external heat source such as a gas boiler. It is this thermal battery that transfers the heat to the cold water.

There are several heat exchanger types available in thermal stores, ranging from simple coils to more complex plate heat exchangers, but they all share the same advantage: mains-pressure hot water with a good flow rate from a safe, vented cylinder. Thermal stores also have other advantages over unvented systems.

■ Thermal stores store water at a much hotter temperature than in an unvented hot water cylinder, which means they hold far more energy. To prevent the water that is delivered to the taps from getting too hot, a thermostatic mixing valve blends hot water, leaving the heat exchanger with cold water from the mains.

■ Thermal stores can accept heat from uncontrollable heat sources such as wood stoves and solid fuel boilers where it is impossible to simply switch the heat off. These cannot be used on unvented systems.

■ A thermal store can also be used as a delivery source for many central heating systems, including underfloor systems. This can be of benefit, especially when uncontrolled heat sources or multiple heat sources are used, as not only are these supplying the thermal store but also the central heating.

■ You do not need to be registered to fit a thermal store. Since these vessels are vented, they are not subject to Document G3 and so Building Control notice is not required.

Recommended design temperatures in hot water systems

There is some form of hot water delivery system in almost all domestic properties in the UK, whether this is from a centralised hot water system or a localised one. However, the overriding concerns with hot water may seem at first glance to be contradictory. *Legionella* and scalding are of major concern to designers, as they present different and conflicting restrictions on the temperatures that can safely be used in the domestic hot water systems we install. It is a fact that hot water temperatures that are not high enough to cause scalding provide ideal conditions for legionella bacteria to multiply, and that hot water temperatures that are high enough to kill legionella bacteria can cause scalding. Because of this, careful temperature control must be exercised both in the storage vessels/heaters and at the point of use.

Legionella pneumophila, the bacterium causing Legionnaires' disease, presents a very real risk, especially in any hot water system that contains a storage vessel. Between 20°C and 45°C, *Legionella* bacteria multiply roughly every 2 minutes. At 60°C *Legionella* bacteria are dead within 32 minutes. However, water at this temperature is likely to cause a partial thickness burn in about 5 seconds.

Legionella bacteria (*legionella pneumophila*)

Bacteria which breed in stagnant water. They can give rise to a lung infection called legionnaire's disease, which is a type of pneumonia.

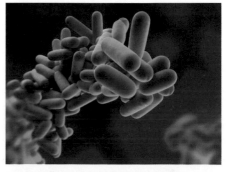

Legionella pneumophila bacteria

The recommended temperatures, therefore, are as follows.

- **Hot water storage vessels** – these should store water at a temperature of 60–65°C, as this temperature offers protection against the reproduction of legionella bacteria.

- **Hot water outflow** – the Water Supply (Water Fittings) Regulations 1999 state that hot water should be distributed at a temperature not less than 55°C and should reach the hot water outlets at 50°C within 30 seconds of the tap being turned on. While this is possible with most hot water storage systems, it is not realistically possible with instantaneous hot water delivery-type systems.

- **Secondary return** – the Health and Safety Executive (HSE) publication 'The control of legionella bacteria in water systems: Approved Code of Practice and guidance' states that: 'The circulation pump is sized to compensate for the heat losses from the distribution circuit such that the return temperature to the calorifier is not less than 50°C.'

The recommended temperatures at point-of-use are as follows.

- **Instantaneous heaters** – most instantaneous hot water heaters have a varying temperature range between 35°C and 55°C but this depends on the flow rate through the heater, with the flow rate being higher at lower temperatures. Typically, power outputs of between 3kW and 12kW are available.

- **Storage heaters** – hot water temperatures typically range up to 75°C for vented-type storage water heaters with a 5- to 10-litre capacity, and up to 80°C with an energy cut-out set at 85°C for unvented types of up to 15 litres capacity.

- **Thermostatic blending valve installations** – the subject of thermostatic blending valves and the control of discharge temperatures requires careful consideration and is discussed separately (see below).

The control of discharge temperatures

The object of any hot water storage system is to store water at the relatively high temperature of 60°C to ensure that it is free from any bacteria, to distribute the water at 55°C, and yet to deliver the water at the hot water outlets at the relatively low temperature of 35–46°C to ensure the safety of the end user. The most efficient way to do this is by the use of thermostatic mixing valves (TMVs).

Thermostatic mixing valves (sometimes known as thermostatic blending valves) are designed to mix hot and cold water to a predetermined temperature, to ensure that the water is delivered to

A typical thermostatic mixing valve

the outlet at a temperature that will not cause injury but is hot enough to facilitate good personal hygiene. There are three methods of installing TMVs.

- **Single valve installations** – this is probably the most common of all TMV installation types. The maximum pipe length to a single appliance is 2m from the TMV to the outlet. Back-to-back installations are acceptable from a single valve provided that the use of one appliance does not affect the other and that both appliances have a similar flow rate requirement, eg two wash basins. Typical installations include:

  - **Baths** – it is now a requirement of Building Regulations Approved Document G3 that all bath installations in new and refurbished properties incorporate the use of a TMV. This would normally be set to a temperature of between 41°C and 44°C, depending on personal comfort levels. Temperatures above this can only be used in exceptional circumstances.

  - **Showers** – these installations usually require a temperature of not more than 43°C. In residential care homes and other medical facilities a temperature of not more than 41°C should be used, according to NHS guidelines.

  - **Wash basins** – careful consideration must be given to wash basin installations, because this is probably the only appliance used in domestic dwellings where the user puts their hands directly into the running water without waiting for the water to get hot. When the water reaches maximum temperature, scalding can occur. Therefore, typical temperatures between 38°C and 41°C can be used, depending on the application. Again, NHS guidelines recommend a temperature of no more than 41°C.

  - **Bidets** – a maximum temperature of 38°C should be used with bidet installations.

  - **Kitchen sinks** – this is probably the area where the user is most at risk. The need to ensure that bacteria and germs are killed and that grease is thoroughly removed dictates that a water temperature of between 46°C and 48°C is used. However, as the kitchen is an area with no published recommendations on hot water temperature, a safe temperature similar to that of wash basins should be considered to lessen the risk of scalding unless notices warning of very hot water are used.

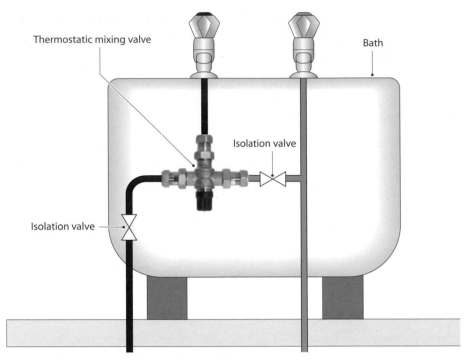

A single thermostatic mixing valve installation

- **Group mixing** – installations in which a number of appliances of a similar type are fed from a single TMV are allowed under certain circumstances. However, installations of this type are not recommended where the occupants are deemed to be high risk, such as in nursing homes. For a group installation, consider the following points:

  - The operation of any one appliance should not affect others on the run.

  - When one TMV is used with a number of similar outlets, the length of the pipework from the valve to the outlets should be kept as short as possible so that the mixed water reaches the furthest tap within 30 seconds.

  - With group shower installations, it is not unusual to see pipe runs in excess of 10m. Pipework runs of this length carry an unacceptable legionella risk. These situations can be dealt with by careful monitoring of the water at the showerheads, and appropriate treatment should legionella be detected, as well as regular and very hot water disinfection when the system is not in use.

- **Centralised mixing** – centralised mixing is very similar to group mixing, but occurs when there are groups of different hot water appliances to be served from a single TMV. The following recommendations should be followed.

  - If the mixed water is re-circulated within the legionella growth temperature range, anti-legionella precautions similar to those recommended for group mixing will need to be implemented

- If the mixed water is to re-circulated at about legionella growth temperature, the recommendations for single TMV installations are appropriate.
- The operation of any one outlet should not affect other outlets.

Typical group installations are as follows.

- Group showers – with the correctly sized TMV, a number of shower outlets may be served at a temperature of between 38°C and 40°C. For safety reasons, the temperature must not exceed 43°C.
- Wash basins – rows of wash basins may be served from a single TMV. Temperatures of between 38°C and 40°C are typical, and should not exceed 43°C for safety reasons.

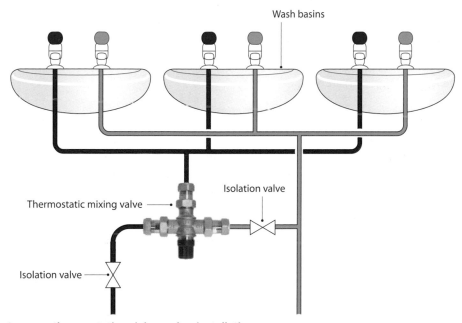

A group thermostatic mixing valve installation

Types of thermostatic mixing valves

Thermostatic mixing valves are certificated under a third-party certification scheme set up and administered by BuildCert. Under the BuildCert scheme, thermostatic mixing valves are certified and approved for use depending on their application. They are divided into two groups:

- **TMV2** – Approved Document G (Sanitation, hot water safety and water efficiency) of the Building Regulations requires that the hot water outlet to a bath should not exceed 48°C. It also states that valves conforming to BS EN 1111 or BS EN 1287 are suitable for this purpose. Similar requirements exist in Scotland.

  TMV2 approval is for domestic thermostatic installations, and uses BS EN 1111 and BS EN 1287 as a basis for thermostatic valve performance testing.

- **TMV3** – these valves are manufactured and tested for the Healthcare and Commercial thermostatic installations and use the NHS specification D08 as a basis for thermostatic valve performance testing.

Table 2 provides a guide to the selection of TMVs for particular applications.

Table 2: TMV selector chart

Environment	Appliance	TMV required by legislation or authoritative guidance?	TMV recommended by legislation or authoritative guidance?	TMV suggested best practice?	Valve type
Private dwelling	Bath	Yes			TMV2
	Basin				
	Shower			Yes	
	Bidet	Yes		Yes	
Housing Association dwelling	Bath	Yes			TMV2
	Basin				
	Shower			Yes	
	Bidet	Yes		Yes	
Housing Association dwelling for the elderly	Bath	Yes			TMV2
	Basin	Yes			
	Shower	Yes			
	Bidet	Yes			
Hotel	Bath			Yes	TMV2
	Basin			Yes	
	Shower			Yes	
NHS nursing home	Bath		Yes		TMV3
	Basin		Yes		
	Shower		Yes		
Private nursing home	Bath		Yes		TMV3
	Basin		Yes		
	Shower		Yes		
Young persons' care home	Bath	Yes			TMV3
	Basin	Yes			
	Shower	Yes			

Environment	Appliance	TMV required by legislation or authoritative guidance?	TMV recommended by legislation or authoritative guidance?	TMV suggested best practice?	Valve type
Schools, including nurseries	**Bath**	Yes, but 43°C max			TMV2
	Basin		Yes		
	Shower	Yes			
Schools for the severely disabled, including nurseries	**Bath**	Yes, but 43°C max			TMV3
	Basin		Yes		
	Shower	Yes			
NHS hospital	**Bath**	Yes			TMV3
	Basin	Yes			
	Shower	Yes			
Private hospital	**Bath**		Yes		TMV3
	Basin		Yes		
	Shower		Yes		

The requirements of Building Regulations Part G for hot water installations

In the past, Building Regulations Approved Document G3 only related to unvented hot water supply systems. In 2010 it was updated to encompass all hot water delivery systems in domestic dwellings. It is divided into four parts.

1 Part 1 of G3 is a new requirement. It states that heated wholesome water must be supplied to any wash basin or bidet that is situated in or adjacent to a room containing a sanitary convenience, to any wash basins, bidets, fixed baths or showers installed in a bathroom and to any sink in an area where food is prepared.

2 Part 2 is an expanded requirement. It states that any hot water system, including associated storage (including any cold water storage cistern) or expansion vessel must resist the effects of any temperature or pressure that may occur during normal use as a consequence of any reasonably anticipated fault or malfunction. This amendment was enforced after the failure of an immersion heater thermostat that caused the collapse of a storage cistern containing water almost at boiling point.

3 Part 3 is also an amended requirement. It states that any part of a hot water system that incorporates a hot water storage vessel must include precautions to ensure that the temperature of the stored water does not exceed 100°C, and that any discharge from such safety devices is safely conveyed to a point where it is visible without constituting a danger to persons in or about the building.

4 Part 4 states that any hot water supply to a fixed bath must include provision to limit the temperature of the discharged water from any bath tap so that it cannot exceed 48°C. This requirement applies to new-builds and property conversions. It is a new requirement that is intended to prevent scalding.

It is interesting to note that Regulation G3 applies to all domestic dwellings including greenhouses, small detached buildings, extensions and conservatories, but only if they are served with hot water supplied from a dwelling.

It should be noted that the local Building Control officer should be informed before commencing any installation of a hot water system.

Different fuels in domestic hot water systems

There are several different fuel sources available for heating the supply of hot water in a dwelling. They can be divided into two categories.

Fossil fuels

Formed by anaerobic decomposition of buried dead carbon-based plants, these fuels are known as hydrocarbons and release a high carbon dioxide content when burnt.

- **Fossil fuels:**
  - **Gas (natural gas)** – probably the most popular way of heating the hot water supply in the UK. It is also the cheapest and cleanest of all fossil fuel types, with CO_2 emissions much lower than with solid fuel for example. However, it is not available to all buildings, as some remote parts of the UK are not connected to the natural gas network.
  - **Gas (liquid petroleum gas, or LPG)** – an alternative gas source for properties not connected to the gas network. It is supplied from a pressurised storage tank positioned some distance from the building. Supplies are replenished by tanker delivery. Liquid propane is preferred to liquid butane, as it boils at a much lower temperature (−45°C) than butane (−4°C). LPG tends to be very expensive.
  - **Electricity** – this fuel can be used to heat water in a variety of ways, such as via immersion heaters, instantaneous water heaters or showers from a 230V single-phase supply.
  - **Oil** – this is another alternative gas source for properties not connected to the gas network. It is supplied from a storage tank positioned some distance from the building. Supplies are replenished by tanker delivery. There are many types of heating oil, with the most common being C2 grade, 28-second viscosity

oil or kerosene. They are very similar to diesel fuel. Appliances tend to be big and quite noisy and they require regular servicing.

- **Solid fuel** – there are many types of solid fuel used to produce energy and provide heating and hot water, although the use of some solid fuels (eg coal) is restricted or prohibited in some urban areas, due to unsafe levels of toxic emissions. In some areas, smokeless coal and coke are the only solid fuels used.

■ **Renewable energy:**

- **Solar thermal** – solar thermal technology utilises heat from the Sun to generate domestic hot water supply to offset the water heating demand from other sources such as electricity or gas.

- **Geothermal** – geothermal energy is heat directly from the Earth. It is a clean, renewable resource. Geothermal heat can be used directly, without involving a power plant or heat pump, for a variety of applications such as space heating and cooling, hot water supply and industrial processes. Its uses for bathing can be traced back to the ancient Romans.

- **Biomass** – the term biomass can be used to describe many different types of solid and liquid fuels. It is defined as any plant matter used directly as a fuel or that has been converted into other fuel types before combustion. Generally, solid biomass is used as heating fuel, including wood pellets, vegetable waste (such as wood waste and crops used for energy production), animal materials/wastes and other solid biomass.

The operating principles of hot water digital showers

Digital showers use state-of-the-art technology to give very accurate control of the showering temperature and flow rate. The shower control and the water mixer have digital intelligence built in.

Digital mixer showers take water from both the hot and cold water supplies and mix them in an electronically controlled mixing valve to accurately reach the desired temperature. An electronic control panel mounted in the showering area provides separate control of the water flow rate and the temperature. A processor box sited remotely adjusts the flow and temperature to the settings selected by the user on the control panel. These are then controlled electronically to provide the desired temperature and flow at the shower.

Accurate control is maintained by the adjustment of separate proportioning valves and pumps or by motorised control of a mixing valve.

A digital shower

Understand the operating principles of components found in hot water systems (LO2)

There are six assessment criteria for this Outcome:

1 Explain the function of safety devices in unvented hot water systems.

2 Explain the method of operation of functional devices in unvented hot water systems.

3 Calculate the diameter of discharge pipework.

4 Specify the requirements of discharge pipework from temperature and expansion relief valves.

5 Specify the layout features for pipework systems that incorporate secondary circulation.

6 Explain balanced and unbalanced supplies in unvented hot water storage systems.

Range	
Safety devices	Control thermostat, overheat thermostat (thermal cut-out), temperature relief valve
Functional devices	Line strainer, pressure-reducing valve, single-check valve, expansion device (vessel or integral to cylinder), expansion relief valve, tundish arrangements, composite valves
Discharge pipework	D1, D2
Requirements	Maximum dimensions, minimum dimensions, material, termination points, fall
Features	Pump type and location, timing devices, prevention of reverse circulation, methods of balancing circuits

The function of safety devices in unvented hot water systems

SmartScreen Unit 304
Presentation 3

With the water inside the unvented hot water storage vessel at a pressure above atmospheric pressure, the control of the water temperature becomes vitally important. This is because as the pressure of the water rises, so the boiling point of the water rises. In simple terms, if total temperature control failure were to occur, the water inside the vessel would eventually exceed 100°C with disastrous consequences. The line graph below demonstrates the pressure/temperature relationship.

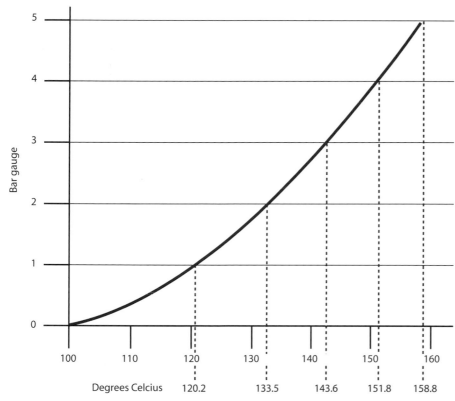

Boiling point/pressure relationship

On the graph you can see that at the relatively low pressure of 1 bar, the boiling point of the water has risen to 120.2°C. If a sudden loss of pressure at the hot water storage vessel were to occur due to vessel fracture, at 120.2°C the entire contents of the cylinder would instantly flash to steam, with explosive results, causing structural damage to the property. Calculating how much steam would be produced illustrates the point further.

$1cm^3$ of water creates $1,600cm^3$ of steam. If the storage vessel contains 200 litres of water and each litre of water contains $1,000cm^3$, then the amount of steam produced would be:

200 x 1,000 x 1,600 = $320,000,000cm^3$ of steam!

Building Regulations Approved Document G3 states that unvented hot water storage systems must have a three-tier level of safety built into the system. This takes the form of three components that are fitted to the storage vessel. The aim of these components is to ensure that the water within the system never exceeds 100°C. These components are as follows.

- **Control thermostat (set to 60°C to 65°C)** – this can take two forms, depending on the type of storage vessel:

- With direct heated vessels this is the immersion heater user thermostat.
- With indirectly heated vessels it is the cylinder thermostat wired to the central heating wiring centre.

  Indirectly fired systems are also controlled, in part, by the boiler thermostat (82°C maximum setting) and the boiler high limit stat, designed to operate, typically, at 90°C).

- **Overheat thermostat (thermal cut-out 90°C maximum, but more usually factory set at between 85°C and 89°C)** – again, this can take two forms:

  - With direct heated systems, it is incorporated into the immersion heater thermostat.
  - With indirectly heated systems, it is a separate component factory-wired into the vessel and designed to operate the motorised valve at the primary hot water coil.

- **Temperature/pressure relief valve (95°C)** – a standard component used on most vessels that is designed to discharge water when the temperature exceeds 95°C. Most types have a secondary pressure relief function.

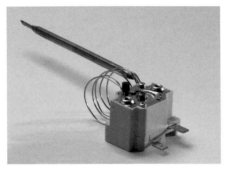

An unvented hot water vessel indirectly heated thermal cut-out

The method of operation of functional devices in unvented hot water systems

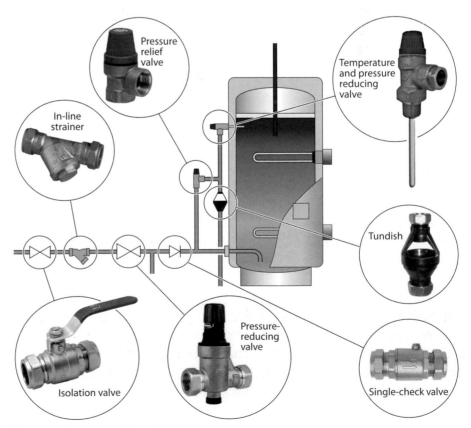

The controls on a modern UVHWSS

The functional controls of an unvented hot water storage system (UVHWSS) are designed to protect the water supply.

■ To avoid contamination, the storage cylinder or vessel must be of an approved material, such as copper or duplex stainless steel, or have an appropriate lining that will not cause corrosion or contamination of the water contained within it. Where necessary it must be protected by a sacrificial anode.

■ A single-check valve (often referred to as a non-return valve) must be fitted to the cold water inlet to prevent hot or warm water from entering the water undertaker's mains supply.

■ A means of accommodating and containing the increase in volume of water as a result of expansion must be installed. This can either be by the use of an externally fitted expansion vessel or via an integral air bubble.

■ An expansion valve (also known as the pressure relief valve) must be installed and should be designed to operate should a malfunction occur with either the pressure-reducing valve or the means of accommodating the expanded water. The expansion valve must be manufactured to BS EN 1491:2000 – Building valves. Expansion valves. Tests and requirements.

A BS 1010 screw-down stop valve

The Water Supply (Water Fittings) Regulations 1999 also state that: 'Water supply systems shall be capable of being drained down and fitted with an adequate number of servicing valves and drain taps so as to minimise the discharge of water when water fittings are maintained or replaced.'

To comply with this requirement, a servicing valve should be fitted on the cold supply close to the storage vessel but before any other control. The valve may be a full-bore spherical plug, a lever action-type isolation valve or a screw-down stop valve to BS 1010. Any drain valves fitted should be manufactured to BS 2879 and should be type A drain valves with a locking nut and an O-ring seal on the spindle.

A lever-action spherical plug isolation valve

The functional controls of an unvented hot water storage system are described below. You will look at each one in turn and identify its position within the system.

In-line strainer

The in-line strainer is basically a filter designed to prevent any solid matter within the water from entering and fouling the pressure-reducing valve and any other mechanical components sited downstream. In modern storage systems, this is incorporated into the composite valve, which is discussed later in this section.

A type A drain-off valve

An in-line strainer

Pressure-reducing valve

Pressure-reducing valves were looked at in detail in Unit 303. However, they are sufficiently important that they warrant an explanation here too.

A pressure-reducing valve

The pressure-reducing valve of an unvented hot water storage system reduces the pressure of the incoming water supply to the operating pressure of the system. In all cases this will be set by the manufacturer and sealed at the factory. The outlet pressure will remain constant even during periods of fluctuating pressures. Should the pressure of the water supply drop below that of the operating pressure of the PRV, it will remain fully open to allow the available pressure to be used.

Replacement internal cartridges are available and are easily fitted without changing the valve body should a malfunction occur. Modern PRVs for unvented hot water storage systems are supplied with a balanced cold connection already fitted.

Single-check valve

A single-check valve (also known as a non-return valve) is fitted to prevent hot water from backflowing from the hot water storage vessel and causing possible fluid category 2 contamination of the cold water supply. The single-check valve also ensures that the expansion of water when it is heated is taken up within the system's expansion components or expansion bubble. Single-check valves are classified as either type EA or EB backflow prevention devices.

A single-check (non-return) valve

In most cases, the check valve will be part of the composite valve, which is discussed later in this section.

Expansion device (vessel or integral to cylinder)

Water expands when heated. Between 4°C and 100°C it will expand by approximately 4%. Therefore 100 litres of water at 4°C becomes 104 litres at 100°C. This expansion of water must be accommodated in an unvented hot water storage system. This can be achieved in one of two ways:

- by the use of an externally fitted expansion vessel, or
- by the use of a purpose-designed internal expansion space, or 'expansion bubble'.

Expansion vessel

An expansion vessel is a cylindrical-shaped vessel that is used to accommodate the thermal expansion of water to protect the system from excessive pressures. It is installed as close to the storage vessel as possible, and preferably higher. There are two basic types.

- **Bladder-type expansion vessel** – also known as a bag-type expansion vessel, this is usually made from steel and contains a neoprene rubber bladder to accept the expanded water. At no time does the water come into contact with the steel vessel. It is contained at all times within the bladder.

The inside of the steel vessel is filled with either air or nitrogen to a predetermined pressure. The initial pressure charge from the manufacturer is usually made with nitrogen to negate the corrosive effects on the steel vessel's interior. A Schrader valve is fitted to allow the pressures to be checked and to allow an air 'top-up' if this becomes necessary. The diagrams show the workings of a bladder-type expansion vessel.

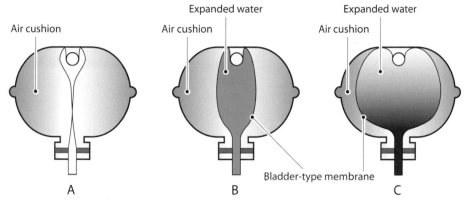

The workings of a bladder (bag)-type expansion vessel

Diagram A shows the bladder in its collapsed state – the only pressure is the air/nitrogen charge compressing the empty bladder. There is no water in the bladder.

Diagram B shows that water under pressure has entered the bladder during the initial cold fill of the storage cylinder, causing the bladder to expand and pressurising the air in relation to the water pressure. The bladder has expanded because the water pressure is greater than the pressure of the air.

Diagram C shows the bladder fully expanded due to the hot water expansion when the system is heated.

A bladder (bag)-type expansion vessel

With some bladder expansion vessels, the bladder is replaceable in the event of its failure. A flange at the base of the vessel holds the bladder in place. By releasing the air and removing the bolts, the bladder can be withdrawn and replaced.

- **Diaphragm-type expansion vessel** – diaphragm expansion vessels are used when the water has been de-oxygenated by the use of inhibitors, or because the water has been heated repeatedly, such as in a sealed central heating system. They must not be used with UVHWSSs because the water is always oxygenated and comes into direct contact with the steel of the vessel.

They are made in two parts, with a neoprene rubber diaphragm separating the water from the air charge. Again, as with the bladder-type expansion vessel, a Schrader valve is fitted to allow top-up and testing of the air pressure. The image below shows the workings of a diaphragm-type expansion vessel.

A diaphragm-type expansion vessel

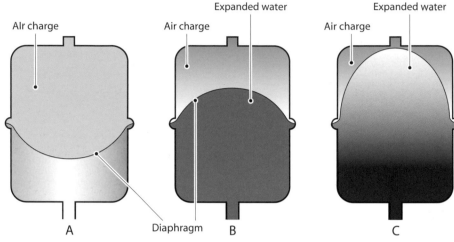

The workings of a diaphragm-type expansion vessel

Internal expansion

With internal expansion, an air pocket is formed as the hot water storage vessel is filled. A floating baffle plate provides a barrier between the air and the water so that there is minimum contact between them in the cylinder. When the water is heated, the expansion pushes the baffle plate upwards in a similar manner to an expansion vessel.

Over a period of time, the air within the air bubble will dissipate as it leaches into the water. When this happens, expansion cannot take place and the pressure relief valve will start to discharge water. However, this will only occur as the water heats up. Once the cylinder is at its full temperature, the pressure relief valve will close and will only begin to discharge water again when expansion is

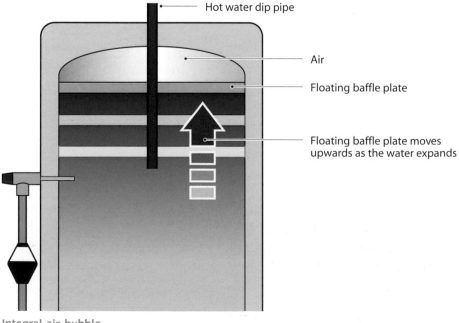

Integral air bubble

taking place. Because of this, manufacturers of bubble-top units and packages recommend that the cylinder is drained down completely and refilled to recharge the air bubble. This should be done on an annual basis or as and when required.

The scientific principles of an expansion vessel

The vessel works on the idea that gases are compressible, but liquids are not. The principle is based on Boyle's law. In this case the gas is air or nitrogen and the liquid is water.

Boyle's law states: 'The volume of a gas is inversely proportional to its absolute pressure provided that the temperature remains constant.' In other words, if the volume is halved, the pressure is doubled.

Mathematically, Boyle's law is expressed as: $P_1V_1 = P_2V_2$

Where:

P_1	=	initial pressure	=	1 bar
V_1	=	initial volume	=	20 litres
P_2	=	final pressure	=	to be found
V_2	=	final volume	=	20 litres – 10 litres of expanded water

So, to find the pressure in the vessel, the formula must be transposed:

$$P_2 = \frac{P_1 \times V_1}{V_2}$$

Therefore:

$$P_2 = \frac{1 \text{ bar} \times 20 \text{ litres}}{10 \text{ litres}}$$

$$= 2 \text{ bar final cold pressure}$$

If, on the initial cold fill of the system, the vessel required, say, 5 litres of water to be taken in, the air pressure to apply to the vessel can be calculated. You can assume a water pressure of 1 bar.

P_1	=	1 bar
V_1	=	20 litres
V_2	=	20 litres – 5 litres = 15 litres
P_2	=	pressure to be calculated
P_2	=	$\frac{P_1 \times V_1}{V_2}$ = $\frac{1 \text{ bar} \times 15 \text{ litres}}{20 \text{ litres}}$ = 0.75 bar

The capacity left in the vessel after the initial fill is 15 litres, with a cold fill pressure of 1 bar, and that 10 litres of water are to expand inside the vessel, so the final pressure of the system will be:

$$P_2 = \frac{P_1 \times V_1}{V_2} \quad = \frac{1 \times 15}{15 - 10} \quad = \frac{15}{5} = 3 \text{ bar}$$

SUGGESTED ACTIVITY

Using the formula $P_1V_1 = P_2V_2$ as shown in the example above, find the initial cold fill pressure of the expansion vessel and the final hot operating pressure of the storage cylinder.

Where:

P_1	=	initial pressure	=	1.5 bar
V_1	=	initial volume	=	18 litres
P_2	=	final pressure	=	to be found
V_2	=	final volume	=	
		18 litres – 9 litres of expanded water		

The initial pressure of the empty 20-litre vessel was 0.75 bar. On initial cold fill 5 litres of water entered the vessel, reducing the capacity to 15 litres. As a result the air was compressed even more when the water expansion took place, and instead of 2 bar final pressure, the pressure when the water was heated was 3 bar.

Pressure relief valve

Often referred to as the expansion relief valve, the pressure relief valve is designed to automatically discharge water in the event of excessive mains pressure or malfunction of the expansion device (expansion vessel or air bubble). It is important that no valve is positioned between the pressure relief valve and the storage cylinder.

The pressure at which the pressure relief valve operates is determined by the operating pressure of the storage vessel and the working pressure of the pressure-reducing valve. The valve is pre-set by the manufacturer and must not be altered.

The pressure relief valve will not prevent the storage vessel from exploding should a temperature fault occur and, as such, is not regarded as a safety control.

A pressure relief valve

Tundish arrangements

The tundish is part of the discharge pipework and is supplied with every unvented hot water storage system. It is the link between the D1 and D2 pipework arrangements. It has three main functions:

- to provide a visual indication that either the pressure relief or temperature relief valve is discharging water as a result of a malfunction
- to provide a physical type A air gap between the discharge pipework and the pressure/temperature relief valves
- to give a means of releasing water through the opening in the tundish in the event of a blockage in the discharge pipework.

A tundish

The tundish must always be fitted in the upright position in a visible place close to the storage vessel. The tundish is looked at in more detail when discharge pipework arrangements are covered later in this section.

Composite valves

These days, it is very rare to see individual controls fitted on an unvented hot water storage system unless it is an early type manufactured in the 1990s. Most manufacturers now prefer to supply composite valves, which incorporate many components into one multi-valve. A typical composite valve will contain:

- a strainer
- a pressure-reducing or pressure-limiting valve, followed immediately by
- a balanced cold take-off, and finally
- a pressure relief valve.

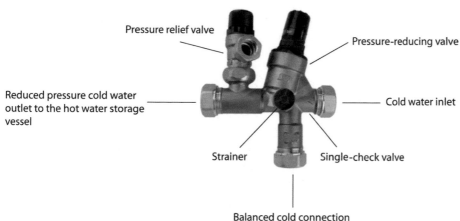

Pressure relief valve

Pressure-reducing valve

Reduced pressure cold water outlet to the hot water storage vessel

Cold water inlet

Strainer

Single-check valve

Balanced cold connection

A typical composite valve

Some composite valves may also contain an isolation valve. With all controls contained in a single valve, making the connection to an unvented hot water storage vessel is a simple matter of connecting the cold supply, without the need to ensure that the controls have been fitted in the correct order.

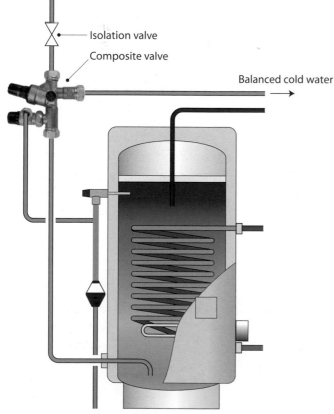

Isolation valve

Composite valve

Balanced cold water

Position of a composite valve

SmartScreen Unit 304

Handout 5

Calculate the diameter of discharge pipework

The discharge pipework from the tundish must not exceed the hydraulic resistance of a 9m straight length of pipe without increasing the pipe size. Where the discharge pipework exceeds 9m, the size of the discharge pipe will require calculation, including the resistance caused by bends and elbows.

Table 3: Diameters of discharge pipework

Valve outlet size	Maximum size of discharge pipe to tundish (D1)	Maximum size of discharge pipe from tundish (D2)	Maximum resistance allowed, expressed as a length of straight pipe without bends or elbows	Resistance created by each bend or elbow
G½	15mm	22mm	Up to 9m	0.8m
		28mm	Up to 18m	1.0m
		35mm	Up to 27m	1.4m
G¾	22mm	28mm	Up to 9m	1.0m
		35mm	Up to 18m	1.4m
		42mm	Up to 27m	1.7m
G1	28mm	35mm	Up to 9m	1.4m
		42mm	Up to 18m	1.7m
		54mm	Up to 27m	2.3m

The table works as follows.

The temperature and pressure relief valves both have ½in BSP outlets. Therefore the D1 pipework, as can be seen from the table, can be installed in 15mm tube. The discharge pipe run is 6m long to the final termination and there are six elbows installed in the run of pipe.

Using the first row in the table, the first option has to be 22mm because the D2 pipework must be at least one pipe size larger than the D1 pipework. The maximum length of 22mm pipe is 9m, but there are six elbows in the run and each of these has a resistance of 0.8m.

$6 \times 0.8 = 4.8m$

If you add the original length of 6m, you get:

$4.8 + 6 = 10.8m$

The length of 22mm discharge pipe, as you have already seen, is 9m so, at 10.8m, 22mm pipe is not large enough for the discharge pipe run. Another pipe size will have to be chosen.

Looking at 28mm, you can see that the maximum run of pipe is 18m, but the 28mm elbows now have a resistance of 1m and there are six of them. Therefore:

6 × 1 = 6m

Add this to the original length of 6m:

6 + 6 = 12m

In this case the discharge pipework is well within the 18m limit, so 28mm discharge pipework can be installed.

Specify discharge pipework from temperature and expansion relief valves

With unvented hot water systems, there is always the possibility, however undesirable, that the pressure relief and temperature relief valves may discharge water. The discharge pipework is designed specifically to remove the discharged water from the building safely. It is therefore very important that it is installed correctly, with the correct size of pipe, and that the pipework is made from the correct material – especially as the discharged water may be near boiling point.

There are three sections to the discharge pipework:

- D1 pipework arrangement
- the tundish
- D2 pipework arrangement.

You have seen the role of the tundish earlier in the unit, so this section concentrates specifically on the D1 and D2 sections of the discharge pipework.

To ensure that there is no damage to the property, the discharge pipework should be positioned in a safe but visible position and should conform to the following.

- The discharge must be via an air break (tundish) positioned within 600mm of the temperature relief valve.
- The tundish must be located within the same space as the hot water storage vessel.
- It should be made of metal or other material capable of withstanding the temperature of the discharged water. The pipe

A customer wishes to have an unvented hot water storage system installed. You have been asked by the site supervisor to size the discharge pipework. The temperature and pressure relief valves both have ½in BSP outlets. Therefore, the D1 pipework can be installed in 15mm tube. The discharge pipe run is 9m long to the final termination and there are five elbows installed in the run of pipe.

What size of discharge pipework should be installed?

should be clearly and permanently marked to identify the type of product and its performance standards.*

- The discharge pipe must not exceed the hydraulic resistance of a 9m straight length of pipe without increasing the pipe size.

- It must fall continuously throughout its entire length with a minimum fall of 1 in 200.

- The D2 pipework from the tundish must be at least one pipe size larger than the D1 pipework.

- The discharge pipe should not connect to a soil discharge pipe unless the pipe material can withstand the high temperatures of discharge water, in which case it should:

  - contain a mechanical seal (such as a Hepworth Hepvo Valve), not incorporating a water trap, to prevent foul air from venting through the tundish in the event of trap evaporation

  - be a separate branch pipe with no sanitary appliances connected to it

  - where branch pipes are to be installed in plastic pipe they should be either polybutylene (PB) to Class S of BS 7291-2:2006 or cross-linked polyethylene (PE-X) to Class S of BS 7291-3:2006

  - be marked along the entire length with a warning that no sanitary appliances can be connected to the pipe.

- The D1 pipework must not be smaller than the outlet of the temperature relief valve.

- The D1 discharge from both the pressure relief and temperature relief valves may be joined by a tee piece provided that all of the points above have been complied with.

- There must be at least 300mm of vertical pipe from the tundish to any bend in the D2 pipework.

*Note: Paragraph 3.9 of Approved Document G3 Guidance Notes specifies metal pipe for the discharge pipework. However, G3 itself states only that hot water discharged from a safety device should be safely conveyed to where it is visible but will not cause a danger to persons in or about the building. Since many types of plastic pipe are now able to withstand the heat of the discharge water, the responsibility for the choice of material rests with the installer, the commissioning engineer and the local Building Control officer to ensure that G3 is complied with. It is also important that if plastic pipes are used, that the type of plastic is clearly indicated for future reference when inspections and servicing are carried out.

The diagram shows some of the requirements described above.

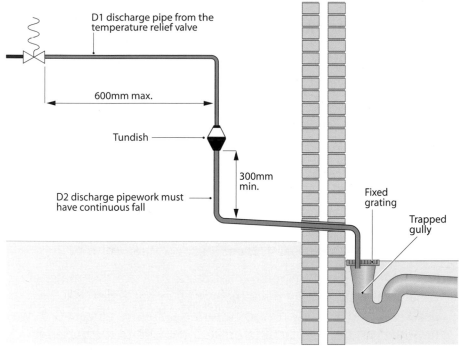

D1 discharge pipe from the temperature relief valve

600mm max.

Tundish

300mm min.

D2 discharge pipework must have continuous fall

Fixed grating

Trapped gully

The layout of the discharge pipework

Correct termination of the discharge pipework

A risk assessment is likely to be needed where any termination point for the discharge pipework is to be considered. This will determine any special requirements in relation to the termination point and its access. Points to consider include:

- areas where the public may be close by or have access to
- areas where children are likely to play or have access to
- areas where the discharge may cause a nuisance or a danger
- termination at height
- the provision for warning notices in vulnerable areas.

Building Regulations Approved Document G3 states that the discharge pipe (D2) from a tundish must terminate in a safe place with no risk to persons in the discharge vicinity, as follows:

- The D2 pipework can be taken to a trapped gully, with the discharge pipe discharging below the gully grate but above the water line.
- Downward discharges at low level up to a maximum 100mm above external surfaces, such as car parks, hard standings and grassed areas, are acceptable provided that a wire cage or similar guard is provided to prevent contact, while maintaining visibility.

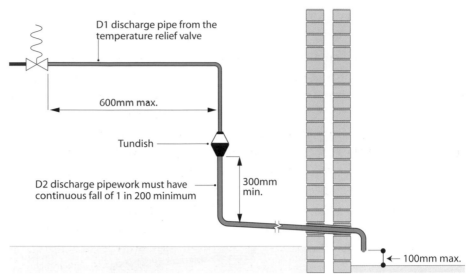

The low-level termination of discharge pipework

- Discharges at high level, onto a flat metal roof or other material capable of withstanding the temperature of the water, may be used provided that any plastic guttering system is at least 3m away from the point of discharge to prevent damage to the guttering.

- Discharges at high level, into a metal hopper and metal downpipe, may be used provided that the end of the discharge pipe is clearly visible. The number of discharge pipes terminating in a single metal hopper should be limited to six to ensure that the faulty system is traceable.

- Discharge pipes that turn back on themselves and terminate against a wall or other vertical surface should have a gap of at least one pipe diameter between the discharge pipe and the wall surface.

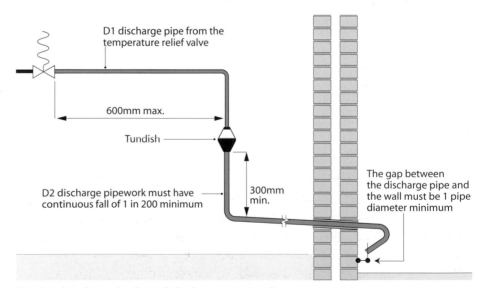

The low-level termination of discharge pipework

Note: The discharge may consist of high temperature water and steam. Asphalt, roofing felt and other non-metallic rainwater goods may be damaged by very-high-temperature hot water discharges.

Termination of the discharge pipework where the storage vessel is sited below ground level

When storage vessels are sited below ground, such as in a cellar, the removal of the discharge becomes a problem because it cannot be discharged safely away from the building. However, with the approval of the local authority and the vessel manufacturer it may be possible to pump the discharge to a suitable external point. A constant temperature of 95°C should be allowed for when designing a suitable pumping arrangement. The pump should include a suitable switching arrangement installed in conjunction with a discharge collection vessel made from a material resistant to high-temperature water. The vessel should be carefully sized in line with the predicted discharge rate and should include an audible alarm to indicate that discharge from either of the pressure or temperature relief valves is taking place.

Specify the layout features for pipework systems with secondary circulation

Secondary circulation is required where the length of any draw-off pipework is excessive. BS EN 806 (and BS 6700) and the Water Supply (Water Fittings) Regulations 1999 give the maximum length a hot water draw-off pipe may travel without a secondary circulation system being installed.

Secondary circulation is a method of returning the hot water draw-off back to the storage cylinder in a continuous loop, to eliminate cold water 'dead legs' by reducing the distance the hot water must travel before it arrives at the taps.

In all installations, secondary circulation must use forced circulation via a bronze or stainless steel-bodied circulating pump to circulate the water to and from the storage cylinder. The position of the pump will depend on the type of hot water system installed.

Secondary circulation installations on unvented hot water storage systems

In most cases, a secondary circulation connection is not fitted on an unvented hot water storage vessel and, unlike with open-vented hot water storage vessels, it is not possible to install a connection on the vessel itself. Where secondary circulation is required, this must be taken to the cold water feed connection using a swept tee just before the cold feed enters the unit. To safeguard against reverse circulation, a non-return valve or single-check valve must be fitted after the circulating pump and just before the swept tee branch. The pump should be fitted on the secondary return, close to the hot water storage vessel.

> **KEY POINT**
>
> What is a dead leg? When a hot tap is opened, a certain amount of cold water is usually drawn off and allowed to run to drain before hot water arrives at the tap. This wasted cold water is known as a dead leg. Under the Water Regulations, dead legs must be restricted to the lengths given in Table 12 of the regulations. If this is not possible, secondary circulation is required.

A bronze-bodied secondary circulation pump

> **KEY POINT**
>
> A secondary circulation pump is very similar in design to a central heating circulator. The difference is that the secondary circulating pump has a body cast from bronze to eliminate discoloured water and subsequent contamination. Bronze, as you know from Level 2, is a non-ferrous metal that does not rust. Central heating circulators have a cast-iron or steel body that would rust if used on a secondary circulation system. Under no circumstances should a central heating circulator be used for this purpose.

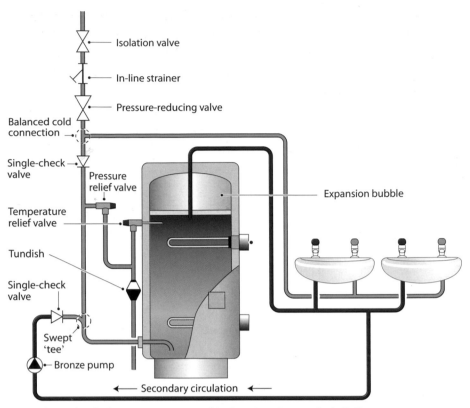

Isolation valve

In-line strainer

Pressure-reducing valve

Balanced cold connection

Single-check valve

Pressure relief valve

Temperature relief valve

Tundish

Single-check valve

Swept 'tee'

Bronze pump

Expansion bubble

Secondary circulation

Secondary circulation on an unvented hot water storage installation

Secondary circulation installations on open-vented hot water storage systems

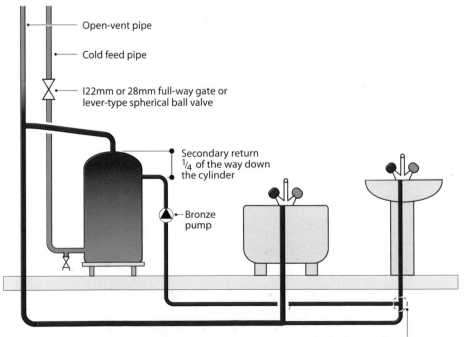

Open-vent pipe

Cold feed pipe

l22mm or 28mm full-way gate or lever-type spherical ball valve

Secondary return ¼ of the way down the cylinder

Bronze pump

Secondary return connection at the furthest appliance

Secondary circulation on an open-vented hot water storage installation

With secondary circulation on open-vented systems, the return pipe runs from the furthest hot tap back to the cylinder where it enters about a quarter of the way down. A circulating pump is placed on the return, close to the hot water cylinder, pumping into the vessel. As with all secondary circulation systems, the pump must be made from bronze or stainless steel to ensure that corrosion does not pose a problem. Isolation valves must be installed either side of the pump so that the pump may be replaced or repaired. The system is shown in the diagram below.

Some open-vented cylinders can be purchased with a secondary return connection already installed on the cylinder. Alternatively, an Essex flange can be used on cylinders where no connection exists.

An Essex flange

Positioning secondary circulation components

As you have seen, the secondary flow (the hot water draw-off) should have a temperature of at least 60°C. The secondary return of the secondary circulation circuit should have a return temperature of 50°C when it reaches the cylinder at the end of the circuit. In this way, the hottest part of the cylinder will always be the top where the hot water is drawn off. If reversed circulation were to occur, the water in the cylinder would never reach the disinfecting temperature of 60°C and so would always present a *Legionella* risk, however slight.

Preventing reversed circulation in secondary circulation systems

By installing a single-check valve on the return and positioning it between the pump and the cylinder, reverse circulation is prevented.

Time clocks for secondary circulation

If secondary circulation is used on hot water systems, it should be controlled by a time clock so that the circulating pump is not running 24 hours a day. The time clock should be set to operate only during periods of demand and should be wired in conjunction with pipe thermostats (also known as aquastats) to switch off the pump when the system is up to the correct temperature and circulation is not required, and to activate the pump when the water temperature drops.

Insulating secondary circulation pipework

If secondary circulation systems are installed, they should be insulated for the entire length of the system. This is to prevent excessive heat loss through the extended pipework as a result of the water being circulated by a circulating pump. The insulation should be thick enough to maintain the heat loss shown in Table 4.

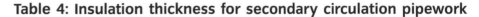

Table 4: Insulation thickness for secondary circulation pipework

Tube/pipe size	Maximum heat loss per metre
15mm	7.89W/m
22mm	9.12W/m
28mm	10.07W/m

Secondary circulation on large open-vented hot water storage systems

The diagram on the next page shows a large domestic hot water system with secondary circulation. There are some significant differences from other secondary circulation systems:

- The hot water vessel includes a shunt pump (component 7 on the diagram). This is to circulate the water within the cylinder to ensure that the stratification (varying temperature) of the water inside is kept to a minimum and to ensure an even heat distribution throughout, thereby preventing the growth of *Legionella* bacteria. Stratification is desirable during the day so that the draw-off water is maintained at its hottest for the longest period of time. Because of this, the shunt pump should only operate during periods of low demand, ie at night.

- The secondary circulation pump (component 5 on the diagram) is installed on the secondary flow and not the secondary return as with other, smaller systems.

- A non-return valve (component 6 on the diagram) is installed on the secondary flow to ensure that reverse circulation does not occur.

- A cylinder thermostat (component 3 on the diagram) is provided to maintain the temperature within the cylinder at a maximum of 60°C.

- A pipe stat (component 2 on the diagram) installed on the secondary flow maintains the temperature at a minimum of at least 50°C.

- A motorised valve (component 4 on the diagram) is installed on the secondary return close to the hot water storage vessel to prevent water being drawn from the secondary return when the pump is not operating.

- Lockshield gate valves (components 9 and 10 on the diagram) are provided to balance the system to ensure even circulation throughout the secondary water system.

- The secondary circulation system, shunt pumps and thermostats are controlled through a control box (component 1 on the diagram).

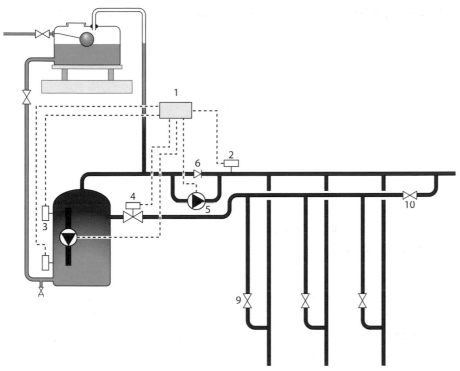

Secondary circulation on a large domestic open-vented hot water storage installation

There are some points to remember about large centralised hot water systems.

- The pipework should be carefully designed to prevent dead legs, as this is a major concern with regard to *Legionella pneumophila.*

- The hot water storage vessel should be capable of being heated to 70°C; again, to kill any *Legionella* bacteria that may be present.

- There should be easy access for draining, cleaning, inspection and maintenance.

- If a shunt pump is installed, the storage vessel should be insulated on its underside to prevent excessive heat loss.

Trace heating instead of secondary circulation

Electric trace heating uses an electric cable that forms a heating element. It is positioned directly in contact with the pipe along the whole length of the pipe. The pipe is then covered in thermal insulation. The heat generated by the element keeps the pipe at a specific temperature.

The operation of the trace heating element should be timed for a period when the hot water system is in most use, ie early in the morning or in the evening. If the pipe is well insulated and installed with a timer, the amount of energy usage will be minimal.

By using trace heating, the additional cost of the extra pipework for the secondary return and its associated pump and running costs are removed.

Trace heating

Balanced and unbalanced supplies in unvented hot water storage systems

Balanced pressure means that both the hot and the cold water are supplied at the same pressure. Most modern mixer taps, shower mixer valves and thermostatic blending valves require a balanced pressure to operate correctly and to ensure that the correct mixing of hot and cold water takes place without causing backflow problems or presenting the danger of scalding.

With unvented hot water systems, balancing the hot and cold water supplies is a relatively easy task to perform. The balanced cold supply must be taken from the cold supply to the unvented storage vessel. It must be connected after the pressure-reducing valve and before the single-check valve, as shown in the diagram. The pressure-reducing valve ensures that the same pressure is supplied to both hot and cold outlets, and the non-return valve ensures that the balanced cold water supply is not affected by backflow from the pressurised hot water storage vessel in the event of a sudden loss of pressure from the cold water supply.

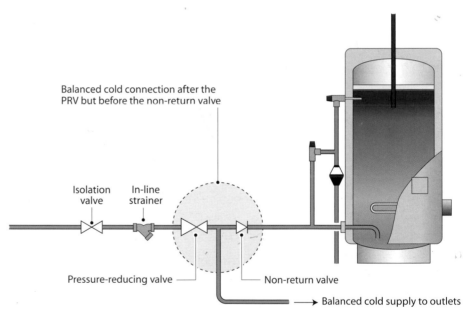

Balanced cold supply connection position

On systems that use a composite valve (a valve which combines the strainer, PRV, non-return valve and pressure relief/expansion valve), a balanced cold connection is usually designed into the valve in the form of a compression fitting connection.

Know the fault-diagnosis and rectification procedures for hot water systems and components (LO3)

There are four assessment criteria for this Outcome:

SmartScreen Unit 304
Worksheet 5

1 Specify the periodic servicing requirements of hot water systems.
2 Interpret documents to identify diagnostic requirements of hot water system components.
3 Describe diagnostic checks on hot water system components.
4 Specify methods of repairing faults in hot water system components.

Range	
Documents	Manufacturers' instructions, industry standards
Diagnostic checks	Pressures, flow rates, levels, correct operation, temperatures, thermostats, pumps, timing devices, expansion and pressure vessels, gauges, controls
Hot water system components	Pumps, expansion/pressure vessels, safety valves, pressure-reducing valves

Specify the periodic servicing requirements of hot water systems

Hot water systems, like other plumbing systems in the home, require a certain amount of periodic maintenance to ensure a continued and efficient operation. Open-vented and unvented systems have different maintenance requirements, as shown in Table 5.

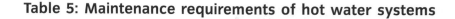

Table 5: Maintenance requirements of hot water systems

Open-vented systems	Unvented systems

• Check the installation for signs of leakage around the storage vessel and associated pipework
 	
• Check the cold water storage cistern in the roof space and clean as necessary. Check that the base and the bearers that the cistern is sitting on are in good condition and fit for purpose
 	
• Adjust the float-operated valve (FOV) to the correct water level. If there have been signs of the water overflowing, the FOV should be re-washered
 	
• Check the operation of the isolation valves and gate valves to ensure that they operate correctly. Advise the customer if they require replacement
 	
• Check and replace the sacrificial anode inside the cylinder as necessary
 	
• Run the heat source and check the temperature of the hot water
 	
• Run the system to 65°C to 70°C to ensure that the cylinder has been disinfected. Do not forget to reset the thermostat to 60°C for safety purposes
 	
• Check the system flow rates using a weir gauge
	• Unvented hot water storage systems should be serviced every 12 months
	
	• Check the installation for signs of leakage around the storage vessel and associated pipework
	
	• Check that the components are approved for use with the storage vessel. Ask the customer if any of the components have been replaced during the lifetime of the vessel
	
	• Check the pressure in the expansion vessel and top up the pressure with a foot pump as necessary
	
	• If the system has an internal expansion bubble, the system should be drained down and refilled to recharge the air
	
	• While the system is drained, remove the in-line strainer (filter) and clean it of any debris
	
	• Check the discharge pipework to ensure that it complies with the regulations. Check the termination point
	
	• Operate the pressure relief valve by twisting the top and holding open for 30 seconds
	
	• Operate the temperature/pressure relief valve by twisting the top and holding open for 30 seconds
	
	• Check the tundish to ensure that water is not discharging from the air gap:
	  • Run the heat source(s) and check the temperature of the hot water. Check the operation of the system thermostats to ensure that they shut down at the desired temperature
	  • Check that the non-self-setting high limit (energy cut-out) thermostat operates to the manufacturer's specification. This is a requirement of the Building Regulations
	  • Run the system to 65°C to 70°C to ensure that the cylinder has been disinfected. Do not forget to reset the thermostat to 60°C for safety purposes

Open-vented systems	Unvented systems
	• Check the static and dynamic pressures of the system to determine if the pressure-reducing valve is operating within the manufacturer's limits
	• Check the temperature of the water at the outlets
	• Check the flow rates using a weir gauge
	• Check that any information and warning notices required by the unvented cylinders are displayed permanently on the storage vessel. This is a requirement of the Building Regulations
	• Complete the servicing log in the BENCHMARK handbook

Identify diagnostic requirements of hot water system components

When attempting to identify faults with hot water systems, the most important document to consult is the manufacturer's instructions. In most cases these will contain a section on fault-finding that will prove an invaluable source of information. Fault-finding using the manufacturer's instructions usually takes three forms:

- known problems that can occur and the symptoms associated with them

- methods by which to identify problems in the form of flow charts. These usually follow a logical, step-by-step approach, especially if the equipment has many parts that could malfunction, such as a pressure-reducing valve or an expansion vessel

- the techniques required for replacement of the malfunctioning component.

A replacement parts list will also be present for those components that can be replaced. When ordering parts, it is advisable to use the model number of the equipment and the parts number from the replacement parts list. This will ensure that the correct part is purchased.

BS 6700 may also be consulted, as it contains important information regarding minimum flow rates required by certain appliances. This should always be used in conjunction with the manufacturer's instructions, and remember: manufacturers' instructions always take precedence over the British Standards and regulations.

Diagnostic checks on hot water system components

Routine checks on components and systems can help to identify any potential problems that may be developing within the system as well as keeping the system operating to its maximum performance and within the system design specification. Checks performed can include the following.

- Checking components for correct operating pressures, temperatures and flow rates. These are important checks, simply because they can indicate whether a component has started to fail and will require replacement, or whether the component will require re-calibration. Components that are pressure and temperature related, such as expansion vessels, pressure-reducing valves and thermostats, are particularly susceptible to failure.

- Cleaning system components (including dismantling and reassembly). Components such as in-line strainers should be checked during periodic maintenance or when there is a noticeable drop-off in flow rate. A blocked strainer will dramatically reduce the flow of water and may well affect the dynamic pressure of the system also. Pressure-reducing valves and composite valves can also be checked and cleaned but usually these contain sealed cartridges where dismantling is not advisable. New cartridges should be installed wherever a blockage in the PRV is suspected.

- Checking for correct operation of system components:
  - Thermostats – these can be checked using a thermometer in the hot water flow once the thermostat has shut off. This will indicate whether the thermostat is operating at the correct temperature.
  - Pumps – these should be checked using the manufacturer's commissioning procedures to ascertain whether the pump is performing as the data dictates. A slight fall in performance is to be expected with age. Check to ensure that there no signs of damage or wear and tear on the pump, or leakage from the pump; that the pump switches on and off at the correct pressure; and that there are no unusual noises or vibrations when the pump is operating.
  - Timing devices – time clocks can be checked to see if they activate at the correct time and that any advance timings, such as 1-hour boost buttons, work correctly. The time display should be checked against the correct time and any alterations to the time made.
  - Expansion and pressure vessels – these should be checked for the correct pressure using a portable bourdon pressure gauge. The type used to check tyre pressures is ideal for this. Any signs of water leakage should be investigated. Always refer to the manufacturer's instructions for the correct charge and pre-charge pressures.

- Gauges and controls – gauges are notorious for requiring replacement or re-calibration as they often display an incorrect pressure. They should be replaced as necessary.

- Checking for correct operation of system safety valves:
  - Temperature relief and expansion/pressure relief valves – these can be checked by twisting the top and holding the valve open for 30 seconds. Always ensure that the valve closes completely and that the water stops without any drips.

Should any components require replacing, they should be replaced with like-for-like components or, if this is not possible, check with the manufacturer that the part is approved for use with the particular storage vessel.

Specify methods of repairing faults in hot water system components

Repair of system components should be undertaken using the manufacturer's servicing and maintenance instructions, as these will contain the order in which the component should be dismantled and reassembled. With all components, there will be occasions when they cannot be repaired and replacement is the only option. Some of the components that may be repaired and/or replaced are described in Table 6.

Table 6: Hot water components

Component	Known faults	Symptoms	Repairs
Pump	• Worn/broken impeller	• Motor working but water not being pumped • No water at the outlets	• No repair possible • Replace the pump
	• Burnt-out motor	• Voltage detected at the pump terminals but pump not working	• No repair possible • Replace the pump
	• Cracked casing	• Water leaking from the pump body	• No repair possible • Replace the pump
	• Faulty capacitor	• Slow starting pump	• Replace the capacitor if possible • Check manufacturer's instructions

Component	Known faults	Symptoms	Repairs
Expansion vessel	• Pressure loss due to faulty Schrader valve	• No pressure in the expansion vessel • Water discharging from the pressure relief valve during water heat-up	• Pump air into the expansion vessel using a foot pump and check the Schrader valve with leak detector fluid • Check for bubbles • Replace Schrader valve as necessary
	• Ruptured bladder/ diaphragm	• Water discharging from the Schrader valve. Water discharging from the pressure relief valve on water heat-up	• It is possible to replace the bladder/ diaphragm of some accumulators • Check the manufacturer's instructions
Expansion (pressure) relief valve	• Water dripping intermittently when the water is being heated	• Usually an indication that the expansion vessel has lost its air charge or internal expansion bubble has disappeared	• Check and recharge the expansion vessel or internal air bubble
	• Water running constantly	• Usually an indication of incorrect pressure due to a malfunction of the pressure-reducing valve	• Check and replace the pressure-reducing valve
Temperature/ pressure relief valve	• Cold water running constantly	• Incorrect pressure due to a malfunction of the pressure-reducing valve • Faulty pressure relief valve	• Check and replace the pressure-reducing valve and the pressure relief valve

Component	Known faults	Symptoms	Repairs
Temperature/ pressure relief valve cntd.	• Hot water running constantly	• Usually a sign of thermostat and high limit stat malfunction	• Isolate the system from the electrical supply and allow to cool before attempting a repair • Check and replace the thermostat and high limit stat as necessary
Thermostats	• Hot water too hot	• System thermostat is not operating at the correct temperature	• Check the temperature of the hot water with a thermometer against the setting on the thermostat • Replace the thermostat as necessary
	• No hot water	• System thermostat is not operating	• Check the thermostat with a GS38 electrical voltage indicator for correct on/off functions • Replace as necessary
High limit thermostat	• No hot water	• Usually an indication that the system thermostat has malfunctioned and the high limit thermostat has activated to isolate the heat source	• Check the main system thermostat and reset the high limit thermostat
Pressure (bourdon) gauge	• Sticking pressure indicator needle	• Gauge not reading the correct pressure and does not move when the pressure is raised or lowered	• No repair possible • Replace the gauge

Note: When replacing or repairing valves and controls, it is important to ensure that the water supply is isolated and the section of pipework is completely drained before beginning to repair or replace the valve.

Table 7: Unvented hot water storage system fault-finding

Fault	Probable causes	Recommended solutions
No hot water flow	Mains cold water off	Check and open isolation valve/ stop valve
	Strainer blocked	Turn off water and clean filter
	Cold water connection incorrectly installed	Check and refit mains cold water as necessary
Poor flow rate	Strainer blocked	Turn off water and clean filter
Water from the hot taps is cold	Immersion heater not switched on	Check and switch on as necessary
	Thermal cut-out (high limit thermostat) has operated	Check and reset by pushing the reset button or replace immersion heater as necessary
	Indirect boiler is not working	Check boiler operation. If a fault is detected, repair boiler
	Indirect boiler thermal cut-out (high limit thermostat) has operated	Reset the boiler cut-out and check the operation of the boiler thermostat
	Motorised valve is not working properly	Check wiring of motorised valve and replace/repair as required
Water discharges from the pressure relief valve	Intermittently: air bubble has reduced or expansion vessel has lost its air charge	Recharge the air bubble by draining down, or check and recharge expansion vessel as necessary
	Continually: • Pressure-reducing valve not working correctly • Pressure relief valve seating damaged	• Check and replace pressure-reducing valve as required • Replace pressure relief valve as required
Water discharges from the temperature relief valve	Hot: thermal control failure	Switch off electrical power. Do not turn off water supply. When the discharge stops, check all thermal controls. Replace as necessary
	Cold: joint failure of the pressure-reducing valve and the pressure relief valve	Check and replace pressure-reducing valve as required. Replace pressure relief valve as required

Replacing an immersion heater

Table 8: Open-vented hot water storage system fault-finding

Fault	Probable causes	Recommended solutions
No hot water flow	Mains cold water off	Check and open isolation valve/stop valve
	FOV stuck in the 'off' position	Check and clean or replace the FOV as necessary
	System airlocked	Drain down the cylinder and refill the system
Water flowing cold	Immersion heater failure	Check and replace the immersion heater
	Heat source not working	Check the boiler for correct operation. Repair as necessary
Water flowing only lukewarm	Cylinder thermostat or immersion heater thermostat set too low	Increase the temperature setting to 60°C and check the correct operation
Poor flow rate Scale build-up in pipes	Cold-feed pipe and/or hot water draw-off blocked with scale	Check the cold feed connection and the hot water draw-off, and de-scale as required
Poor pressure	Cold water feed/storage cistern too low	Raise the cistern to increase the distance to the hot water outlets
Very hot water discharging into the cold water feed/storage cistern from the open vent	Failure of the immersion heater thermostat and energy cut-out	Check and test the thermostat and energy cut-out and replace

Thermostats

Faulty thermostats are usually indicated by one of two symptoms:

- excessive hot water
- no hot water.

The type of thermostat on the system will depend on whether the system is directly or indirectly heated. The use of manufacturers' instructions when diagnosing faults with thermostats and high limit thermostats is recommended.

Below is a fault-finding chart to use to determine the cause of excessive hot water in an indirectly heated hot water storage system.

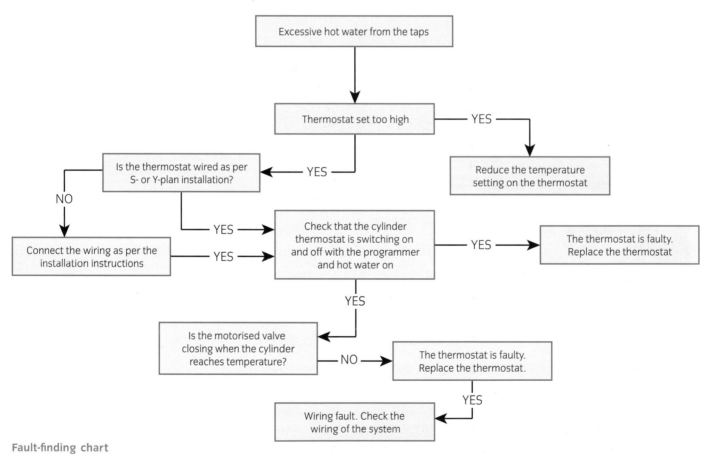

Fault-finding chart

Know the commissioning requirements of hot water systems and components (LO4)

There are five assessment criteria for this Outcome:

1 Interpret documents required to carry out commissioning tasks.
2 State the commissioning checks required on hot water systems.
3 Explain how to balance a secondary circulation system during commissioning activities.
4 State the actions to be taken when commissioning procedures identify faults.
5 Describe information required on a commissioning record for a hot water system.

SmartScreen Unit 304
Presentation 2 and Handout 7

Range	
Commissioning checks	Soundness, flushing, check operating pressures, check temperatures and flow rates, check for correct operation of system components – thermostats, pumps, timing devices, expansion and pressure vessels, gauges and controls, check for correct operation of system safety valves, temperature relief, expansion relief
Actions	Investigate, rectify, re-check
Information	Type of appliance, location, pressures, flow rates, temperature, installation information (who, when), maintenance requirements, components
Documents	Manufacturers' instructions, industry standards

Testing and commissioning of hot water systems is probably the most important part of any installation, as it is at this stage that the system design is finally put into operation. For an installation to be successful, it has to comply with both the manufacturer's installation instructions and the regulations in force. It also has to satisfy the design criteria and flow rates that have been calculated, and the customer's specific requirements.

Testing and commissioning together perform a vital role, and their importance cannot be overstated. Correct commissioning procedures and system set-up often make the difference between a system working to the specification and failing to meet the required demands.

This part of the unit looks at the correct methods of testing and system commissioning.

Documents required to carry out commissioning tasks

Inadequate commissioning, system set-up, system flushing and maintenance operations can affect the performance of any hot water system, irrespective of the materials that have been used in the system installation. Building debris and swarf (pipe filings) can easily block pipes, and these can also promote bacteriological growth. In addition, excess flux used during the installation can cause corrosion and may lead to the amount of copper that the water contains exceeding the permitted amount for drinking water. This could have serious health implications and, in severe cases, may cause corrosion of the pipework, fittings and any storage vessel installed.

It is obvious, then, that correct commissioning procedures must be adopted if the problems stated are to be avoided. There are four documents that must be consulted:

- The Water Supply (Water Fittings) Regulations 1999
- BS 6700 and BS EN 806 (in conjunction with BS 8558)
- Building Regulations Approved Document G3
- the manufacturer's instructions for any equipment and appliances.

Commissioning checks required on hot water systems

Before soundness testing a hot water system, you should perform visual inspections of the installation. These should include the following.

- Walk around the installation. Check that you are happy that the installation is correct and meets installation standards.
- Check that all open ends are capped off and all valves are isolated.
- Check that all capillary joints are soldered and that all compression joints are fully tightened.
- Check that enough pipe clips, supports and brackets are installed and that all pipework is secure.
- Check that the equipment, eg unvented hot water storage cylinder, shower boosting pumps, expansion vessels and subsequent safety and functional controls, is installed correctly

and that all joints and unions on and around the equipment are tight.

- Check that the pre-charge pressure in the expansion vessel is correct and in accordance with the manufacturer's data.

- Check that any cisterns installed on open-vented hot water storage systems are supported correctly and that FOVs are provisionally set to the correct water level.

- Check that all appliance isolation valves and taps are off. These can be turned on and tested when the system is filled with water.

- Check that the D1 and D2 discharge pipework complies with the building regulations and that it terminates in a safe but visible position.

Initial system fill

The initial system fill is always conducted at the normal operating pressure of the system. The system must be filled with fluid category 1 water direct from the water undertaker's mains cold water supply. It is usual to conduct the fill in stages so that the filling process can be managed comfortably. There are several reasons for this.

- Filling the system in a series of stages allows operatives time to check for leaks stage by stage. Only when the stage being filled is leak free should the next stage be filled.

- Open-vented systems – airlocks from cistern-fed open-vented systems are less likely to occur as each stage is filled slowly and methodically. Any problems can be assessed and rectified as the filling progresses without the need to isolate the whole system and initiate a full drain-down. Allowing cisterns to fill to capacity and then opening any gate valves is the best way to avoid airlocks. This ensures that the full pressure of the water is available and the pipes are running at full bore. Trickle filling can encourage airlocks to form, causing problems later during the fill stage.

- Unvented systems – before an unvented hot water storage system is filled, the pressure at the expansion vessel (if fitted) should be checked with a bourdon pressure gauge to check the pre-charge pressure. Unvented hot water storage systems should be filled with all hot taps open. This is to ensure that pockets of air at high pressure are not trapped within the storage vessel, as this can cause the system to splutter water, even after the system has filled. Water should be drawn from every hot water outlet to evacuate any air pockets from the system. The taps can be closed when the water runs freely without spluttering. The temperature and pressure relief valves should be opened briefly to ensure their correct operation and to test the discharge pipework arrangement.

- When the system has been filled with water, it should be allowed to stabilise to full operating pressure. Any FOVs should be allowed to shut off. The system will then be deemed to be at normal operating pressure.

Once the filling process is complete, another thorough visual inspection should take place to check for any possible leakage. The system is then ready for pressure testing.

Soundness testing hot water systems

The procedure for soundness testing hot water systems is described in BS 6700. There are two types of test:

- for testing metallic pipework installations
- for testing plastic pipework systems.

Both of these test procedures are covered in detail in Unit 303.

Flushing procedures for hot water systems and components

This subject was covered in detail in Unit 303, but differs slightly where appliances and equipment are installed on hot water systems.

As with cold water installations, the flushing of hot water systems is a requirement of the British Standards. All systems, irrespective of their size, must be thoroughly flushed with clean water direct from the water undertaker's mains supply before being taken into service. This should be completed as soon as possible after the installation has been completed to remove potential contaminants such as flux residues, PTFE, excess jointing compounds and **swarf**. Simply filling a system and draining down again does not constitute a thorough flushing. In most cases, this will only move any debris from one point in the system to another. In practice, the system should be filled and the water run at every outlet until the water runs completely clear and free of any discoloration. It is extremely important that any hot water storage vessels and cold water storage cisterns are drained down completely.

It is generally accepted that systems should not be left charged with water once the flushing process has been completed, especially if the system is not going to be used immediately, as there is a very real risk that the water within the system could become stagnant. In practice, it is almost impossible to effect a complete drain-down of a system, particularly a large system where long horizontal pipe runs may hold water. This in itself is very detrimental, as corrosion can often set in and this can also cause problems with water contamination. It is therefore recommended, to minimise the risk of

Swarf

Fine chips of stone, metal, or other material produced by a machining operation.

corrosion and water quality problems, to leave systems completely full and flush through at regular intervals of no less than twice weekly, by opening all terminal fittings until the system has been taken permanently into operation. If this is the case, provision for frost protection must be made.

Taking flow rate and pressure readings

Once the hot water system has been filled and flushed, the heat source should be put into operation and the system run to its operating temperature. Thermostats and high limit thermostats should be checked to ensure that they are operating at their correct temperatures. When the system has reached full operating temperature and the thermostats have switched off, the flow rates, pressures and water temperatures can then be checked against the specification and the manufacturer's instructions. This can be completed in several ways.

- Flow rates can be checked using a weir gauge. This is sometimes known as a weir cup or a weir jug, and the method of use is simple. The gauge has a slot running vertically down the side of the vessel, which is marked with various flow rates. When the gauge is held under running water, the water escapes out of the slot. The height that the water achieves before escaping from the slot determines the flow rate. Although the gauge is usually accurate, excessive flow rates will cause a false reading because the water will evacuate out of the top of the gauge rather than the side slot.

- System pressures (static) can be checked using a bourdon pressure gauge at each outlet or terminal fitting. Bourdon pressure gauges can also be permanently installed on either side of a boosting pump to indicate both inlet and outlet pressures.

- Both pressure (static and running) and flow rate can be checked at outlets and terminal fittings using a combined pressure and flow rate meter.

- The temperature should be checked using a thermometer at the hot water draw-off to ensure that it is at least 60°C but does not exceed 65°C. Each successive hot water outlet, moving away from the storage vessel, should be temperature checked to ensure that any thermostatic mixing valves are operating at the correct temperature and that the hot water reaches the outlet within the 30-second limit. If a secondary return system is installed, then the circulating pump should be running when the tests are conducted and the temperature of the return checked just before it re-enters the cylinder to ensure that the temperature is no less than 10°C lower than the draw-off, or 50°C minimum.

Checking hot water flow rates

Checking the hot water temperature using an infrared thermometer

Balancing a secondary circulation system during commissioning activities

Large secondary circulation systems should contain bronze lockshield valves on every return leg of the hot water secondary circuit. These should be fitted as close to the appliances as possible, and are used to balance flow rates to each leg in the system, so that:

- heat loss through the circuit is kept to a minimum
- the temperature of each leg is constant, and
- the temperature of the return at the cylinder is not less than 50°C.

Correct balancing is achieved by opening the valves on the longest circuits, and then successively closing the lockshield valves a little at a time, working towards the cylinder until the flow rates through all circuits are equal. The flow rate should be balanced so that all the circuits achieve the same temperature at the same time. This is especially important for systems that operate through a time clock.

What to do when commissioning procedures identify faults

Commissioning is the part of the installation where the system is filled and run for the first time. It is at this point that you will see if it works as designed. Occasionally problems will be discovered when the system is fully up and running, such as systems that do not meet correct installation requirements.

- **Systems that do not meet the design specification** – problems such as incorrect flow rates and pressures are quite difficult to deal with. If the system has been calculated correctly and the correct equipment has been specified and installed to the manufacturer's instructions, then problems of this nature should not occur. However, if the pipe sizes are too small in any part of the system, flow rate and pressure problems will develop almost immediately downstream of where the mistake has been made. In this instance, the drawings should be checked and confirmation sought from the design engineer that the pipe sizes are correct before any action is taken. It may also be the case that too many fittings or incorrect valves have been used, causing pipework restrictions.

  Another cause of flow rate and pressure deficiency is the incorrect set-up of equipment such as boosting pumps and accumulators. In this instance, the manufacturer's data should be consulted and set-up procedures followed in the installation instructions. It is here that mistakes are often made. If problems still continue, the manufacturer's technical support team should be contacted for advice. In a very few cases, the equipment specified may be at fault and will not meet the design specification. If this is the case the equipment must be replaced.

- **Poor installation techniques** – installation is the point at which the design is transferred from the drawing to the building.

- **Noise** – incorrectly clipped pipework can often be a source of frustration within systems running at high pressures because of the noise that it can generate. Incorrect clipping distances and, often, lack of clips and supports can put strain on the fittings and cause the pipework to reverberate throughout the installation, even causing fitting failure and leakage. To prevent these occurrences, the installation should checked as it progresses and any deficiencies brought to the attention of the installing engineer. Upon completion, the system should be visually checked before flushing and commissioning begins.

- **Leakage** – water causes a huge amount of damage to a building and can even compromise the building structure. Leakage from pipework, if left undetected, causes damp, mould growth and an unhealthy atmosphere. It is therefore important that leakage is detected and cured at a very early stage in the system's life.

Damage to ceiling casued by damp

It is almost impossible to ensure that every joint on every system installed is leak free. Manufacturing defects on fittings and equipment and damage sometimes cause leaks. Leakage due to badly jointed fittings and poor installation practice is common, especially on large systems where literally thousands of joints have to be made until the system is complete. This can often be avoided, though, by taking care when jointing tubes and fittings, using recognised jointing materials and compounds, and using manufacturers' recommended jointing techniques.

A plumber's nightmare! A badly designed plumbing system makes fault-finding almost impossible

Legionella pneumophila in hot water systems

According to the Health and Safety Executive, the instances of Legionnaires' disease derived from hot water supply have diminished over recent years as a result of better installation techniques and more awareness of sterilisation methods. However, large hot water systems can often be complex in their design, so they still present a significant risk of exposure. The environments in which legionella bacteria proliferate are as follows:

- at the base of the cylinder or storage vessel where the cold feed enters and cold water mixes with the hot water within the vessel. The base of the storage vessel may well eventually contain sediments, which support the growth of *Legionella* bacteria
- the water held in a secondary circulation system between the outlet and the branch to the secondary circulation system, as this may not be subject to the high temperature sterilisation process.

In general, hot water systems should be designed for safe operation by preventing or controlling conditions that allow the growth of legionella bacteria. They should, however, permit easy access for cleaning and disinfection. The following points should be considered:

- Materials such as natural rubber, hemp, linseed oil-based jointing compounds and fibre washers should not be used in domestic water systems. Materials and fittings acceptable for use in water systems are listed in the directory published by the Water Research Centre.
- Low-corrosion materials (copper, plastic, stainless steel, etc) should be used where possible.

Defective components and equipment

Defective components cause frustration and cost valuable installation time. If a component or piece of equipment is found to be defective, do not attempt a repair, as this may invalidate the manufacturer's warranty. The manufacturer should be contacted first, as they may wish to send a representative to inspect the component prior to replacement. The supplier should also be contacted to inform them of the faulty component. In some instances, where it can be proven that the component is defective and was not a result of poor installation, the manufacturer may reimburse the installation company for the time taken to replace the component.

The information required on a commissioning record for a hot water system

Commissioning records such as BENCHMARK certificates for hot water systems should be kept for reference during maintenance and repair and to ensure that the system meets the design specification.

Typical information that should be included on the record is as follows:

- the date, the time and the name(s) and ID numbers of the commissioning engineer(s)
- the location of the installation
- the amount of hot water storage and cold water storage (if any)
- the types and manufacturers of equipment and components installed
- the type of pressure test carried out, and its duration
- the incoming static water pressure
- the flow rates and pressures at the outlets
- the expansion vessel pressure
- whether temperature and pressure relief valves have been fitted
- the results of tests on the discharge pipework.

The BENCHMARK certificate should be signed by the operative and the customer and kept on file in a secure location.

The BENCHMARK logo

Be able to install and inspect hot water systems (LO5)

There are five assessment criteria for this Outcome:

1 Install components and final pipework connections to unvented cylinders.
2 Position and fix safety relief pipework from unvented cylinders to termination point.
3 Carry out commissioning checks.
4 Inspect faults on unvented storage cylinder components.
5 Inspect faults on hot water shower pumps.

Range	
Pipework	Line strainer, pressure-reducing valve, check valve (may be installed in a combination valve) expansion vessel, storage cylinder
Commissioning checks	Unvented cylinders, shower pumps
Storage cylinder components	Expansion vessel, pressure

This Outcome is part of the practical assessment that will be conducted within the workshop environment.
There are three practical tasks.

- **Task HW1** – install and commission an unvented hot water system. This task may be combined with tasks CH1, 2, 3 and 4 from Unit 306. Photographic evidence and a written summary are requirements for this task. You must:
  - ensure that you have all necessary documentation before beginning the task
  - confirm the suitability of the UVHWSS cylinder supplied by your assessor
  - select the appropriate fittings and confirm the suitability of the cold water supply
  - prepare the work area
  - install the cylinder and control components into a previously prepared position to manufacturer's instructions and industry regulations using safe practices
  - install the relief pipework to manufacturer's instructions and industry regulations using safe practices
  - visually check that the system is ready for the commissioning procedure to begin

- hydraulically test the installation with a hydraulic test bucket in line with BS EN 806 (BS 6700)
- thoroughly flush the system
- commission and performance test the system
- check the correct operation of safety controls and relief pipework
- complete a BENCHMARK/commissioning certificate.

- **Task HW2** – identify and rectify two faults on unvented hot water installations. The task involves identifying faults that have been previously installed by your assessor. You must:
  - confirm that you have all the documentation required to correctly identify the faults
  - visually inspect the installation to ensure that it meets current standards and regulations
  - successfully diagnose and rectify the faults
  - reinstate and commission the system in accordance with the manufacturer's instructions, including heat testing where appropriate
  - complete any fault-diagnosis documentation.

- **Task HW3** – fault identification and rectification on pumped shower installations. The task involves identifying faults that have been previously installed by your assessor. You must:
  - check the installation to confirm compliance with industry requirements and manufacturer's instructions
  - successfully diagnose the fault
  - rectify the fault and replace any part in a safe manner, eg by electrical isolation
  - reinstate the supplies to the system and commission the system including heat testing (where appropriate)
  - complete fault-identification records for the work activity.

Good luck with your assessment!

Conclusion

In this unit, you will have seen that hot water is a very complex subject. It is obvious that careful consideration must be given to the customer's requirements if the system that you fit is to meet their specific needs. The system choice is often the result of the calculations you make to determine flow rates, pipe sizes and quantity of hot water required. This unit gives you the knowledge needed to install good, well-thought-out and planned hot water storage systems, as well as an insight into the complexities of good hot water system design.

Test your knowledge questions

1 Briefly describe a centralised system of hot water supply.

2 What is the purpose of the open vent on an open-vented system?

3 Briefly describe an unvented hot water system.

4 Name the two categories of unvented hot water storage systems.

5 Name the two types of localised hot water systems.

6 Which Approved Document governs hot water systems?

7 Complete the following table.

Maximum recommended lengths of uninsulated hot water pipes	
Outside diameter (mm)	**Max. length (m)**
12	
Over 12 and up to 22	
Over 22 and up to 28	
Over 28	

8 Name the type of unvented cylinder shown.

9 Name the components and controls in the correct order in an unvented hot water storage system.

10 Which gas law do expansion vessels make use of?

11 At what temperature should water be stored in a hot water storage vessel?

12 On a secondary circulation system, what is the non-return valve for?

13 Which temperature components make up the three-tier level of safety protection in an unvented hot water system?

14 Name the functional controls of an unvented hot water storage system.

15 What is the purpose of trace heating on domestic hot water systems?

16 Which piece of equipment is used to measure flow rates?

17 What action should be taken before commissioning of a hot water system takes place?

18 Which three elements go together to make up the discharge pipework?

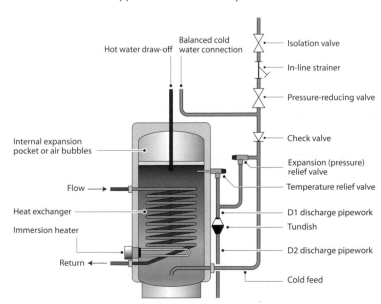

Assessment checklist

What you now know (Learning Outcome)	What you can now do (Assessment criteria)	Where this is found (Page number)
1. Understand the types of hot water systems	1.1 Explain the working principles of different types of centralised hot water supply systems used in buildings.	284–301
	1.2 Identify types of localised hot water supply systems used in buildings.	301–303
	1.3 Compare thermal stores and unvented hot water storage systems.	303–305
	1.4 State the recommended design temperatures within hot water systems.	305–311
	1.5 Describe the requirements of Part G of the Building Regulations for hot water installations.	311–312
	1.6 Evaluate the use of different fuels in domestic hot water systems.	312–313
	1.7 Explain the operating principles of hot water digital showers.	313
2. Understand the operating principles of components found in hot water systems	2.1 Explain the function of safety devices in unvented hot water systems.	314–315
	2.2 Explain the method of operation of functional devices in unvented hot water systems.	316–323
	2.3 Calculate the diameter of discharge pipework.	324–325
	2.4 Specify the requirements of discharge pipework from temperature and expansion relief valves.	325–329
	2.5 Specify the layout features for pipework systems that incorporate secondary circulation.	329–333
	2.6 Explain balanced and unbalanced supplies in unvented hot water storage systems.	334
3. Know the fault-diagnosis and rectification procedures for hot water systems and components	3.1 Specify the periodic servicing requirements of hot water systems.	335–337
	3.2 Interpret documents to identify diagnostic requirements of hot water system components.	337
	3.3 Describe diagnostic checks on hot water system components.	338–339
	3.4 Specify methods of repairing faults in hot water system components.	339–344

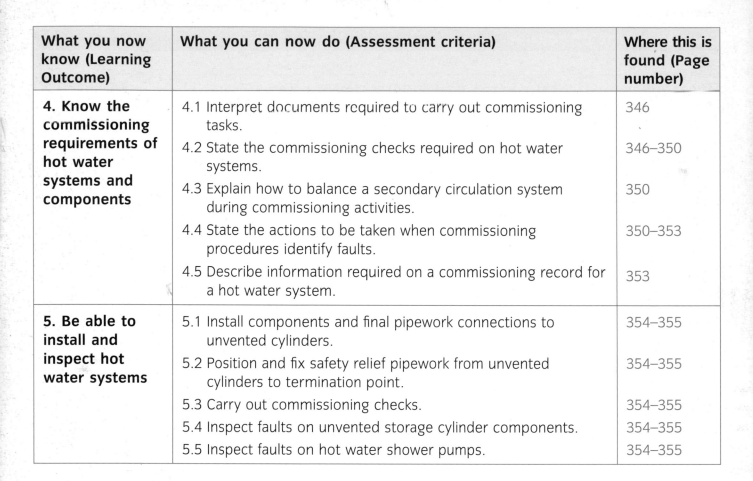

What you now know (Learning Outcome)	What you can now do (Assessment criteria)	Where this is found (Page number)
4. Know the commissioning requirements of hot water systems and components	4.1 Interpret documents required to carry out commissioning tasks.	346
	4.2 State the commissioning checks required on hot water systems.	346–350
	4.3 Explain how to balance a secondary circulation system during commissioning activities.	350
	4.4 State the actions to be taken when commissioning procedures identify faults.	350–353
	4.5 Describe information required on a commissioning record for a hot water system.	353
5. Be able to install and inspect hot water systems	5.1 Install components and final pipework connections to unvented cylinders.	354–355
	5.2 Position and fix safety relief pipework from unvented cylinders to termination point.	354–355
	5.3 Carry out commissioning checks.	354–355
	5.4 Inspect faults on unvented storage cylinder components.	354–355
	5.5 Inspect faults on hot water shower pumps.	354–355

This combination unit provides the learner with knowledge, understanding and skills associated with complex central heating systems and associated controls. Learners will be introduced to complex systems including their design considerations, boiler types, control systems, underfloor heating and de-commissioning and commissioning requirements. Learners will also carry out installation of pipework and controls used in complex central heating systems with testing and commissioning procedures.

There are six Learning Outcomes (LOs) for this unit and each LO will be discussed in turn. There are 56 Guided Learning Hours for this unit. The learner will:

1 Understand complex domestic heating system layouts and controls.
2 Understand the layouts and operating principles of sealed systems.
3 Understand the types of boilers used in domestic central heating systems.
4 Understand the types of heat emitters used in underfloor heating systems.
5 Know how to de-commission, commission and fault-find on central heating systems.
6 Be able to de-commission, install, commission and fault-find on sophisticated central heating systems and their components.

Carbon footprint

The amount of carbon dioxide released into the atmosphere as a result of the activities of a particular individual, organisation or community.

Central heating is a vast and complex subject. There are now more options with regard to sources of heat, pipe materials and heat emitters than ever before. Environmentally friendly technology and the re-emergence of underfloor heating mean that the customer can now afford to be selective about the system they have installed in their property. The advent of heat pumps and solar systems, with the associated savings on fuel and running costs, has dramatically lowered the **carbon footprint** of domestic properties. No longer does the customer have to rely on appliances that burn carbon-rich fuels such as gas and oil. Zero-carbon and carbon-neutral fuels have revolutionised domestic heating, while advances in technology have lowered the cost of the energy-saving appliances that were previously only available to a select few.

Unit 306/606 builds on the knowledge of central heating systems you gained at Level 2. It investigates new and exciting technology that has the potential to dramatically cut the cost of heating our homes while, at the same time, reducing our carbon emissions. It also looks at new controls and components that can transform an existing wasteful installation into an energy-efficient system.

Understand complex domestic heating system layouts and controls (LO1)

There are eleven assessment criteria for this Outcome:

1 Identify documents relating to central heating design and installation.
2 Describe pipework layouts for complex central heating systems.
3 Describe the working principles of key components in a complex central heating system.
4 Explain boiler interlock.
5 Compare the relationship of positive and negative pressures in relation to feed, vent and pump positions.
6 Describe the wiring arrangements required for S- and Y-plan heating systems and components.
7 Identify alternative methods of wiring arrangements.
8 Describe the procedures for safely isolating supplies.
9 Describe testing of wiring in domestic heating systems.
10 Describe the working principles of low loss headers.
11 Explain the effect bore diameter of tube has on heat loads.

Range	
Documents	Manufacturers' instructions, Part L of the Building Regulations, Central Heating System Specifications (CHeSS) 2008, Domestic Building Services Compliance Guide 2010, Part P of the Gas Safety (Installation and Use) Regulations 1998
Central heating system	S-plan, Y-plan, zoned
Key components	Motorised valves, room stat, programmable room stat, cylinder stat and overheat stat, frost stat, pipe stat, timer, programmer, compensator, boiler energy management programmer, pump, auto bypass, feed and expansion cistern, automatic air vent, air separator
Alternative methods	Plug-and-play, wireless
Procedures	Prove device, isolate, test all phases, prove device, lock off isolator or remove fuse
Testing	Earth continuity, short-circuit, resistance to earth, continuity, polarity, fuse rating, voltage

Documents relating to central heating design and installation

The installation of central heating systems is governed strictly by various regulations, standards (the British Standards) and recommendations.

The Regulations

The following are the important regulations that relate to the installation of central heating systems:

- The Building Regulations:
  - Approved Document L 2010 – Conservation of fuel and power
    - a) L1A – Conservation of fuel and power in new dwellings
    - b) L1B – Conservation of fuel and power in existing dwellings

- Approved Document F – Ventilation
- Approved Document J – Combustion appliances and fuel storage systems
- Approved Document P – Electrical safety – Dwellings
  - The Water Supply (Water Fittings) Regulations 1999
  - The Gas Safety (Installation and Use) Regulations 1998
  - BS 7671 – The IET Wiring Regulations 17th edition.

Approved Documents L1A and L1B

The main requirement of Approved Document L is for a boiler interlock. A boiler interlock is a series of controls (cylinder thermostats, programmable room thermostats, programmers and time switches) that prevents the boiler from cycling when there is no demand for heat. In addition note the following.

- Every home must be divided into a least two heating zones, using a thermostat controlling a motorised valve.
- If the house is less than $150m^2$, these can be controlled by the same time clock or programmer.
- If the house is larger than $150m^2$, each zone must be controlled by its own time clock/programmer.
- Living and sleeping areas (zones) must be controlled at different temperatures by means of a thermostat.
- Every radiator should be fitted with a thermostatic radiator valve, unless the radiator is being used as the reference radiator for a thermostat situated elsewhere in the room.

These requirements apply every time a home is built.

Where existing installations are concerned, the requirements of Document L are made retrospectively. In other words, if an existing system does not comply with the regulations, the system must be updated:

- every time a home has an extension or change of use
- every time more than one individual component, such as a boiler, is replaced in a heating system.

Simple boiler servicing is exempt from this, but the recommendation is made that radiator thermostats should be fitted when the system is drained down.

The British Standards

The following British Standards relate to the installation of central heating systems:

- **BS EN 12828:2003** – Heating systems in buildings. Design for water-based heating systems

- **BS EN 12831:2003** – Heating systems in buildings. Method for calculation of the design heat load
- **BS EN 14336:2004** – Heating systems in buildings. Installation and commissioning of water-based heating systems
- **BS EN 1264-1:1998** – Floor heating. Systems and components. Definitions and symbols
- **BS EN 1264-2:2008** – Water-based surface-embedded heating and cooling systems. Floor heating. Proving methods for the determination of the thermal output using calculation and test methods
- **BS EN 1264-3:2009** – Water-based surface-embedded heating and cooling systems. Dimensioning
- **BS EN 1264-4:2009** – Water-based surface-embedded heating and cooling systems. Installation
- **BS EN 442:2003** – Specification for radiators and convectors

The recommendations

The following recommendations relate to the installation of central heating systems.

- **Domestic Building Services Compliance Guide 2010** – a free document that can be downloaded from http://www. planningportal.gov.uk/buildingregulations/approveddocuments/ partl/bcassociateddocuments9/compliance, it offers practical assistance when designing and installing to the Building Regulations requirements for space heating and hot water systems, mechanical ventilation, comfort cooling, fixed internal and external lighting, and renewable energy systems.
- **Central Heating System Specifications (CHeSS) 2008** – this publication offers compliance advice and best practice for the installation of central heating systems.
- **Chartered Institution of Building Services Engineers (CIBSE) Domestic Heating Design Guide 2007** – this was produced to help heating engineers specify and design wet central heating systems.

Manufacturers' technical instructions

Central heating systems and components must be installed, commissioned and maintained strictly in accordance with the manufacturer's instructions.

If the manufacturer's instructions are not available or have been misplaced, most manufacturers now provide the facility to download the instructions from the company website.

SmartScreen Unit 306

Presentation 1 and Handout 1

Pipework layouts for complex central heating systems

Domestic central heating systems fall into two different categories, based on the way the system is filled with water and the pressure at which they operate.

- Low-pressure, open-vented central heating systems are fed from a feed and expansion cistern in the roof space. These can be both modern fully pumped systems and existing gravity hot water/pumped heating installations.

- Sealed, pressurised central heating systems are fed directly from the mains cold water supply and incorporate an expansion vessel to take up the expansion of water as a result of it being heated. These are generally more modern fully pumped and combination boiler systems and are discussed later in the unit.

The water in low-pressure open-vented central heating systems is kept below 100°C. For existing systems the flow water from the boiler is usually about 80°C, and the return water temperature is usually 12°C to 15°C lower.

Circulation of the water can be either by gravity circulation to the heat exchanger in the hot water cylinder and pumped heating to the heat emitters, or by means of a fully pumped system in which both the hot water heat exchanger and heat emitters are heated using a circulating pump. Fully pumped systems have the advantage that system resistance created by the pipework, fittings and heat emitters can be overcome much more easily, and this enables the system to heat up quicker, giving the occupants a much more controllable system.

Sealed heating systems operate at a higher pressure, with modern systems incorporating condensing boilers operating at a slightly higher temperature of 82°C for the flow temperature with a return temperature 20°C lower, at 62°C.

In both cases the difference between the flow and return temperatures is the amount of heat lost to the heated areas.

The illustration on the next page shows the development of central heating, from the open-vented one-pipe system through to the more modern sealed combination boiler systems and fully pumped systems using system boilers.

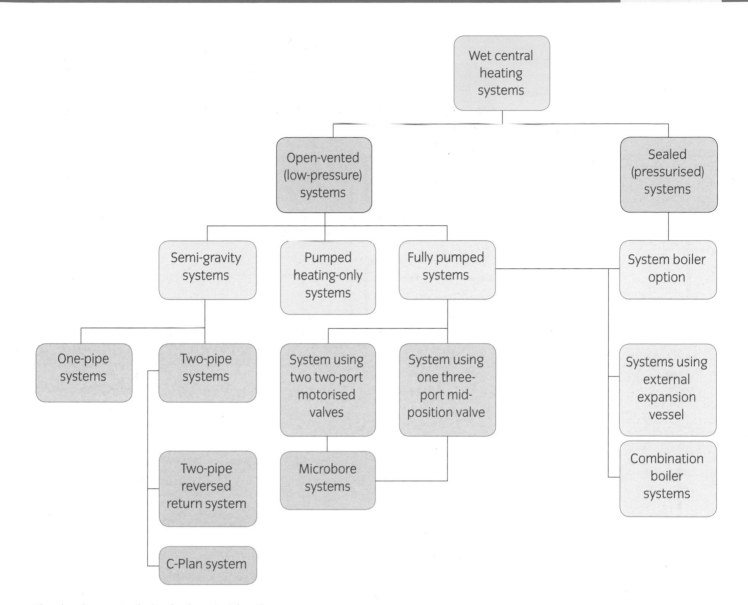

The development of plumbed central heating

Fully pumped systems with two- or three-port zone valves (S-plan/S-plan plus)

The S-plan system has two two-port motorised zone valves to control the primary and heating circuits separately, by the cylinder and room thermostats respectively. This system is recommended for dwellings with a floor area greater than 150m^2 because it allows the installation of an additional two-port zone valve to zone the upstairs heating circuit separately from the downstairs circuit (S-plan plus). A separate room thermostat and possibly a second time clock/programmer would also be required for upstairs zoning. A system bypass is required for overheat protection of the boiler.

A two-port zone valve

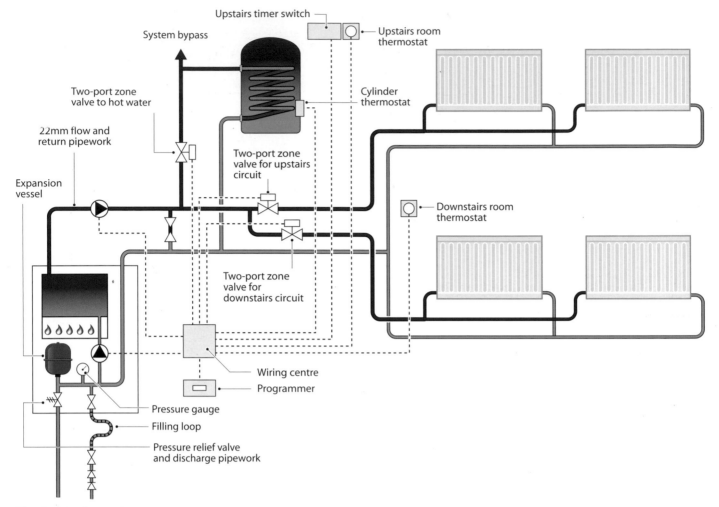

Upstairs timer switch

System bypass

Upstairs room thermostat

Two-port zone valve to hot water

Cylinder thermostat

22mm flow and return pipework

Two-port zone valve for upstairs circuit

Expansion vessel

Downstairs room thermostat

Two-port zone valve for downstairs circuit

Wiring centre

Programmer

Pressure gauge

Filling loop

Pressure relief valve and discharge pipework

The S-plan plus system

Fully pumped systems with three-port mid-position valve (Y-plan) or three-port diverter valve (W-plan)

The three-port mid-position valve (Y-plan) or diverter valve (W-plan) controls the flow of water to the primary (cylinder) circuit and the heating circuit. The valve reacts to the demands of the cylinder thermostat or the room thermostat.

The system contains a system bypass fitted with an automatic bypass valve, which simply connects the flow pipe to the return pipe. The bypass is required when all circuits are closed either by the motorised valve or the thermostatic radiator valves as the rooms reach their desired temperature. The bypass valve opens automatically as the circuits close to protect the boiler from overheating by allowing water to circulate through the boiler, keeping it below its maximum high temperature. This prevents the boiler from 'locking out' on the overheat energy cut-out.

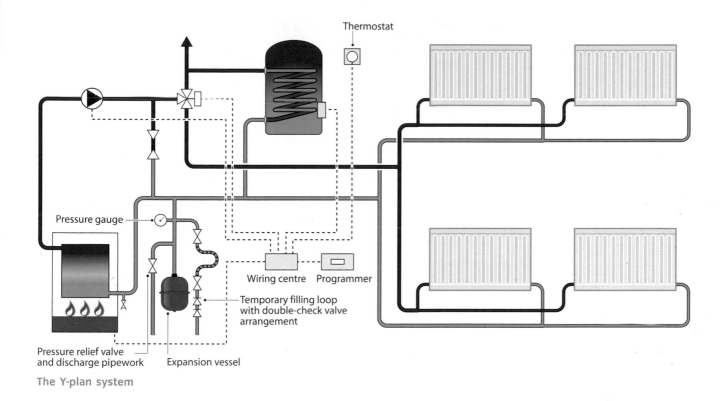

Thermostat

Pressure gauge

Wiring centre   Programmer

Temporary filling loop
with double-check valve
arrangement

Pressure relief valve
and discharge pipework   Expansion vessel

The Y-plan system

Key components in a complex central heating system

No matter how good the central heating design or how accurate the calculations, the system requires proper control to be effective, efficient and economical to run. The types of controls that are added to a system can greatly improve its performance. Even older systems can benefit from the addition of modern and effective controls.

This part of the unit looks at the various components and controls for central heating systems, their functions and how they fit into modern systems.

SmartScreen Unit 306
Presentation 2

Feed and expansion cistern

Open-vented systems contain a feed and expansion cistern, which fulfils two important functions.

- It is the means by which water enters the system for filling and top-up.
- It allows space for the system water to expand to when it is heated.

Generally, the size of the cistern will depend on the size of the system, but for most domestic systems an 18-litre cistern is recommended. The bigger the system, the more water it will contain and so the water expansion will be greater. The water level in the cistern should, therefore, be set at a low level.

The cistern must be located at the highest part of the system and must not be affected by the operation of the circulating pump. For fully pumped systems, the cistern must be at least 1m above the highest part of the pumped primary flow to the heat exchanger in the hot water storage cylinder. For gravity systems, the minimum height of the cistern can be calculated by taking the maximum operating head of the pump and dividing it by three.

The cold feed for the system for most domestic properties is 15mm. The cold-feed pipe should not contain any service or isolation valves. This is to ensure that there is a supply of cold water in the event of overheating and leakage, preventing the system from boiling. Should the valve be inadvertently closed, a dangerous situation could develop, especially if the vent was also blocked, as the pressure would build up in the system, raising the boiling point of the water to dangerous levels. Both the cistern and any float-operated valve (FOV) it may contain must be capable of withstanding hot water at a temperature close to 100°C.

Primary open safety vent

The purpose of the open-vent pipe is safety. The open vent is installed to:

- provide a safety outlet should the system overheat due to a component failure, and
- ensure that the system always remains at atmospheric pressure, limiting the boiling point to 100°C.

In a fully pumped system, the height of the open vent should be a minimum of 450mm from the water level in the cistern to the top of the open-vent pipe. This is to allow for any pressure surges created by the circulating pump. The minimum size of pipe for the open vent is 22mm and this, like the cold-feed pipe, should not be fitted with any valves.

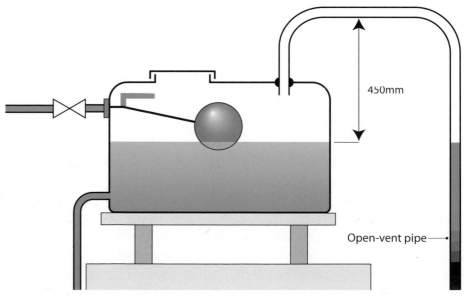

450mm

Open-vent pipe

The height of the open-vent pipe

The function of central heating system components

At Level 2, you investigated some of the more common controls, such as room thermostats, cylinder thermostats, time clocks and programmers. Here some of the more advanced controls that can be installed alongside these are investigated, and the working principles of other common components are recapped.

Zone control valves

Zone control of multiple spaces within a dwelling is achieved by the use of motorised valves activated and controlled by a time clock/programmer/room thermostat arrangement, or a programmable room thermostat that will perform the same function.

The most common types of motorised valves are two-port zone valves and three-port mid-position and diverter valves. The method of use of these valves will depend on the pipework layout and installer/end user preference.

- Three-port valves provide separate hot water and heating circuits. Zoning of the living spaces can be achieved by the inclusion of additional two-port valves on the individual space circuits, ie upstairs and downstairs circuits. Three-port valves include a mid position which allows shared flow (see page 366).

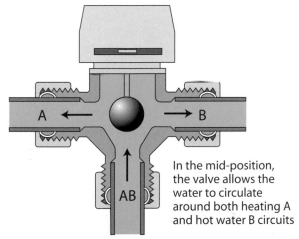

In the mid-position, the valve allows the water to circulate around both heating A and hot water B circuits

A mid-position valve showing the heating connection (A), hot water connection (B) and the common or primary connection (AB)

- Two-port motorised zone valves are probably the most common of all zone arrangements used. They provide zoning of individual circuits and are used where more than one zone is required. A separate zone valve is used for each zone (see page 365).

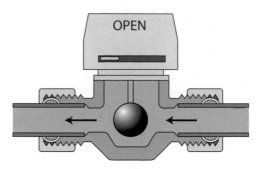

The two-port zone valve open

Valves of 22mm can be used on boilers up to around 20kW. 28mm or larger should be used where the system is greater than 20kW.

Electrical central heating controls

Modern central heating systems cannot function without electrical controls. They are required at every stage of operation from switching the system on to shutting it down when the required temperature has been reached. They provide both functional operation and safety and are a requirement of Building Regulations Document L: Conservation of fuel and power. There are many different types:

- time clocks and programmers
- room thermostats
- cylinder thermostats
- frost thermostats and pipe thermostats
- motorised valves.

Before looking at the various controls, consider the implications of Document L, which was updated in October 2010. The main points are described in Table 1.

Table 1: The main points of Approved Document L1

Controls for new dwellings Part L1, England and Wales; Part J, Scotland	Controls for existing dwellings Part L1, England and Wales.
1) Provide installations that are fully pumped. 2) Provide independent time and temperature control to both the heating and hot water circuits. Part J, Scotland requires the installation of seven-day timing. 3) Provide control systems with a boiler interlock. Note 1 4) Install an automatic bypass valve if a bypass is fitted. Note 2	1) When a new installation is fitted, the controls should be as for a new dwelling. When replacing the boiler and/or the hot water vessel, the opportunity to improve the controls should be considered. To be confident that the requirements of the Building Regulations are met this would entail the following: 2) Provide fully pumped installations with a boiler interlock. Note 1 3) Install an automatic bypass valve, if a bypass is fitted. Note 2

Controls for new dwellings Part L1, England and Wales; Part J, Scotland	Controls for existing dwellings Part L1, England and Wales.

5) Split the heating circuit into zones using either:
  a) room thermostats or programmable room thermostats in all zones
  b) a room thermostat or programmable room thermostat in the main zone and thermostatic radiator valves (TRVs) on all radiators in the other zones.
6) Ensure installation of a cylinder thermostat and a zone valve to control stored hot water. Note 3
7) Provide time control by the use of either:
  a) a full programmer with separate timing to each circuit
  b) two or more separate timers providing timing control to each circuit
  c) programmable room thermostat(s) to the heating circuit(s), with separate timing of the hot water.
8) For dwellings with a total usable floor space greater than 150m², then:
  a) the heating circuit should be split into a minimum of two zones plus a hot water service zone, and
  b) each zone should be separately timed by the use of a multi-channel programmer or multiple heating programmers or programmable room thermostats.
9) Any boiler management control system that meets the specified zoning, timing and temperature requirements is a wholly acceptable alternative.
	4) Provide time and temperature control to both the heating and hot water circuits, for fully pumped installations.
	4.1) Separate the space heating system into zones and:
	  a) if a room thermostat is already fitted, fit TRVs on at least all radiators in the sleeping areas and check if a new room thermostat or programmable room thermostat should be fitted
	  b) if there is no room thermostat, install either a room thermostat or programmable room thermostat, and fit TRVs on at least all radiators in the sleeping areas.
	4.2) Provide time control by the use of either:
	  a) a full, standard or mini programmer
	  b) one or more separate time switches
	  c) programmable room thermostat(s).
	5) For semi-gravity installations, the recommended option is to convert it to fully pumped. The controls should then be as detailed above.
	5.1) If either the new boiler or hot water storage vessel can only be used on a pumped circuit, the installation must be converted to fully pumped. Install controls as detailed above.
	5.2) If the new boiler or storage vessel is designed for gravity hot water, or it is impractical to convert to a fully pumped installation, provide the following controls:
	  a) a cylinder thermostat and zone valve to control the hot water and provide a boiler interlock. Note 3
	  b) a room or programmable room thermostat
	  c) a programmer or time switch
	  d) TRVs on at least all radiators in the bedrooms.
	6) Any boiler management control system that meets the specified zoning, timing and temperature requirements is a wholly acceptable alternative.

Note 1: Boiler interlock is achieved by the correct use of the room thermostat(s) or programmable room thermostat(s), the cylinder thermostat and zone valve(s) in conjunction with the timing device(s). These should be wired such that when there is no demand from the heating or hot water both the pump and boiler are switched off. The use of TRVs alone does not provide interlock.

Note 2: Although the regulations may not always require a bypass to be fitted, the performance of any installation with multi-zoning or TRVs is improved by the use of an automatic bypass valve.

Note 3: The use of non-electric hot water controllers does not meet this requirement.

The above table is taken from the TACMA Guide to Building Regulations Part L1A. To comply with the requirements, the correct electrical controls must be fitted.

Time clocks and programmers

Time clocks are the simplest of all central heating timing devices. They are only suitable for switching on one circuit, such as the

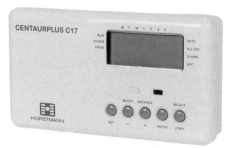

A central heating time clock

A digital programmer

A room thermostat

A cylinder thermostat

heating circuit, so are ideally suited for combination boiler installations. Both mechanical and digital time clocks are available.

Programmers are two-way time clocks, which are able to switch on both heating and hot water at various times throughout the day. There are three basic types:

- A mini-programmer, which allows the heating and hot water circuits to be on together, or hot water alone, but not heating alone. These are ideally suited to C-plan and C-plan plus systems.

- A standard programmer, which uses the same time settings for space heating and hot water.

- A full programmer, which allows the time settings for space heating and hot water to be fully independent. Some will allow seven-day programming of both heating and hot water so that the two circuits can be used individually or both together.

Programmers are often fitted to the front fascia of the boiler and integrated into the boiler design. However, this is not always convenient, especially if the boiler is sited in a garage or roof space.

Room thermostats

A room thermostat senses air temperature. It is simply a temperature-controlled switch that connects or breaks an electrical circuit when either calling for heat or shutting the circuit down when the correct temperature has been reached. Most room thermostats contain a very small heater element called an accelerator, which 'tops up' the heat to the room thermostat by 1°C or 2°C, smoothing out the temperature cycle and preventing the boiler from cycling when it isn't required.

Programmable room thermostats allow different temperatures to be set for different days of the week. They also provide a night setback feature, allowing a minimum temperature to be maintained at night. Some units also allow the time control of the hot water cycle.

Cylinder thermostats

A cylinder thermostat is a simple control of stored hot water temperature, usually strapped to the side of the hot water cylinder about a quarter of the way up. It is used with a motorised valve to provide close control of water temperature and should be set to 55°C.

Frost thermostats and pipe thermostats

The purpose of the frost thermostat is to stop the boiler and any other vulnerable parts of the system from freezing in extremely cold weather. It is wired into the system to override all other programmers and thermostats. Frost thermostats should be set to between 3°C and 5°C and should be placed close to the vulnerable parts of the system, especially if they are fitted in unheated garages and roof spaces.

Frost thermostats are much more effective when installed alongside a pipe thermostat.

A pipe thermostat is strapped to vulnerable pipework, and senses water temperature. It is designed to override all other controls when the temperature of the water is close to 0°C, and works in conjunction with the frost thermostat. The pipe thermostat and frost thermostat should be wired in series.

A frost thermostat

System design and control

Now that you have seen the controls and the system layouts, you need to look at how the controls work together to ensure efficient operation of the systems. This unit concentrates on fully pumped systems, as these are the systems that you will need to use for new installations.

A pipe thermostat

Table 2a: The Y-plan system

Three-port valve	• The flow from the boiler must be connected to the AB port, which is marked on the valve • The A port must be connected to the heating circuit • The B port must be connected to the hot water circuit • The valve must not be installed upside down, as leakage of water could penetrate the electric actuator
Time control	• Must be provided by a programmer that allows individual use of hot water and heating circuits
Heating circuit	• Must have a room thermostat positioned in the coolest room away from heat sources and cold draughts • Should be wall mounted at 1.5m from floor level • The room thermostat controls the three-port valve • All radiators must have thermostatic radiator valves fitted

Hot water circuit	• The hot water temperature must be controlled by a cylinder thermostat placed a third up from the base of the cylinder • The cylinder thermostat controls the three-port valve
Bypass	• An automatic bypass valve is required
Frost/pipe thermostat	• Must be provided where parts of the system are in vulnerable positions

Table 2b: The S-plan system

The S-plan gives better overall control of the system and this improves system efficiency.

Two-port zone valves	• A single zone valve must be installed on the hot water circuit controlled by a cylinder thermostat • The heating circuit must contain one or more (if the system is to be zoned) two-port zone valves, controlled by individual room thermostats
Time control	• This must be provided by a programmer that allows individual use of hot water and heating circuits • A second time clock may be required if the system is zoned
Heating circuit	• One or more room thermostats control downstairs and upstairs heating circuits • These should be installed 1.5m from floor level
Hot water circuit	• The hot water temperature must be controlled by a cylinder thermostat placed a third up from the base of the cylinder
Bypass	• An automatic bypass valve is required
Frost/pipe thermostat	• Must be provided where parts of the system are in vulnerable positions

Advanced controls: weather compensation, delayed start and optimum start

Domestic central heating systems can benefit from more advanced controls, especially when a condensing boiler is fitted. **Condensing boilers** respond to lower flow and return temperatures better than non-condensing appliances. Advanced controls enhance system efficiency.

Condensing boiler

A boiler that extracts all usable heat from the combustion process, cooling the flue gases to the dew point. The collected water is then evacuated from the boiler via a condensate pipe.

- **Weather compensation** – this control uses an externally fitted temperature sensor fitted on a north- or northeast-facing wall so as not to be in the direct path of solar radiation. As the external temperature rises, the weather compensator reduces the circulation temperature of the flow from the boiler to compensate for the warmer outside temperature. Similarly, the reverse occurs if the weather gets colder.

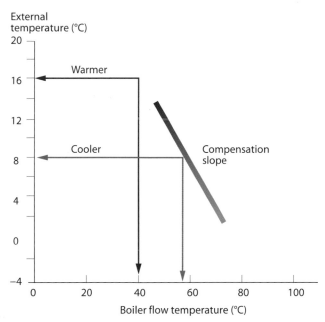

A weather compensation graph

A weather compensator

- **Delayed start** – here, the end user sets the time to bring on the heating, taking into account the time it would normally take to warm the dwelling – ie, most people would set the heat to come on at 5pm if they were due to arrive home from work at 6pm. A delayed start unit will, at the time the heat is due to come on, compare the current indoor temperature to that required by the room thermostat. It will then delay the start of the boiler firing if required. The benefits are that during milder weather when the heat requirement is less, energy will be saved. Room thermostats with a delayed start function are now available.

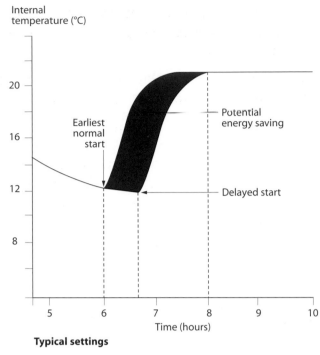

Typical settings

Maximum heat-up period, eg 6am to 8am
Normal occupancy period, eg 8am to 10am

Delayed/optimum start function

An optimum start controller

■ **Optimum start** – with optimum start, the end user sets the required occupancy times and the required room temperature and the controller calculates the necessary heat-up time so that the rooms are at the required temperature irrespective of the outside temperature. The idea is based around comfort rather than energy savings.

Domestic boiler management systems (home automation systems)

A boiler management system (BMS) is an electronic controller that provides bespoke control solutions for domestic central heating systems.

Standard functions of BMS controls include real-time temperature and boiler/controls monitoring, room temperatures (known as set points) and time schedule adjustment, optimisation, and night setback control.

The system remembers key points, such as how quickly the building heats up or cools down, and makes its own adjustments so that energy savings can be made. If it is very cold outside at, say, 2am, the BMS will switch the heating on at 4.15am to allow the building to be at the correct temperature by the time the user has set the heating to come on – say 7am – irrespective of the time that the user has set the time for the heating to activate. On milder nights, the heating may not come on until 6.15am, but it will still reach its set point by 7am.

It will also learn how well your house retains heat and may shut down early if it calculates that your set point will still be maintained at your 'off' time of, say, 10pm.

These systems provide a cost-effective means of monitoring system efficiency and can reduce heating costs by up to 30%.

Electronic sensors are fitted to the flow and return pipework and an external temperature sensor is fitted for weather compensation. The information is used to accurately vary the system output according to demand. This helps to significantly reduce fuel wastage caused by temperature overshoot, heat saturation of the heat exchanger, unnecessary boiler cycling and flue gas losses while maintaining internal comfort levels and reducing CO_2 emissions.

Selection of system and control types for single-family dwellings

The installation of an effective system of central heating controls has a major effect on the consumption of energy and the effectiveness of the system. Choosing the right controls will lead to:

- improved energy efficiency
- reduced fuel bills
- lower CO_2 emissions.

The establishment of a minimum standard of heating controls is vital if the heating system is to achieve satisfactory efficiencies when the system is in use. The efficiency of the boiler is only part of the story. For the boiler to achieve these efficiencies, at least a minimum standard of controls must be installed.

A good system of controls must:

- ensure that the boiler does not operate unless there is demand – this is known as 'boiler interlock'
- only provide heat when it is required to achieve the minimum temperatures.

There are two levels of controls for domestic properties, and these are set out in Central Heating System Specification (CHeSS) CE51 2008:

1 **Good practice** – this set of controls achieves good energy efficiency in line with Approved Document L 2010. This is described in detail in the CHeSS document:
   a **HR7** – good practice for systems with a regular and a separate hot water storage:
   i) full programmer
   ii) room thermostat

     iii) cylinder thermostat

     iv) boiler interlock (see Note 1)

     v) TRVs on all radiators, except in rooms with a room thermostat

     vi) automatic bypass (see Note 2).

b HC7 – good practice for systems using a combination boiler or combined primary storage unit:

     i) time switch

     ii) room thermostat

     iii) boiler interlock (see Note 1)

     iv) TRVs on all radiators, except in rooms with a room thermostat

     v) automatic bypass valve (see Note 2).

2 Best practice – this standard uses enhanced controls to further increase energy efficiency in line with Approved Document L1A/B 2010. This is described in detail in the CHeSS document:

a HR8 – best practice for systems with a regular and a separate hot water storage:

     i) programmable room thermostat, with additional timing capability for hot water

     ii) cylinder thermostat

     iii) boiler interlock (see Note 1)

     iv) TRVs on all radiators, except in rooms with a room thermostat

     v) automatic bypass valve (see Note 2)

     vi) more advanced controls, such as weather compensation, may be considered.

b HC8 – best practice for systems using a combination boiler or combined primary storage unit:

     i) programmable room thermostat

     ii) boiler interlock

     iii) TRVs on all radiators, except in rooms with a room thermostat

     iv) automatic bypass valve (see Note 2)

     v) more advanced controls, such as weather compensation, may be considered.

Note 1 (from CHeSS): Boiler interlock is not a physical device but an arrangement of the system controls (room thermostats, programmable room thermostats, cylinder thermostats, programmers and time switches) which ensure that the boiler does not fire when there is no demand for heat. In a system with a combination boiler this can be achieved by fitting a room thermostat. In a system with a regular boiler this can be achieved by correct wiring interconnection of the room thermostat, cylinder thermostat and motorised valve(s). It may also be achieved by more advanced controls, such as a boiler energy manager. TRVs alone are not sufficient for boiler interlock.

Note 2 (from CHeSS): An automatic bypass valve controls water flow in accordance with the water pressure across it, and is used to maintain a minimum flow rate through the boiler and to limit circulation pressure when alternative water paths are closed. A bypass circuit must be installed if the boiler manufacturer requires one, or specifies that a minimum flow rate must be maintained while the boiler is firing. The installed bypass circuit must then include an automatic bypass valve (not a fixed-position valve).

Care must be taken to set up the automatic bypass valve correctly, in order to achieve the minimum flow rate required (but not more) when alternative water paths are closed.

KEY POINT

A free copy of the CHeSS specifications can be downloaded from http://www.centralheating.co.uk/system/uploads/attachments/0000/0157/CE51_CHeSS_WEB_FINAL_JULY_081.pdf.

Explain boiler interlock

A boiler interlock is not a single control device but the interconnection of all of the controls on the system, such as room thermostats, cylinder thermostats and motorised valves. The idea behind a boiler interlock is to prevent the boiler firing up when it is not required, which is a problem with older systems. A boiler interlock can also be achieved by the use of advanced controls, such as a building management system (BMS), which was in the past reserved for larger systems but is now available for domestic properties.

Positive and negative pressures in relation to feed, vent and pump positions

The position of the open-vent pipe, the cold-feed pipe and the circulating pump to a fully pumped system is an important part of the system design. If these components are positioned on the system incorrectly, it will not work properly. They may even induce system corrosion as a result of constant aeration of the system water.

The open-vent and the cold-feed pipe should be positioned on the flow from the boiler on the suction side of the circulating pump, with a maximum of 150mm between them. This is called the neutral point, as the circulating pump acts on both the feed pipe and the open-vent pipe with equal suction. If they are any further apart, the neutral point becomes weak and the pump will act on the feed pipe with a greater force than the open-vent pipe. This creates an imbalance which leads to a lowering of the water in the feed and expansion cistern. When the pump switches off, the water returns to its original position. The constant see-sawing motion aerates the water, creating corrosion within the system.

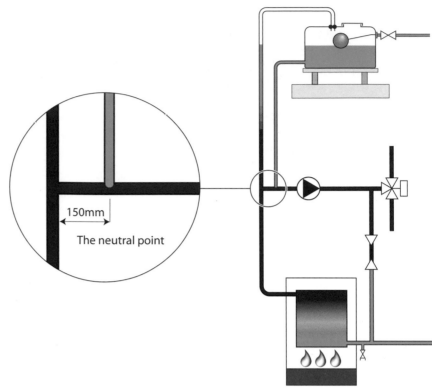

150mm

The neutral point

Position of the cold-feed and open-vent pipes

Circulating pumps

Circulating pumps must also be positioned with care to avoid design faults that could lead to problems with corrosion by aeration of the water as a result of water movement in the feed and expansion cistern. This occurs when water is either pushed up the cold-feed pipe and the open-vent pipe or is circulated between the cold-feed pipe and the open-vent pipe.

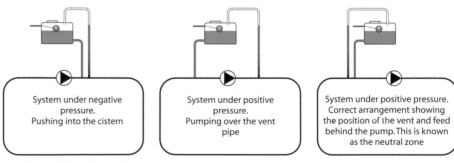

System under negative pressure. Pushing into the cistern

System under positive pressure. Pumping over the vent pipe

System under positive pressure. Correct arrangement showing the position of the vent and feed behind the pump. This is known as the neutral zone

Position of the circulating pump

A circulating pump, or hydronic central heating circulator, is a simple electric motor with a fluted water wheel-like **impeller** that circulates the water around the system by centrifugal force. The faster the impeller rotates, the greater the circulation through the system. For quiet operation of the system, the flow rate should not exceed 1 litre per second (1l/s) and 1.5 litres per second for microbore systems. Most domestic circulating pumps have three speeds which correspond to varying circulatory pressures, or heads. Domestic circulating pumps have either a 6m or a 10m head.

Impeller

A rotor used to increase the pressure and flow of a fluid.

The central heating circulating pump

Air separators

The use of an air separator helps in the positioning of the open vent and cold feed by ensuring that the neutral point is built in to the system. The positioning of the pipework on an air separator creates a turbulent water flow in the separator body and this helps to remove air from the system, which makes the system quieter in operation and significantly reduces the risk of corrosion.

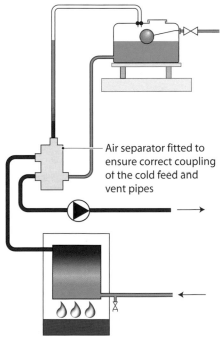

Air separator fitted to ensure correct coupling of the cold feed and vent pipes

The use of an air separator

An air separator

Wiring arrangements for S- and Y-plan heating systems and components

This part of the unit covers some of the more common wiring diagrams for domestic central heating systems:

SmartScreen Unit 306
Handout 3

- pumped heating systems with combination boilers
- fully pumped systems incorporating three-port valves – mid-position and diverter valves

- fully pumped systems incorporating two two-port valves
- fully pumped systems incorporating hot water and multiple space heating zones
- fully pumped systems incorporating weather compensation, optimum start or delayed start controllers
- application of frost thermostats and boilers with pump overrun facility.

Pumped heating systems from a combination boiler

The wiring diagram shows a typical combination boiler wired to separate upstairs and downstairs zone control through a wiring centre. In this particular drawing the system would have an external time clock or programmer and not a boiler-integrated model.

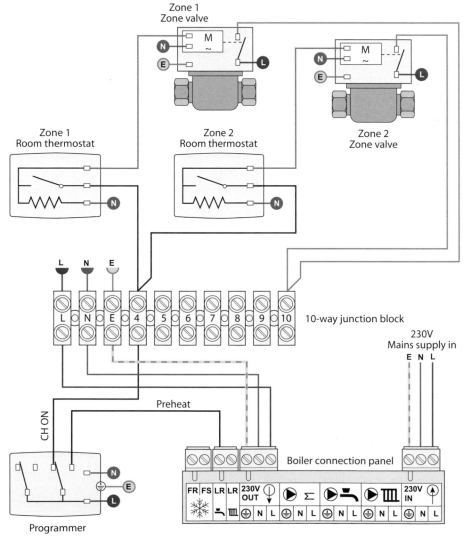

A pumped heating system with combination boiler

Fully pumped systems incorporating three-port valves: mid-position (Y-plan) and diverter (W-plan)

The two diagrams look similar but show two very different systems. The mid-position valve (Y-plan) allows water to be circulated to both heating and hot water circuits simultaneously because the valve will sit in the mid position. However, the diverter valve (W-plan) will allow flow either to the hot water or central heating circuits. The W-plan system is known as a hot water priority system. In other words, the hot water storage cylinder must be satisfied before the central heating circuit will operate. The W-plan is identifiable by the fact that there are only three wires (L/N/E) from the diverter valve.

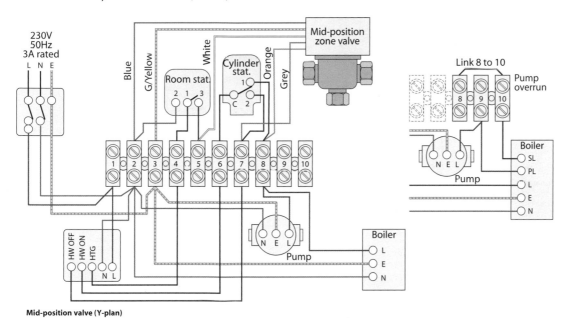

Mid-position valve (Y-plan)

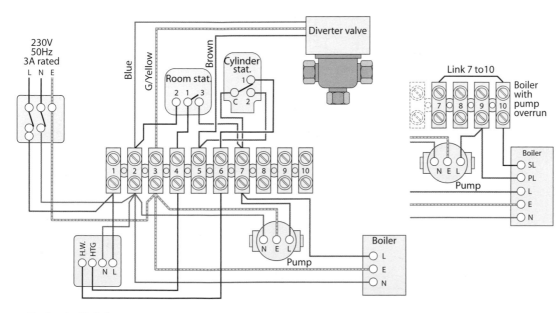

Diverter valve (W-plan)

A fully pumped system incorporating three-port valves – mid position and diverter

Fully pumped systems incorporating two two-port valves (S-plan)

This system is immediately identifiable by the fact that it has two two-port motorised zone valves. Some systems that include a system boiler will not require an external pump as shown in the diagram, and this can be omitted.

S-PLAN

If using a 6-wire 28mm or 1" BSP V4043H on either circuit, the white wire is not needed and must be made electrically safe.

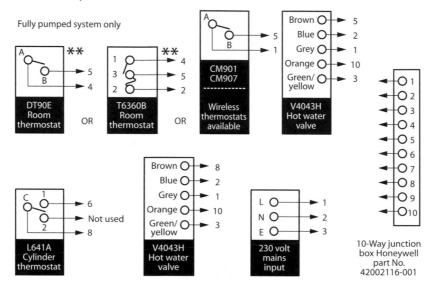

When circuit is wired as above: Complete wiring will be as diagram below.

NOTE:

It is recommended that either the 10-way junction box or Sundial Wiring Centre should be used to ensure first-time, fault-free wiring.

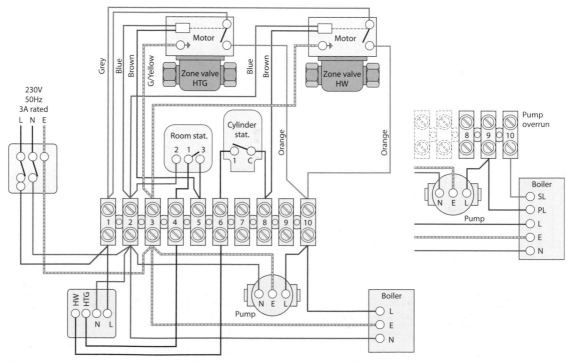

A fully pumped system incorporating two two-port valves (S-plan)

Fully pumped systems incorporating hot water and multiple space heating zones (S-plan plus)

To comply with Building Regulations Document L1, this system is the recommended system for all new-build domestic dwellings and refurbishments. It incorporates multiple heating zones with separate time/temperature control. Two heating zones are shown, but more can be added if required.

S-PLAN PLUS

If using a 6-wire 28mm or 1" BSP V4043H on either circuit, the white wire is not needed and must be made electrically safe.

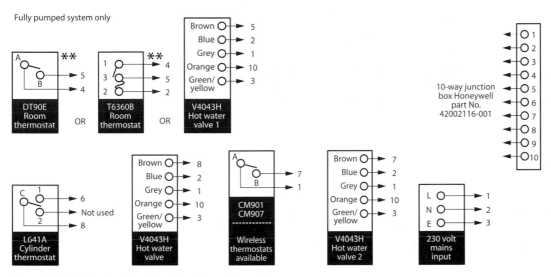

When circuit is wired as above: Complete wiring will be as diagram below.

NOTE: It is recommended that a 10-way junction box should be used to ensure first-time, fault-free wiring.

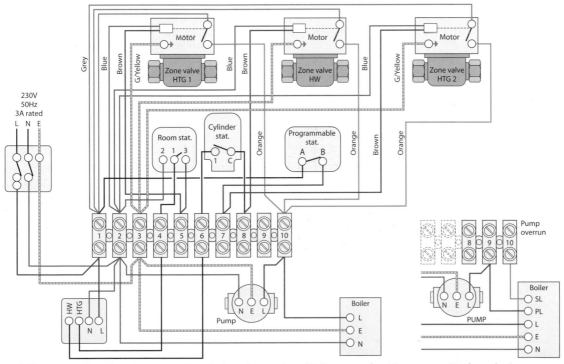

A fully pumped system incorporating hot water and multiple space heating zones (S-plan plus)

Fully pumped systems incorporating weather compensation, optimum start controllers or delayed start controllers

This diagram shows a fully pumped S-plan system including weather compensation. The red dashed line shows the alteration that must be made to a standard S-plan wiring arrangement to accommodate the weather compensator.

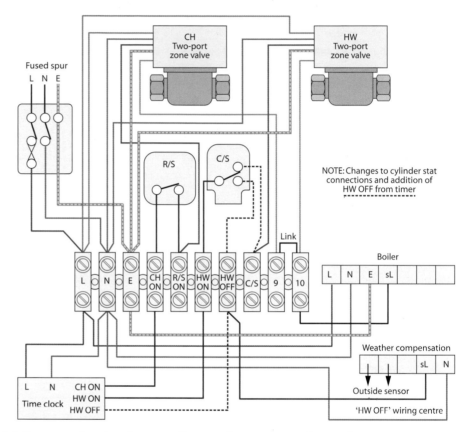

A fully pumped system incorporating weather compensation, optimum start or delayed start controllers

Application of frost thermostats and boilers with pump overrun

Pump overrun is a feature required by some boilers to dissipate the heat after the burner has finished firing. Pump overrun allows the hot water to be circulated away from the boiler, preventing the operation of the energy cut-out. It often occurs when both the hot water and heating circuits have reached temperature and the zone valves have closed. Water is then circulated through the automatic bypass valve to keep the boiler below the high limit temperature.

The diagrams on page 385 indicate the wiring arrangement of a pump overrun facility.

Frost thermostats

Frost thermostats should always be fitted in accordance with a pipe thermostat. The frost thermostat will sense the temperature of the air, while the pipe thermostat will sense the temperature of the water. Both thermostats have to be closed before the boiler will fire. When the water temperature reaches near freezing point, the boiler will fire preventing the system from freezing. Frost thermostats will override all other controls to ensure the safety of the system.

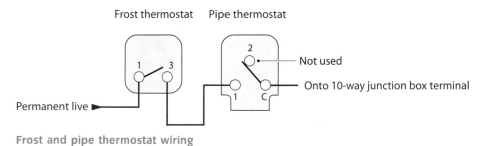

Frost and pipe thermostat wiring

Alternative methods of wiring arrangements

The easiest method of wiring a domestic central heating system is using a plug-and-play-type system. Controls are simply connected to the system using a special low-voltage plug and socket system. Controls such as the room thermostat and motorised valve are plugged into a base unit which is mounted in a convenient location close to them.

Procedures for safely isolating electricity supplies

The Electricity at Work Regulations 1989 forbid anyone from working on or near live equipment unless there are good reasons why it cannot be switched off, such as live working required by a manufacturer's instructions. In all other instances, the circuit should be properly isolated before work commences.

SmartScreen Unit 306
Handout 4 and Worksheet 2

You must do the following before starting work.

1   Check with the occupiers of the property that it is convenient and safe for you to isolate the electricity supply. You should also advise them of the length of time for which the power is likely to be off, and give an estimated time for reinstatement of the supply.

2   Identify the electrical equipment that is to be worked on, and how that equipment is isolated from the supply. This can be done by opening the circuit breakers, isolating switches or removing the fuses. On domestic systems, by far the best method of isolation is to switch off the main switch on the consumer unit and lock it in the 'off' position, making the whole of the system dead. It may be

possible to 'lock off' individual circuits using proprietary locking devices with a padlock and a warning label. A suitable label is shown below.

Warning!

Electrical circuit isolated
DO NOT attempt to switch on

Contact telephone number _____

A warning sign

3 The next stage is to isolate and prove the installation is dead using a suitable voltage-indicating device or test lamp. A GS38 type voltage-indicating device and proving unit is suitable for this purpose. A proprietary socket tester can also be used to prove the circuit is dead if a socket is located on the circuit to be isolated.

4 If there are people in the property, they must be informed that they must not, under any circumstances, attempt to switch the circuit or the supply on.

Where shuttered sockets are to be tested, an alternative test lamp can be used in the form of a socket tester. These can be plugged into a socket and a series of LED lights will illuminate to indicate the power, earthing and polarity of the wiring to the socket.

A Safe Isolation Procedure is carried out as follows.

1 Confirm with the occupier of the property that it is safe and convenient to isolate the circuit to be worked on.

2 Select a suitable voltage-indicating device (refer to GS38) like the one shown in the photograph. This may be a purpose-designed test lamp or a socket tester.

3 Identify, at the consumer unit or other distribution point, the circuit that is to be isolated and open the circuit breaker or remove the fuse.

4 Prove that the test instrument is working correctly on a known voltage or proving unit.

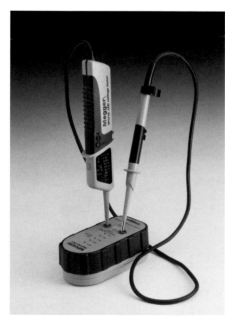

Proving the voltage-indicating device is working correctly

5 Confirm that the circuit is dead by testing between line (live) and earth, line and neutral, and earth and neutral at the point of work. It is not unknown for circuit breakers and fuses to be incorrectly identified.

6 Re-prove that the test instrument is working correctly on a known voltage or proving unit.

7 If the test instrument is still working correctly, it can be assumed that the circuit is dead.

8 Secure the means of isolation using a suitable device and padlock like the one shown below. If the fuse has been removed, keep it with you. Do not leave it lying around near the consumer unit.

9 Apply the warning label to the consumer unit, indicating that work is taking place and that the circuit must not be switched on

A circuit breaker locked off and with warning notice applied

Checking to see if the circuit is dead

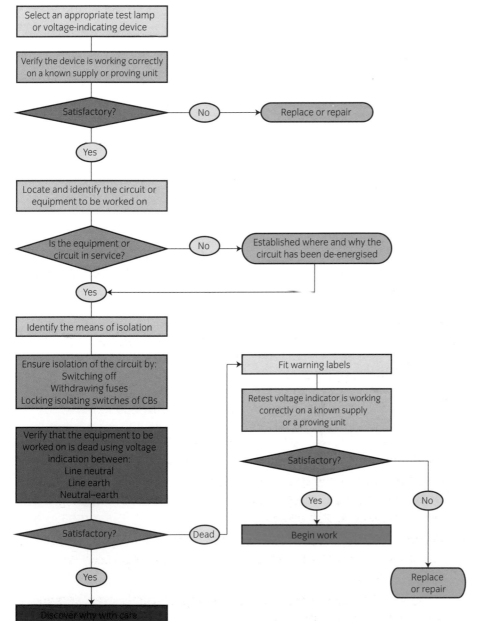

Drop chart of the isolation procedure

Testing of wiring in domestic heating systems

When work on the circuit has been completed successfully, the installation can be visually inspected. Some of the points to look for are listed below.

- Conductor identification – is the correct colour in the correct place?
- Connection of the conductors – are all conductors securely connected and tight?
- Routing of cables – are all cables routed in safe zones?
- Correct cable used – is the cable used big enough to carry the load?
- Termination of cables – does the outer sheathing go all the way into the fitting?
- Clipping and trunking of cables – is the cable correctly positioned, supported and protected?
- Switches, sockets and equipment – have the switches and sockets been fitted securely, and all switch and socket fronts fitted correctly?

Dead testing

Now that the installation has been inspected and any installation faults found have been rectified, testing of the installation can begin. The following tests need to be performed in order:

1 earth continuity
2 short circuit
3 insulation resistance
4 polarity.

Pre-test checks

Before performing any tests, it is important that the test equipment is checked to ensure that it is safe to use, will not cause harm to people, animals or the property and that it reads exactly what it is supposed to.

Calibration

Calibration of the test equipment is too often overlooked, but it is a crucial part of the testing process. In simple terms, it is a measure of how 'true' the readings given by the test equipment are. The readings the equipment gives must be in line with the tables in BS 7671 – The IET Wiring Regulations 17th edition, to ensure complete compliance and, therefore, safety.

If the reading given by test equipment is checked against the corresponding table of minimum values and found to be over, the test is deemed safe. However, what would happen if the test equipment calibration had drifted and the reading was under? A situation would arise in which a circuit that should have failed passes. The consequences would be that:

- BS 7671 has not been complied with
- the Electricity at Work Regulations 1989 have been disregarded
- there is a risk of electric shock and fire.

This is why it is important that test equipment is tested and calibrated regularly.

Earth continuity test

STEP 1 Once the system has been safely isolated, set the test meter to Ω (Ohms) range, then touch the tips of the probes together, which should give a reading of less than 1Ω.

STEP 2 Connect a lead from the multi-meter to a suitable and accessible earth point and then connect the other lead to at least five earth points on the appliance.

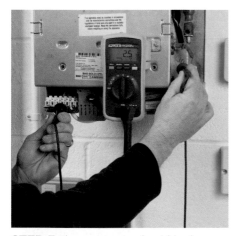

STEP 3 The resistance should be less than 1Ω.

STEP 4 If any reading is more than 1Ω, check the connection points and earth wires on the system.

Short-circuit test

While the appliance is disconnected from the mains supply, set the boiler thermostat and all other switches such as the selector switch and room thermostat to the 'on' position.

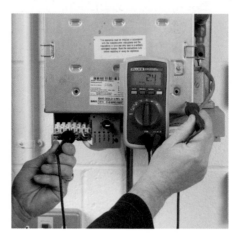 

The test probes are connected to live and neutral with a reading of over 20Ω on the meter

Short circuit

A short circuit is an overcurrent which is the result of a fault between two conductors which have a different potential under normal operating conditions. This situation could occur because of damage from a nail impacting conductor or from poor system design or modification.

Set the multimeter to the Ω (Ohms) range and then connect the test probes across the live and neutral terminals on the boiler connection strip. There should be a resistance reading of more than 20Ω. If there is a reading of less than 20Ω you will need to investigate this, as it is possible that there is a fault across the system. Any reading of less than 1Ω indicates a dead zone, and all components that carry current should be disconnected and individually tested for **short circuit**.

Resistance to earth

Resistance to earth

Opposition of a conductor to a current flow.

The test probes are connected to live and earth and the meter reading is 'OL'

To test for **resistance to earth**, as with the short-circuit test, ensure that the appliance is electrically disconnected and that all of the switches on the system are set to the 'on' position, then set the meter to the Ω (Ohms) range. This time connect the test leads to the to the earth and live contacts on the appliance connection strip. There should be a reading of 'OL', which means 'infinity' or 'open circuit'. Sometimes, because of the electronics found in newer appliances, there may be a resistive reading on the meter. If there is any reading less than 2MΩ then any fault on the appliance should be rectified before the appliance is reconnected to the supply.

Polarity

BS 7671 – The IET Wiring Regulations 17th edition states in Section 10.3.4 that it is important to confirm that overcurrent devices and single-pole controls are in the line conductor and that the polarity of socket outlets and similar accessories is correct.

The purpose of a polarity test is to see whether all the conductors are fitted to the right terminal location on an appliance or on a fitting on a system. This is a live test. The appliance must be connected to the mains and the meter is set to the Vac range. The leads are then connected to the appliance terminal strip. For correct polarity, the readings should be as follows.

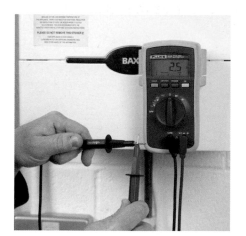

Live to neutral: 230Vac (approx.)

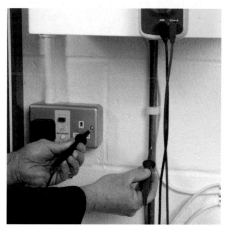

Live to earth: 230V (approx.)

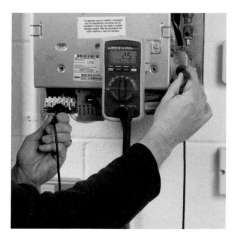

Neutral-earth = 0.247 (0–15V) OK

If a reading of 0–15Vac is discovered on live to neutral or live to earth, this means there is a fault with the polarity and this should be investigated and rectified.

A plug-in tester can determine the polarity of the conductor located on a socket outlet.

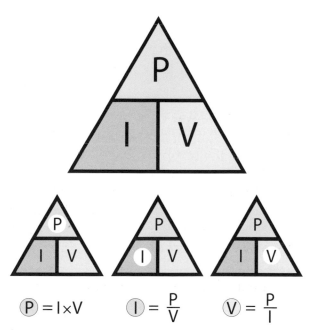

$$P = I \times V \qquad I = \frac{P}{V} \qquad V = \frac{P}{I}$$

The power triangle

SUGGESTED ACTIVITY

When looking at a boiler installation on site, identify where the appliance can be safely isolated from. If it is safe to do so, remove the cartridge fuse from the spur face and calculate the load of the installation to see if it is the correct size to protect the appliance. Use the power triangle for your calculations. (A power triangle is used to calculate the size of a fuse or miniature circuit breaker.)

A plug-in tester

- When handling electrical testing equipment at work visually inspect it to ensure that there is no damage to the probe leads, and check that the probe tips are not more than 4mm. Investigate to see whether there is an **HTC fuse** in one of the probes, and make a note of any labels to identify that the device has been calibrated within the time designated by the manufacturer's instructions.

HTC fuse

This is a special type of fuse used on printed circuit boards.

- When safely isolating a circuit, and once the fused spur has been locked off, locate the consumer unit to see if the appliance you are about to work on has a dedicated radial circuit and is correctly labelled.

Example

If a gas boiler installation, including pump, zone valves and its own components such as a fan and control system, came to a total of 300 watts, P, the amps, I, would add up to 1.3. Therefore a 3-amp fuse would suffice.

By using the equation, the following fuse size can be calculated:

$300P \div 230V = 1.3 \ I$

Earth loop impedance testing

You might have heard electricians talk about earth loop impedance testing, but it must be stressed that this is a live test and so you will never undertake such a test. However, it is very important to understand what this is for.

This test is performed at the end, or termination point, of the circuit. For a central heating circuit this would be the switched fuse spur to the system – everything past that point is classified as equipment and is not subject to the test. A test meter is connected to the live, or phase, conductor and the earth conductor with the electricity turned on. This is carried out at the consumer unit to the property.

The test measures the resistance, or impedance, in Ohms in a loop from the test point right the way back to the transformer, down the line conductor in the substation, and back up the earth and circuit protective conductor to the test point.

The point of the test is to ensure that resistance in the circuit is as low as possible so that, in the event of a fault to earth, the circuit protective device will operate correctly. In domestic circuits, this means a reading of around 1Ω or less.

The working principles of low loss headers

For a boiler to work at its maximum efficiency, the water velocity passing through the heat exchanger needs to be maintained within certain parameters. This is especially important for condensing boilers, as they rely on a defined temperature drop across the flow and return before the condensing mode begins to work effectively. Installation of a low loss header allows the creation of two separate circuits. These are shown in the diagram.

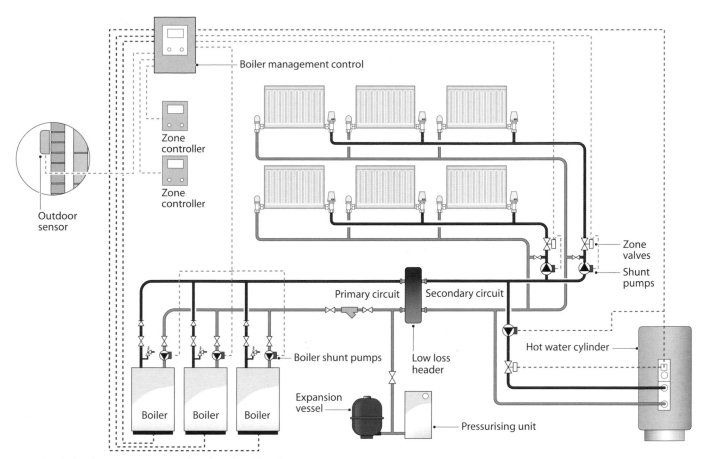

A multiple boiler installation with a low loss header

- **The primary circuit** – the flow rate within the primary circuit can be maintained at the correct flow rate for the boilers, so that the maximum efficiency of the boilers is maintained regardless of the demand placed on the secondary circuit. Each boiler has its own shunt pump so that equal velocity through the boilers is maintained.

- **The secondary circuits** – the secondary circuits allow for varying flow rates demanded by the individual balanced zones or circuits. Each zone would be controlled by a shunt pump set to the flow rate for that particular zone. A two-port motorised zone valve, time clock and room thermostat controls each zone independently and these are often fitted in conjunction with other controls such as outdoor temperature sensors. In some cases the flow rates through each secondary circuit will exceed that required by the boilers. In other cases, the opposite is true and the boiler flow rate will be greater than the maximum flow rate demanded by the secondary circuits, especially where multiple boiler installations are concerned.

Water velocity is just part of the problem. Water temperature is also important. There are two potential problems here.

- If the difference in temperature between the flow and return is too great, it puts a huge strain on the boiler heat exchangers

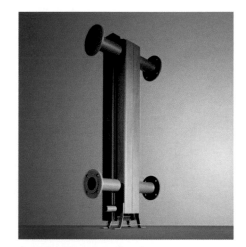

A typical low loss header

because of expansion and contraction. This is known as 'thermal shock'.

■ For a condensing boiler to go into condensing mode, the return water temperature must be in the region of 55°C. In some instances, temperature sensors are fitted to the low loss header to allow temperature control over the primary circuit.

Installation of a low loss header

The low loss header is ideal for use with systems that have a variety of different heat emitters. It is the perfect place for installing an automatic air valve for removing unwanted air from the system. Drain points can also be fitted for removing any sediment that may collect in the header. Both of these features are usually fitted as standard on most low loss headers.

The effect of bore diameter on heat loads

Pipework for central heating systems must be capable of carrying the heat load that is to be supplied to the heat emitters. Each size of tube will have a maximum amount of heat it can carry, but this is dependent on other factors such as the velocity of the water and the temperature difference between the flow and the return pipework.

Table 3: Heat-carrying capacity (kW) of BS EN 1057 copper tube

Diameter of tube (mm)	Wall thickness (mm)	Heat capacity (kW)			
		Temp. drop 11°C		Temp. drop 20°C	
		0.3m/s minimum	1.5m/s maximum	0.3m/s minimum	1.5m/s maximum
6	0.6	0.24	1.24	0.45	2.25
8	0.6	0.49	2.48	0.90	4.51
10	0.7	0.80	4.00	1.45	7.27
12	0.8	1.17	5.86	2.13	10.60
15	0.7	2.00	10.00	3.63	18.10
22	0.9	4.42	22.14	8.05	40.20
28	0.9	7.43	37.17	13.50	67.50

Table 3 shows the amount of heat that can be carried by copper tube in sizes from 6mm to 28mm at velocities between 0.3m/s and 1.5m/s. The temperature drop refers to the temperature difference between the flow and return pipes. With open-vented systems, the temperature drop across the flow and return is normally around 11°C. Sealed systems using condensing boilers, however, operate at slightly higher temperatures and this allows smaller pipework and heat emitters to be fitted with a temperature drop of 20°C across the flow and return pipes. The optimum velocity for central heating is around 1m/s. Any greater than that and the system tends to be noisy. Conversely, velocities of less than 0.3m/s can lead to excess sludging of the pipework and radiators as a result of corrosion and the formation of black oxide sludge.

Understand the layouts and operating principles of sealed systems (LO2)

There are five assessment criteria for this Outcome:

1 Identify components required for sealed central heating systems.
2 Describe the safety hazards associated with sealed central heating systems.
3 Describe the advantages of sealed central heating systems.
4 Explain layout requirements for sealed system components.
5 Calculate the size of expansion vessels for sealed central heating systems.

SmartScreen Unit 306
Presentation 1

	Range
Components	Expansion valve, expansion vessel, temperature relief valve, pressure gauge, filling loop, check valves
Safety hazards	Explosion, high pressure, production of steam
Advantages	Less pipework, fewer components, can come as a package installed inside boiler, less installation, quicker filling, fewer airlocks

Sealed heating systems are those that do not contain a feed and expansion cistern but are filled with water directly from the mains cold water supply via a temporary filling loop. Large systems are filled via an automatic pressurisation unit. The expansion of water is taken up by the use of an expansion vessel and the open vent is replaced by a pressure relief valve which is designed to relieve the excess pressure by releasing the system water and discharging safely to a drain point outside of the dwelling. This is vital, as the water may be in excess of 80°C. A pressure gauge is also included so that the pressure can be set when the system is filled and periodically checked for rises and falls, as these could indicate a potential component malfunction. The system is usually pressurised to around 1 bar.

Components required for sealed central heating systems

Sealed systems do not contain a feed and expansion cistern nor an open-vent pipe. Instead, these systems incorporate the following components:

- an external expansion vessel fitted to the system return
- a pressure relief valve
- a temporary filling loop (the system is filled via this) or a CA disconnection device
- a pressure gauge.

The expansion vessel

The expansion vessel (often incorrectly called the pressure vessel) is a key component of the system. It replaces the feed and expansion cistern on the vented system and allows the expansion of water to take place safely. It comprises a steel cylinder which is divided in two by a neoprene rubber diaphragm.

The vessel is installed on to the return because the return water is generally 20°C cooler than the flow water, and so this does not place as much temperature stress on the expansion vessel's internal diaphragm as the hotter flow water would. If installation of the vessel on the flow is unavoidable, it should be placed on the suction side of the circulating pump in the same way as the cold feed and open-vent pipe on the open-vented system.

On one end of the expansion vessel is a Schrader air pressure valve. Air is pumped into the vessel to 1 bar pressure and this forces the neoprene diaphragm to virtually fill the whole of the vessel.

On the other end is a ½in male BSP (British standard pipe) thread, and this is the connection point to the system. When mains-pressure cold water enters the heating system via the filling loop and the system is filled to a pressure of around 1 bar, the water forces the diaphragm backwards away from the vessel walls, compressing the air slightly as the water enters the vessel. At this point, the pressure on both sides of the diaphragm is 1 bar. As the water is heated, expansion takes place. The expanded water forces the diaphragm backwards, compressing the air behind it still further and, because water cannot be compressed, the system pressure increases.

The expansion vessel with filling loop, pressure relief valve and pressure gauge

On cooling, the water contracts, the air in the expansion vessel forces the water back into the system and the pressure reduces to its original pressure of 1 bar. Periodically, the pressure in the vessel may require topping up. This can be done by removing the cap on the Schrader valve and pumping the vessel up to its original pressure with a foot pump. (The operation of expansion vessels was discussed in more detail in Unit 304, see pages 318–322.)

The pressure relief valve

The pressure relief valve (also known as the expansion valve) is installed on the system to protect against over-pressurisation of the water. Pressure relief valves are usually set to 3 bar pressure. If the water pressure rises above the maximum pressure that the valve is set to, the valve opens and discharges the excess water pressure safely to the outside of the property through the discharge pipework.

Pressure relief valves are most likely to open because of lack of room in the system for expansion due to a malfunction with the expansion vessel. This can be caused by:

- the diaphragm in the expansion vessel rupturing, allowing water both sides of the diaphragm, or
- the vessel losing its charge of air.

A pressure relief valve

The filling loop

The filling loop is an essential part of any sealed system and should contain an isolation valve at either end, and a double-check valve on the mains cold water supply side of the loop. The filling loop is the means by which sealed central heating systems are filled with water. Unlike open-vented systems, sealed systems are filled directly from the mains cold water via a filling loop. The connection of a heating system to the mains cold water supply constitutes a cross-connection between the cold main (fluid category 1) and the heating system (fluid category 3), which is not allowed under the Water Supply (Water Fittings) Regulations 1999. The filling loop must protect the cold water main from backflow, and this is done in two ways.

- The filling loop includes a type EC verifiable double-check valve.
- The filling loop must be disconnected after filling, creating a type AUK3 air gap for protection against backflow.

The filling loop is generally fitted to the return pipe close to the expansion vessel, and may even be supplied as part of the expansion vessel assembly.

The filling loop

Permanent filling connections to sealed heating systems

It is possible to permanently connect sealed heating systems to the mains cold water supply by using a type CA backflow prevention device. The type CA backflow prevention device, when used with a pressure-reducing valve, can be used instead of a removable filling loop to connect a domestic heating system direct to the water undertaker's mains cold water supply. This is possible because the water in a domestic heating system is classified as a fluid category 3 risk. A CA device can also be installed on a commercial heating system, but only when the boiler is rated up to 45kW. Over 45kW, the water in the system is classified as a fluid category 4 risk, and so any permanent connection would require a type BA **RPZ valve**. An example of a CA backflow prevention device is shown on the next page.

The device contains an integral tundish to remove any discharge should a backflow situation occur. Under normal operation, the valve should not discharge water. However, the valve may discharge a small amount of water if the supply pressure falls below 0.5 bar or 11% of the downstream pressure.

RPZ valve

An RPZ valve is a reduced pressure zone valve – a backflow protection device used to protect a category 1 fluid from fluid category 4 contamination.

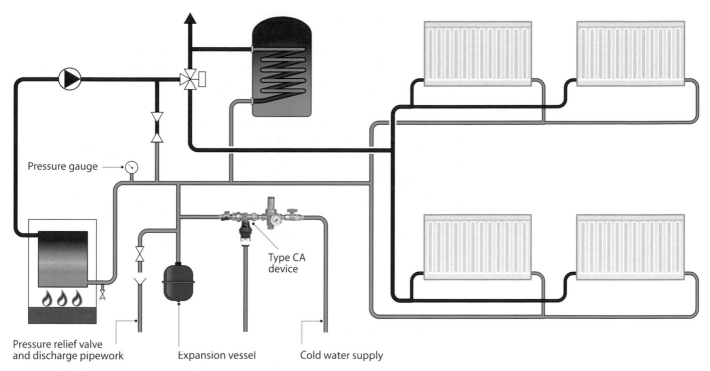

Pressure gauge

Type CA
device

Pressure relief valve
and discharge pipework

Expansion vessel

Cold water supply

A sealed system with CA backflow prevention device

The pressure gauge

This is to allow the correct water pressure to be set within the system. It also acts as a warning of component failure or an undetected leak should the pressure begin to inexplicably rise or fall.

Hazards associated with sealed central heating systems

The greatest hazard with sealed heating systems is that they operate at pressures above atmospheric pressure. Most systems operate at around 1 bar. With the water inside the central heating system at a pressure above atmospheric pressure, the control of the temperature of that water becomes vitally important. This is because as the pressure of the water rises, so the boiling point also rises. In simple terms, if total temperature control failure were to occur at the boiler, the water inside the system would eventually exceed 100°C with disastrous consequences. The line graph on the next page demonstrates the pressure/temperature relationship.

On the graph you can see that at the relatively low pressure of 1 bar, the boiling point of the water has risen to 120.2°C! If a sudden loss of pressure at the hot water storage vessel were to occur due to vessel fracture, at 120.2°C the entire contents of the cylinder would instantly flash to steam with explosive results, causing structural damage to the property. Calculating how much steam would be produced illustrates the point further.

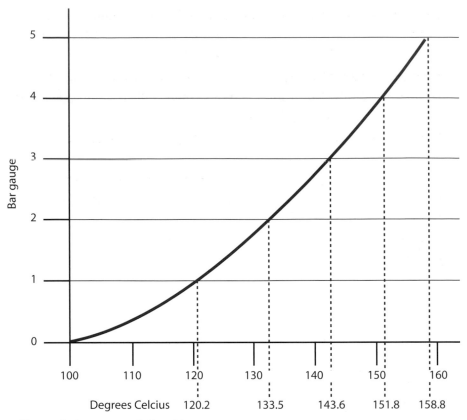

Boiling point/pressure relationship

$1cm^3$ of water creates $1,600cm^3$ of steam. If the system contains 200 litres of water and each litre of water contains $1,000cm^3$, then the amount of steam produced would be:

200 x 1,000 x 1,600 = $320,000,000cm^3$ of steam!

A vital component of the sealed heating system is the pressure relief valve, which operates when the system becomes over-pressurised and discharges the system water through discharge pipework to relieve excess pressure. However, unlike unvented hot water storage systems, they do not contain any form of temperature relief valve. Temperature control of the boiler is provided by a control thermostat and energy cut-out device or high limit thermostat to limit the temperature within the boiler to 85°C.

The advantages of sealed central heating systems

There are several advantages to sealed heating systems.

- **Less pipework** – because the need for a feed and expansion cistern is removed, less pipework is often used.
- **Smaller pipework** – sealed heating systems operate at a slightly higher temperature than vented systems, which means that the heat delivery for the size of pipe is increased. This means that smaller pipework can often be used.

- **Higher heat emitter temperatures** – because sealed systems operate at a slightly higher temperature, this means that slightly smaller heat emitters may be used in some cases.
- **Greater range of heat emitters** – the slightly higher operating temperature offers a greater range in heat emitters, including fan convectors and skirting heating.
- **Less installation time** – sealed systems are often easier to install and this results in less installation time.
- **Quicker filling** – sealed systems fill much quicker than vented systems because the filling water is coming straight from the mains cold water supply.
- **Fewer airlocks** – filling the system from the mains cold water supply eliminates the problems of airlocks.
- **Sealed system components inside the boiler** – system boilers are supplied with all necessary sealed system components already installed as part of the boiler within the boiler casing. This simplifies the system installation.
- **Fewer components** – if a system boiler is used, the problems of siting and installing the expansion vessel and its associated components are eliminated.

Layout requirements for sealed system components

There are several types of sealed heating system:

- sealed systems with an external pressure vessel
- system boilers that contain all necessary safety controls
- combination boilers.

All fully pumped systems, such as those with two or three two-port zone valves (S-plan or S-plan plus) or a three-port mid-position valve (Y-plan) or a three-port diverter valve (W-plan), can be installed as sealed systems or they can be purpose-designed 'heating only' systems using a combination boiler with instantaneous hot water supply.

Fully pumped systems with two or three two-port zone valves (S-plan and S-plan plus)

The S-plan system has two two-port motorised zone valves to control the primary and heating circuits separately, by the cylinder and room thermostats respectively. This system is recommended for dwellings with a floor area greater than 150m^2, because it allows the installation of an additional two-port zone valve to zone the upstairs heating circuit from the downstairs circuit (S-plan plus). A separate room thermostat and possibly a second time clock/programmer would also be required for upstairs zoning.

A system bypass is required for overheat protection of the boiler.

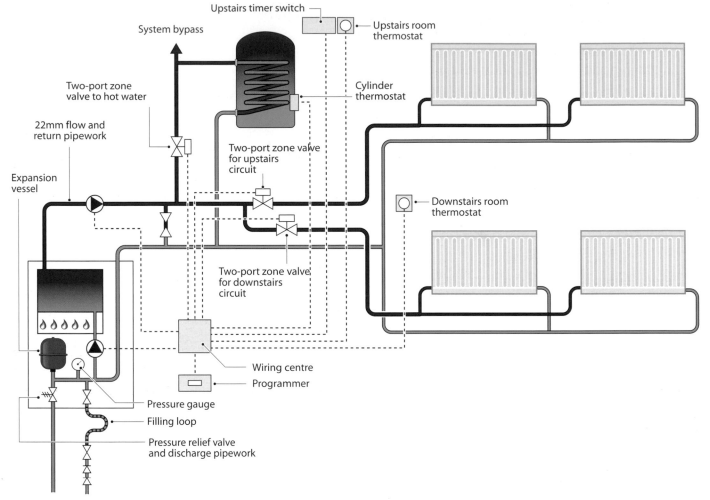

- Upstairs timer switch
- System bypass
- Upstairs room thermostat
- Cylinder thermostat
- Two-port zone valve to hot water
- 22mm flow and return pipework
- Two-port zone valve for upstairs circuit
- Expansion vessel
- Downstairs room thermostat
- Two-port zone valve for downstairs circuit
- Wiring centre
- Programmer
- Pressure gauge
- Filling loop
- Pressure relief valve and discharge pipework

The S-plan plus system (sealed system boiler)

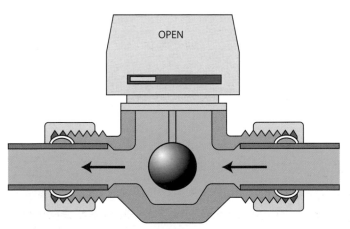

OPEN

The two-port zone valve

Fully pumped systems with three-port mid-position valve (Y-plan) or a three-port diverter valve (W-plan)

The three-port mid-position valve (Y-plan) or diverter valve (W-plan) controls the flow of water to the primary (cylinder) circuit and the heating circuit. The valve reacts to the demands of the cylinder thermostat or the room thermostat.

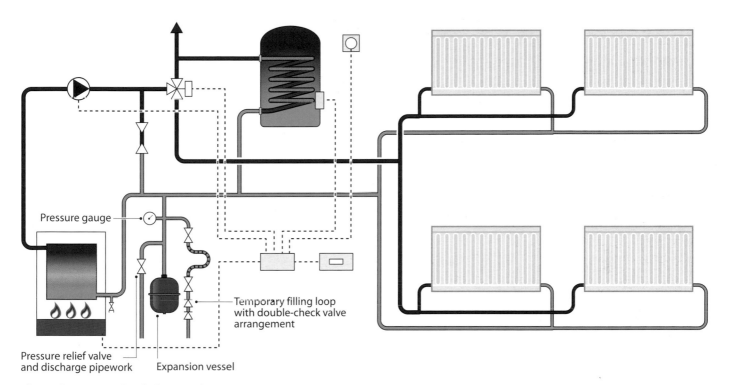

The Y-plan system (sealed system)

The system contains a system bypass fitted with an automatic bypass valve which simply connects the flow pipe to the return pipe. The bypass is required when all circuits are closed either by the motorised valve or the thermostatic radiator valves as the rooms reach their desired temperature. The bypass valve opens automatically as the circuits close to protect the boiler from overheating by allowing water to circulate through it, thereby keeping the boiler below its maximum high temperature. This prevents the boiler from locking out on the overheat energy cut-out.

Combination boilers

In recent years, combination boilers have become one of the most popular forms of central heating in the UK. A combination boiler provides central heating and instantaneous hot water supply from a single appliance. Modern combination boilers are very efficient and they contain all the safety controls – ie expansion vessel and pressure relief valve – of a sealed system. Most 'combis' also have an integral filling loop.

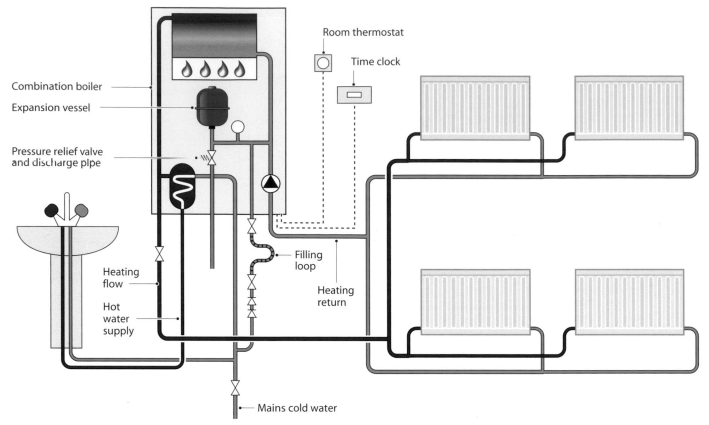

Combination boiler

Expansion vessel

Pressure relief valve and discharge pipe

Room thermostat

Time clock

Filling loop

Heating flow

Heating return

Hot water supply

Mains cold water

A sealed system with a combination boiler

Calculate the size of expansion vessels for sealed central heating systems

There are several methods for sizing expansion vessels. All methods must take into account the volume of cold water in the system and the amount by which it will expand in order to reach its design temperature. The CIBSE method is shown below.

If the system volume is known, expansion vessels can be sized using the following formula:

$$V = \frac{eC}{1 - \frac{p_1}{p_2}}$$

V = total volume of the expansion vessel

C = total volume of water in the system in litres

P_1 = fill pressure in bars absolute (gauge pressure + 1 bar)

P_2 = setting of the pressure relief valve + 1 bar

e = expansion factor that relates to the maximum system requirements

Expansion factor e	Temperature °C
0.0324	85
0.0359	90
0.0396	95
0.0434	100

e can be found from the formula:

$$e = \frac{d_1 - d_2}{d_2}$$

Where:

d_1 = density of water at filling temperature kg/m^3

d_2 = density of water at operating temperature kg/m^3

Example

A sealed central heating system has a total water volume of 600 litres. The pressure of the water main is 1.5 bar and the pressure relief maximum pressure is 6 bar. The system is designed to operate at a maximum temperature of 80°C, which means the expansion factor will have to be calculated. The fill temperature of the water is 10°C.

Calculate the expansion factor using:

$$e = \frac{d_1 - d_2}{d_2}$$

Calculate the expansion vessel volume using:

$$V = \frac{eC}{1 - \frac{p_1}{p_2}}$$

Calculate the expansion factor e

The temperature of the fill water is 10°C with a density of 999.8kg/m^3. The maximum operating temperature is 80°C with a density of 972kg/m^3. Therefore the e factor is:

$$\frac{999.8 - 972}{972} = \textbf{0.0286}$$

SUGGESTED ACTIVITY

A sealed central heating system has a total water volume of 400 litres. The pressure of the water main is 1 bar and the pressure relief maximum pressure is 4 bar. The system is designed to operate at a maximum temperature of 85°C, which means the expansion factor will have to be calculated. The fill temperature of the water is 4°C.

Calculate the expansion factor using:

$$e = \frac{d_1 - d_2}{d_2}$$

Calculate the expansion vessel volume using:

$$V = \frac{eC}{1 - \frac{p_1}{p_2}}$$

SmartScreen Unit 306

Presentation 3

Calculate the expansion vessel volume

V =	total volume of the expansion vessel
C =	600 litres
P_1 =	1.5 + 1
P_2 =	6 + 1
e =	0.0286

Therefore, the expansion vessel volume is:

$$V = \frac{eC}{1 - \frac{p_1}{p_2}}$$

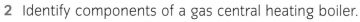

$$\frac{0.0286 \times 600}{1 - \frac{2.5}{7}} = \frac{17.16}{1 - 0.357} = \frac{17.16}{0.643} = 26.68 \text{ litres or } 4.44\%$$

So, the expansion vessel volume is: 26.68 litres or 4.44% of total system volume.

Understand the types of boilers used in domestic central heating systems (LO3)

There are five assessment criteria to this Outcome:

1 Identify fuel sources for central heating.
2 Identify components of a gas central heating boiler.
3 Describe the operating principles of boilers.
4 Describe different flueing arrangements for boilers.
5 Explain the reason for pump overrun on boilers.

Range	
Fuel sources	Natural gas, LPG, oil, solid fuel, heat pumps
Components	Heat exchanger, water-to-water heat exchanger, diverter valve, gas valve, condense trap, air pressure switch
Boilers	Combination, condensing, non-condensing, traditional, systems boiler
Flueing	Open flue, balanced flue, fanned draft flues

Fuel sources for central heating

Even now, in this time of zero- and low-carbon fuels, traditional fuels are still used for domestic central heating systems. Today there is an eclectic mix of fuels available, ranging from coal to biomass and heat pumps.

Solid fuel

There are three main types of solid fuel that can be used for central heating systems. These are:

- coal
- coke
- biomass.

Coal

Coal

This is a fossil fuel created from the remains of plants that lived and died between 100 and 400 million years ago, when large areas of the Earth were covered with huge swamps and forest bogs.

The energy that we get from coal comes from the energy that the plants absorbed from the Sun millions of years ago, in a process called photosynthesis. When plants die this stored energy is usually released during the decaying process, but when coal is formed the process is interrupted, preventing the release of the trapped solar energy.

As the Earth's climate changed and the vegetation died, a thick layer of rotting vegetation built up that was covered with water, silt and mud, stopping the decaying process. The weight of the water and the top layer of mud compressed the partially decayed vegetation under heat and pressure, squeezing out the remaining oxygen and leaving rich hydrocarbon deposits. What once had been plants gradually fossilised into a combustible carbon-rich rock we call coal.

Coal is classified into four main types, depending on the amount of carbon, oxygen and hydrogen present. The higher the carbon content, the more energy the coal contains.

Table 4: Types of coal

Coal type	Heat content kW/kg	Carbon content %	Description
Lignite	2.2–5.5	25–35	The lowest grade of coal, lignite is crumbly and has high moisture content. Most lignite is used to produce electricity.

Coal type	Heat content kW/kg	Carbon content %	Description
Sub-bituminous	5.5–8.3	35–45	Sub-bituminous coal typically contains less heating value than bituminous coal, but more moisture.
Bituminous	7–10	45–86	Bituminous coal is formed by added heat and pressure on lignite. Made of many tiny layers, bituminous coal looks smooth and sometimes shiny. It has two to three times the heating value of lignite. Bituminous coal is used to generate electricity and is an important fuel for the steel and iron industries.
Anthracite	10	86–97	Anthracite was created where additional pressure combined with very high temperatures inside the Earth. It is deep black and looks almost metallic due to its glossy surface.

Coal is still used for domestic and industrial central heating boilers and for steam and electricity generation.

Coke

Coke is produced by heating coal in coke ovens to around 1,000°C. During this process, the coal gives off methane gas and coal tar, both of which can be cleaned and re-used. Coke burns clearly and without a flame and gives out a lot of heat, but it has to be mixed with coal as it will not burn by itself.

Coke is a smokeless fuel that is valued in industry because it has a calorific (heat) value higher than any form of natural coal. It is widely used in steel making and in certain chemical processes, but can also be used in many domestic boilers and room heaters.

Biomass

The term biomass can be used to describe many different types of solid and liquid fuel. It is defined as any plant matter used directly as a fuel or that has been converted into other fuel types before combustion. When used as a heating fuel, it is generally in the form of solid biomass including wood pellets, vegetal waste (such as wood waste and crops used for energy production), animal materials/wastes and other solid biomass.

Other fuels

There are three other fuels that are commonly used for central heating systems. These are:

- fuel oil
- natural gas
- liquid petroleum gas (LPG).

Fuel oil (kerosene, or C2 grade 28-second viscosity oil to BS 2869)

A simple definition for fuel oil is a liquid by-product of crude oil, which is produced during petroleum refining. There are two main categories under which it is classified: distillate oils, such as diesel fuel, and residual oils, which include heating kerosene. Distillate fuel oil is generally used for home heating.

Around 95% of domestic boilers burning fuel oil in domestic properties use kerosene, which is also known generically as C2 grade 28-second viscosity oil. This is the preferred oil fuel grade for domestic heating, due to its clean combustion. Modern oil central heating boilers only require a single annual service, if being used with an atomising pressure jet burner. It is the only oil grade that can be used with balanced or low-level flues.

Container of kerosene

Kerosene has very good cold weather characteristics and remains fluid beyond minus 40°C, although it does tend to thicken slightly during extremely cold weather.

Kerosene is a high-carbon fuel and is clear or very pale yellow in colour. Newer boilers have a label inside the casing with information on nozzle size and pump pressure that show that the boiler has been set up to use kerosene. They may also reference the British Standard for kerosene, BS 2869 Grade C2.

Natural gas

Gas is a light hydrocarbon fuel found naturally wherever oil or coal has formed. Despite its association with global warming, it is one of the cleanest, safest and most useful of all energy sources, being used for a variety of industrial and domestic applications such electricity generation, heating and cooking. All gas, whether natural or liquid petroleum gas, produces carbon dioxide, nitrogen and water vapour as by-products of the combustion process.

Natural gas is a mixture of gases found naturally in coal and oil deposits, and because of this, it has reasonably high carbon content. It consists of over 90% methane, but also contains other flammable gases such as ethane, propane and butane. It is a major source of fuel and is found in many parts of the world such as Russia, the USA, Saudi Arabia and the North Sea.

Before natural gas is introduced to the gas pipe network, it must be rigorously processed and cleaned to remove the waxy oil deposits it contains. These deposits are called naphthas.

Natural gas has no smell, so a chemical with a distinctive rotten-egg smell, called mercaptan, is added to it so it can be detected if it leaks. It has a specific gravity of 0.8, which means it is lighter than air.

Natural gas has many industrial and domestic uses, including providing the base ingredient for products such as plastics, fertiliser, anti-freeze and fabrics, as well as electricity generation, cooking and heating. In fact, industry is the largest consumer of natural gas, accounting for over 40% of natural gas usage.

Table 5: Typical composition of natural gas

Methane	CH_4	70–90%
Ethane	C_2H_6	0–20%
Propane	C_3H_8	
Butane	C_4H_{10}	
Carbon dioxide	CO_2	0–8%
Oxygen	O_2	0–0.2%
Nitrogen	N_2	0–5%
Hydrogen sulphide	H_2S	0–5%

Liquid petroleum gas (LPG)

LPG is generally processed from petroleum refining, or it can be extracted from natural gas. It is stored in liquid form by compressing it, to reduce the gas's volume by 274 times. In other words, 1 litre of liquid makes 274 litres of gas. There are two types of commercially available LPG:

- propane
- butane.

Table 6: Characteristics of LPG

Gas	Chemical symbol	Specific gravity	Boiling point	Characteristics and uses
Propane	C_3H_8	1.5	−45°C	Propane is the most widely available of all LPGs. It is used for cooking, heating equipment including boilers and fires, cars and many industrial applications Propane is 1½ times heavier than air Propane is available in a range of bottle sizes and can also be stored in bulk for domestic and commercial use
Butane	C_4H_{10}	2.0	−4°C	Butane has a slightly higher calorific value than propane, but its use is limited to camping and portable equipment because of its relatively high boiling point Butane is available in a range of bottle sizes. It is twice as heavy as air

As with natural gas, LPG has no smell, so a stenching agent called ethanethiol (also known as ethyl mercaptan) is added. The calorific value of LPG is about 2.5 times higher than that of mains gas, so more heat is produced from the same volume of gas.

Heat pumps

A heat pump is an electrical device with reversible heating and cooling capability. It extracts heat from one medium at a low temperature (the source of heat) and transfers it to another at a high temperature (called the heat sink), cooling the first and warming the second. A heat pump works in the same way as a refrigerator, by moving heat from one place to another. Heat pumps can provide space heating, cooling, water heating and air heat recovery. There are several different types:

- ground source heat pumps
- air source heat pumps
- water source heat pumps.

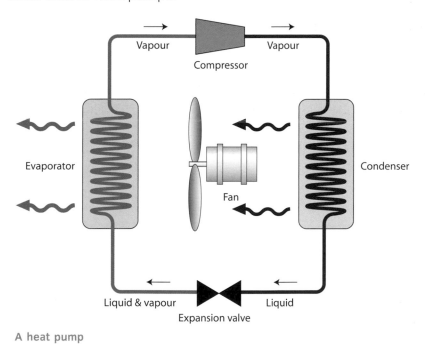

A heat pump

Components of a gas central heating boiler

This part of the unit looks at some of the components of central heating boilers. It is impossible to look at all the components in this book, so some of the more common ones are covered:

- heat exchanger
- water-to-water heat exchanger
- diverter valve

- gas valve (multi-functional control)
- condense trap
- air pressure switch.

The position of these components is shown on the cutaway diagram of a combination boiler below.

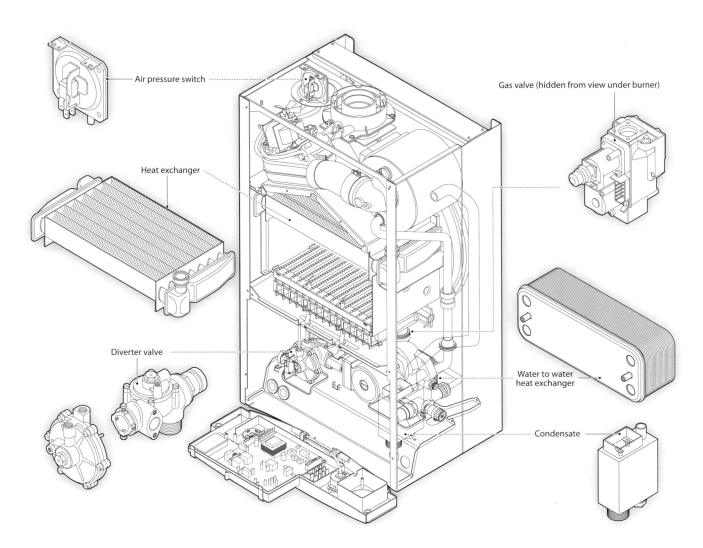

Air pressure switch

Gas valve (hidden from view under burner)

Heat exchanger

Diverter valve

Water to water heat exchanger

Condensate

Combination boiler cutaway

Heat exchanger

Heat exchangers transfer the heat from the burner into the heat transfer medium. With boilers, the heat transfer medium is water. Common materials include cast iron, stainless steel, aluminium and copper.

The heat exchanger is often located above the burner, and the products of combustion pass through it to heat the water, although with modern condensing boilers the heat exchanger is circular, with the burner passing through the middle. Here, a gas/air mixture is blown by a fan through the burner, where it is ignited, heating the heat exchanger. Irrespective of the method, the products of combustion are then directed into the flue system.

The cooler return water typically enters at the bottom of the heat exchanger and the heated flow water exits from the top.

Water-to-water heat exchanger

Often called plate heat exchangers, these are constructed from a number of corrugated plates – the more plates that are sandwiched together, the better the heat transfer. The plates provide two separated circuits for water to flow through. One circuit is connected to the primary system and is heated by the water in the primary heat exchanger. The other circuit is connected to the mains cold water supply.

As the cold water flows through the heat exchanger, it is warmed instantly by the heated plates from the primary circuit, thereby providing instant hot water to the taps.

Diverter valve

In a combination boiler, the diverter valve is an electrically operated valve that provides a means of changing the direction of the flow of heated water produced by the primary heat exchanger.

When the boiler is in heating mode, the diverter valve allows water to flow to the heating circuit. When a hot tap is opened, a flow switch activates the diverter valve to alter the flow of water through the plate heat exchanger to warm the flowing water to the hot tap.

Gas valve (multi-functional control)

The gas valve, or to give its correct name, the multi-functional control, is a valve that incorporates many components in one unit:

- a filter
- a thermoelectric safety shut-off valve – this shuts down the boiler in the event of pilot failure (if applicable)
- a gas regulator – this regulates the gas to the correct pressure for the burner
- a solenoid valve – this opens when the appliance calls for heat
- a gas inlet pressure test point
- a gas burner pressure test point
- a pilot flame adjustment screw (if applicable).

There are many designs of multi-functional control, and most are based on two specific methods of operation:

- controls fitted to boilers that have a permanent pilot flame and incorporate a thermo-electric flame failure device (thermo-couple)
- controls that utilise some form of electronic ignition and flame-ionisation flame failure system.

A typical diverter valve

A multi-functional control

A condensate trap

An air pressure switch

Condense trap

A condense trap, or condensate trap, is fitted to condensing boilers to collect the condensate that forms as a part of the boiler's operation, and to allow a path for the condensate to evacuate from the boiler to the outside of the building via a drain or soakaway. Condense traps are usually made from moulded plastic, and often incorporate a siphon to maintain some water in the trap at all times. The body of the trap has an inlet from the boiler heat exchanger/combustion chamber and an outlet to the drain.

Air pressure switch

The air pressure switch uses pressure differential caused by the running of the flue fan to activate a small micro-switch. Inside the air pressure switch is a small diaphragm that moves due to the movement of air caused when the fan starts. When the diaphragm moves, it operates the small micro-switch that activates the ignition sequence on the boiler. The air pressure switch is usually connected to the fan and the flue by rubber tubing.

The operating principles of boilers

So far this unit has looked at the different central heating systems and their layouts. The next section investigates in detail the different appliances you can use to generate the heat required to warm the systems, and the different fuels they use.

To recap, boilers used for central heating systems are generally heated by one of three different fuel types:

- solid fuel
- gas
- oil.

Table 7: Suitability of fuel types for different appliances

	Open flue	Room-sealed (natural draught)	Room-sealed (fan-assisted)	Freestanding/ independent boiler	Wall-mounted	Condensing	Non-condensing (traditional)	System boiler	Cooker	Open fire with high-output back boiler	Room heaters
Solid fuel	✓	✗	✗	✓	✗	✗	✓	✗	✓	✓	✓
Gas	✓	✓	✓	✓	✓	✓	✓	✓	✓	✗	✗
Oil	✓	✗	✓	✓	✓	✓	✓	✓	✓	✗	✗

Solid-fuel appliances

Solid-fuel appliances are still used in rural areas of the UK where access to piped fuel supply is difficult. Solid fuel is available in many different forms, including:

- coal
- coke
- anthracite
- biomass wood pellets (carbon neutral).

Independent solid-fuel boilers (freestanding)

Domestic open-flued independent solid fuel boilers are designed to provide both domestic hot water and central heating in a whole range of domestic premises, from the very large to the very small.

There are two main types of independent boilers for domestic use.

- **Gravity feed boilers** – often called hopper-fed boilers, these appliances incorporate a large **hopper**, positioned above the firebox, which can hold two or three days' supply of small-sized anthracite. The fuel is fed automatically to the fire bed as required, and an in-built, thermostatically controlled fan aids combustion. This provides a rapid response to an increase in demand. They are available in a wide range of sizes and outputs.

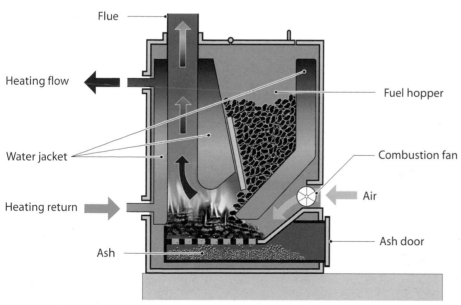

A gravity-fed boiler

The main danger with this type of boiler is the risk of fire in the hopper. The fuel feed to the fire bed needs to be regulated with care.

- **Batch feed boilers** – these are 'hand fired' appliances that require manual stoking. They also require much more refuelling

KEY POINT

The legal requirements for the installation of solid fuel and oil heat-producing appliances, such as boilers, cookers and room heaters, are covered in Building Regulations Document J – Heat-producing appliances. The legal requirements for the installation of gas appliances are given in the Gas Safety (Installation and Use) Regulations 1998. In all cases, manufacturers' instructions must always be followed when installing heat-producing appliances of any kind.

The governing bodies for the different fuels used with heating appliances are:
- gas – Gas Safe www.gassaferegister.co.uk
- oil – Oftec www.oftec.org
- solid fuel – Hetas www.hetas.co.uk.

KEY POINT

The dew point is the temperature at which the moisture within a gas is released to form water droplets. When a gas reaches its dew point, it is cool enough that it can no longer hold the water it contains, and this is released in the form of dew, or water droplets.

Hopper

A container

than hopper-fed boilers. They can, however, be less expensive to run in some cases, and will often operate without the need for an electricity supply, thereby providing hot water and central heating during power failure.

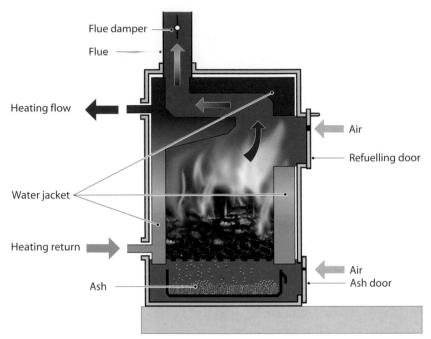

A batch-fed boiler

Gas central heating boilers

Gas central heating boilers are the most popular of all central heating appliances. Over the years there have been many different types, from large multi-sectional cast-iron domestic boilers to small low-water-content condensing types. Both natural gas (those that burn a methane-based gas) and LPG (those that burn propane) types are available.

Central heating boilers can be categorised as follows:

- traditional (non-condensing)
- cast-iron heat exchanger
- low-water-content heat exchanger
- combination (non condensing)
- condensing (system and combination).

Radiator connections are classified by their abbreviations:

- **TBOE** – Top, Bottom, Opposite End (used on heat-sink radiators with solid fuel systems and one-pipe systems)
- **BBOE** – Bottom, Bottom, Opposite End (the usual method of radiator connection)
- **TBSE** – Top, Bottom, Same End (used with some one-pipe systems).

Table 8: The different kinds of gas boilers and their flue arrangements

	Energy efficient	Cast-iron heat exchanger	Low-water-content	Open-vented system	Sealed (pressurised) system	Open flue	Room-sealed (natural draught)	Room-sealed (fan-assisted)	Wall-mounted	Freestanding
Traditional boiler	✗	✓	✓	✓	✓	✓	✓	✓	✓	✓
Condensing boiler	✓	✗	✓	✗	✓	✗	✗	✓	✓	✓
System boiler	✓	✗	✓	✗	✓	✗	✗	✓	✓	✗
Combination boiler	✓	✗	✓	✗	✓	✗	✓	✓	✓	✓

Traditional boilers (non-condensing)

Traditional non-condensing boilers have been around for many years and in many different forms. This part of the unit looks at some of the boilers you may come across when working on the many existing systems there are installed.

Boilers with cast-iron heat exchangers

For many years, boilers were made with cast-iron heat exchangers. They were often very large and heavy, even for small domestic systems. Some heat exchangers were made from cast iron which was cast in a single block, while older types were made up of cast-iron sections bolted together. The more sections a boiler had, the bigger the heat output.

Fuel efficiency was, typically, 55% to 78% with much wasted heat escaping through the flue. Most traditional boilers were fitted on open-vented systems, but sealed (pressurised) systems could also be installed with the inclusion of an external expansion vessel and associated controls (see pages 398–399).

Cast-iron boilers can be either freestanding (floor mounted) or wall mounted using a variety of flue types:

- open flue
- room-sealed (natural draught)
- fan-assisted room-sealed (forced draught).

A traditional open flue gas boiler

Table 9: Advantages and disadvantages of cast-iron heat exchangers

Advantages of cast-iron heat exchangers	Disadvantages of cast-iron heat exchangers
• Long lasting, with a typical life of 20 to 30 years • Very robust	• Heavy • Not energy efficient • Do not comply with Document L • Noisy • Very basic boiler controls

Boilers with low-water-content heat exchangers

Low-water-content heat exchangers were usually made from copper tube with aluminium fins, or lightweight cast iron. They were an attempt to reduce the water content of the heating system, thus speeding up heating times and improving efficiency. Typical efficiencies for this type of boiler were around 82%.

The boilers were always wall mounted, very light in weight and, as a consequence, often quite small in size, designed for fully pumped S- and Y-plan heating systems only. They were the first generation of central heating boilers to use a high temperature limiting thermostat (or energy cut-out) to guard against overheating, and often used a basic printed circuit board to initiate a pump overrun to dissipate latent heat. This is explained in greater detail later in the unit.

Low-water-content boilers can be found with a variety of flue types:

- open flue
- room-sealed (natural draught)
- fan-assisted room-sealed (forced draught).

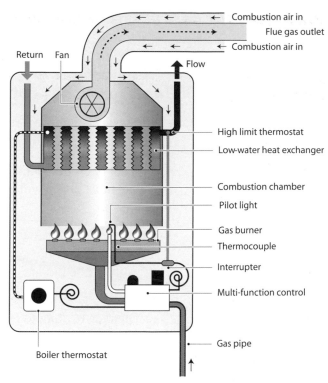

A fan-assisted low-water-content boiler

Table 10: Advantages and disadvantages of low-water-content boilers

Advantages of low-water-content boilers	Disadvantages of low-water-content boilers
• Light in weight • Often a cheaper appliance • Relatively quick water heating times	• Not energy efficient • Do not comply with Document L • Could be very noisy • Relatively short working life • High-maintenance compared with other boilers

Combination boilers (non-condensing)

Combination boilers that supply instantaneous hot water as well as central heating have been around for many years. Early models, although wall mounted, were very large. Most had a sealed (pressurised) heating system but some were the low-pressure open-vented type. Hot water flow rates were often poor in comparison with modern condensing types.

Early combination boilers can be found with a variety of flue types:

■ open flue

■ room-sealed (natural draught)

■ fan-assisted room-sealed (forced draught).

Table 11: Advantages and disadvantages of non-condensing boilers

Advantages of non-condensing combination boilers	Disadvantages of non-condensing combination boilers
• Instantaneous hot water supply • Sealed system means no feed and expansion cistern required in the roof space	• Not energy efficient • Very large in size • Do not comply with Document L • Can be very noisy • High-maintenance compared with other boilers • Poor hot water flow rates • Difficult to maintain

Condensing boilers

The latest addition to the gas central heating family is the condensing boiler. These boilers work in a very different way from traditional boilers.

Natural gas, when it is combusted, contains CO_2, nitrogen and water vapour. As the flue gases cool, the water vapour condenses to form water droplets. It is this process that condensing boilers use.

With a condensing boiler, the flue gases first pass over the primary heat exchanger, which extracts about 80% of the heat. The flue gases, which still contain 20% of latent heat, are then passed over a secondary heat exchanger where a further 12–14% of the heat is extracted. When this happens, the gases cool to their dew point, condensing the water vapour inside the boiler as water droplets, which are then collected in the condense (condensate) trap before being allowed to fall to drain via the condensate pipe. The process gives condensing boilers their distinctive 'plume' of water vapour during operation, which is often mistaken for steam.

Modern condensing boilers are around 93% efficient, releasing only 7% of wasted heat in the cooler flue gases into the atmosphere.

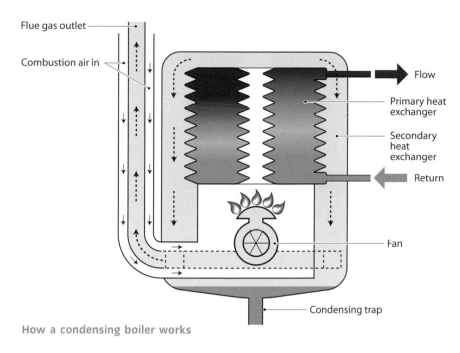

Flue gas outlet

Combustion air in

Flow

Primary heat exchanger

Secondary heat exchanger

Return

Fan

Condensing trap

How a condensing boiler works

Table 12: Advantages and disadvantages of condensing boilers

Advantages of condensing boilers	Disadvantages of condensing boilers
• Document L compliant • Very high efficiency • Sealed (pressurised) system gives better heating flow rates • System corrosion can be reduced • Very quiet in operation • Can be used with all modern fully pumped heating systems (system boilers) • No feed and expansion cistern required in the roof space • Very good flow rate on hot water supply (condensing combination boilers)	• High-maintenance compared with other boilers • Siting of the condense pipework can often prove difficult • Do not work if the condense line freezes during cold weather • Use more gas when not in condensing mode

Oil-fired central heating appliances

Oil-fired appliances are popular where access to mains gas is difficult. They offer a viable alternative to gas appliances. Most oil-fired appliances use C2 grade 28-second viscosity oil (kerosene).

KEY POINT

Dressing a radiator involves getting the radiator ready for hanging by putting in the valves, the air release valve and the plug. The process is as follows.

1 Carefully remove the radiator from its packing. Inside the packing you will find the hanging brackets, the air release valve and the plug and, often, small U-shaped pieces of plastic to be placed on the brackets where the radiator fits. These are designed to prevent the radiator from rattling.

2 Take out the factory-fitted plugs. Be careful, especially if you are working in a furnished property, as the radiator often contains a small amount of water from when it was tested at the factory.

3 Split the valves at the valve unions and wrap **PTFE tape** around the valve tail. Between 10 and 15 wraps will ensure that the joint between tail and radiator does not leak. This may seem a lot of PTFE, but the ½in female sockets on the radiators are notoriously slightly oversized, and this leads to leaks.

4 Make the tail into the radiator using a radiator spanner.

PTFE tape

PTFE stands for polytetrafluoroethylene. It is a tape used to make leak-free joints in copper and low carbon steel installations.

Table 13: The different kinds of oil boilers and their flue arrangements

	Energy efficient	Cast-iron heat exchanger	Open-vented system	Sealed (pressurised) system	Open flue (forced draught)	Open flue (natural draught)	Room-sealed (fan-assisted)	Wall-mounted	Freestanding	Pressure jet burners	Vaporising burners
Traditional boilers	✗	✓	✓	✓	✓	✗	✓	✓	✓	✓	✗
Condensing boilers	✓	✓	✗	✓	✓	✗	✓	✓	✓	✓	✗
System boilers	✓	✓	✗	✓	✓	✗	✓	✓	✗	✓	✗
Combination boilers	✓	✓	✗	✓	✓	✗	✓	✓	✓	✓	✗
Cookers	✗	✓	✓	✗	✗	✓	✗	✗	✓	✗	✓

As can be seen from the table above, oil-fired appliances are available in a variety of different types, which generally use two different firing methods:

- pressure jet or atomising burners
- vaporising burners.

Pressure jet or atomising burners

A typical oil pressure jet burner

Pressure jet burners use an oil burner that mixes air and fuel. An electric motor drives a fuel pump and an air fan. The fuel pump forces the fuel through a fine nozzle, breaking the oil down into an oil mist. This is then mixed with air from the fan and ignited by a spark electrode. Once it is lit, the burner will continue to burn as long as there is a supply of air and fuel in the correct ratio.

Oil pressure jet-type boilers are installed on all modern oil-fired central heating systems including condensing system boilers, condensing combination boilers and wall-mounted types.

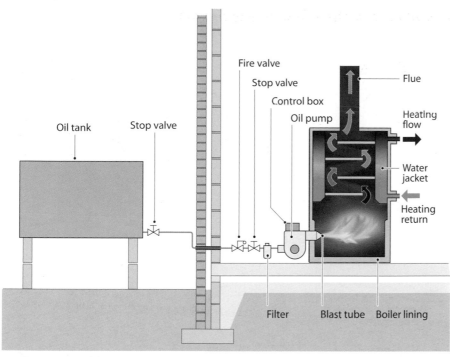

A pressure jet oil burner installation

A pressure jet oil burner combination boiler

Table 14: Advantages and disadvantages of pressure jet oil burner boilers

Advantages of pressure jet oil burner-type boilers	Disadvantages of pressure jet oil burner-type boilers
• Document L compliant • Very high efficiency • Sealed (pressurised) system gives better heating flow rates • Can be used with all modern fully pumped heating systems (system boilers) • No feed and expansion cistern required in the roof space (system and combination types) • Very good flow rate on hot water supply (condensing combination boilers)	• High maintenance compared with gas boilers • Noisy in operation • Needs an oil tank for fuel storage

Vaporising burners

Vaporising burners work on gravity oil feed. There is no pump. The oil flows to the burner, where a small oil heater warms the oil until vapour is given off, and it is the vapour that is then ignited by a small electrode. As the oil burns, vapour is continually produced, which keeps the burner alight.

They are generally only used in oil-fired cookers.

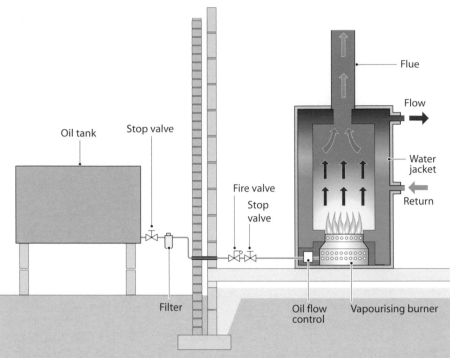

A vaporising oil burner cooker

A vaporising oil burner installation

Table 15: Advantages and disadvantages of vaporising oil burner boilers

Advantages of vaporising oil burner type boilers	Disadvantages of vaporising oil burner type boilers
• Very quiet in operation	• Very limited use (cookers only)

Flueing arrangements for boilers

All central heating appliances need a flue to remove the products of combustion safely to the outside. The basic intention is to produce an updraught, whether by natural means or by the use of a fan, to eject the fumes away from the building. There are two flue types:

- open flues
- room-sealed (balanced) flues.

Table 16: Flue types used on different appliances

	Open flue (natural draught)	Open flue (forced draught)	Room-sealed (natural draught)	Room-sealed (fan-assisted)
Solid fuel boilers	✓	✓	✗	✗
Gas boilers	✓	✓	✓	✓
Pressure jet oil burners	✗	✓	✗	✓
Vaporising oil burners	✓	✗	✗	✗

Open flues

The open flue is the simplest of all flues. Because heat rises, it relies on the heat of the flue gases to create an updraught. There are two different types:

- natural draught
- forced draught.

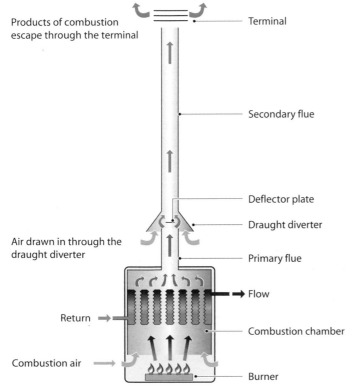

Products of combustion escape through the terminal

Terminal

Secondary flue

Deflector plate

Draught diverter

Air drawn in through the draught diverter

Primary flue

Flow

Return

Combustion chamber

Combustion air

Burner

The operation of an open flue

With a boiler with this type of flue, air for combustion is taken from the room in which the boiler is located. The products of combustion are removed by natural draught vertically to atmosphere, through a suitable terminal. The room must have a route, for combustion air, direct from the outside. This is usually supplied through an air brick on an outside wall. All natural draught open-flue appliances work in this way. The material from which the flue is made, however, will differ depending on the type of fuel used.

Occasionally, an open flue may be forced draught. This is where a purpose-designed fan is positioned either before the combustion chamber or close to the primary flue. The fan helps to create a positive updraught by blowing the products of combustion up the flue. Forced draught open flues are not suitable for all open-flue types and it will depend on the boiler manufacturer and the boiler/ flue design.

Room-sealed (balanced) flues

This boiler draws its air for combustion directly from the outside through the same flue assembly used to discharge the flue products. This boiler is inherently safer than an open-flue type, because there is no direct route for flue products to spill back into the room. There are two basic types:

- natural draught
- fan-assisted (forced draught).

Room-sealed natural draught

Natural draught room-sealed appliances have been around for many years, and there are still many thousands in existence. The basic principle is very simple – both the combustion air (fresh air in) and the products of combustion (flue gases out) are situated in the same position outside the building. The products of combustion are evacuated from the boiler through a duct that runs through the combustion air duct, one inside the other.

The boiler terminals were either square or rectangular and quite large in size. Terminal position was critical to avoid fumes going back into the building through windows and doors.

Room-sealed fan-assisted (forced draught)

Fan-assisted room-sealed appliances work in the same way as their natural draught cousins, with the products of combustion outlet positioned in the same place (generally) as the combustion air intake. However, there are two distinct differences.

- The process is aided by a fan, which ensures the positive and safe evacuation of all combustion products as well as any unburnt gas which may escape.

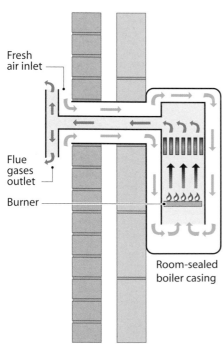

Fresh air inlet

Flue gases outlet

Burner

Room-sealed boiler casing

The operation of a natural draught room-sealed boiler

- The flue terminal is circular, much smaller and can be positioned in many more places than its predecessors.

There are two very different versions of the fan-assisted room-sealed boiler.

- The fan is positioned on the combustion products outlet from the heat exchanger. This creates a desirable negative pressure within the casing.

- The fan is positioned on the fresh air inlet blowing a mixture of gas and air to the burner. This creates a positive pressure within the boiler casing. Nearly all condensing boilers use this principle.

The reason for pump overrun on boilers

Boilers with a low water content use a pump overrun system to keep the pump running for a short period after the boiler has shut down. It is required to dissipate any latent heat build-up in the water in the heat exchanger, as this could trip the energy cut-out, resulting in boiler lock-out.

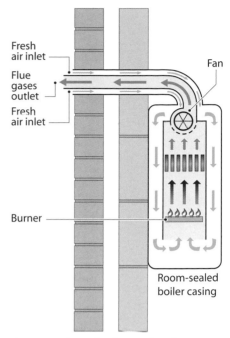

The operation of a fan-assisted room-sealed boiler

Understand the types of heat emitters used in underfloor heating systems (LO4)

There are five assessment criteria to this learning Outcome:

1 Justify selection of heat emitters used in plumbing systems.
2 Describe components required for installation with a range of heat emitters.
3 Describe the design considerations for underfloor heating.
4 Describe the advantages of underfloor heating systems.
5 Describe the operating principles of underfloor heating systems.

SmartScreen Unit 306
Presentation 4

Range	
Heat emitters	Panel radiators, column radiators, low surface temperature radiators, fan convectors, skirting convectors, towel warmers, towel radiators, underfloor heating
Design considerations	Room volume, retrofit, heat-up times, floor covering, floor matting, aesthetics, usable floor area

Range	
Components	Thermostatic radiator valves, manual valves, lockshield valves, combined lockshield and drain-off, blending valves, UFH manifold, UFH control units
Advantages	Lower operating temperature, warms from the floor up, even distribution of heat, operates effectively with condensing boilers, aesthetic appeal

Selecting heat emitters used in plumbing systems

So far in this unit, you have looked at heating systems and the appliances that drive them. Here, you will look at the methods of getting the heat into a room or dwelling. There are many different types of heat emitters available, including the following:

- panel radiators
- column radiators
- low-surface-temperature radiators (LSTs)
- fan convectors:
  - wall mounted
  - kick space
- tubular towel warmers
- towel warmers with integral panel radiators
- skirting heating.

Panel radiators

Modern panel convectors/radiators are designed to emit heat by convection and radiation – 70% of the heat is convected. They have fins (often called convectors) welded to the back of the radiator which swarm the cold air that passes through them, creating warm air currents that flow into the room. This dramatically improves the efficiency of the radiator. Steel radiators that do not have fins rely on radiant heat alone, and this leads to cold spots in the room. Positioning of the radiator is therefore critical. The radiator should be sited on a clear wall with no obstructions, such as a window sill, above it. If this is not possible, enough space should be left between the top of the radiator and the obstruction to allow the warm air to circulate. It is recommended by radiator manufacturers that radiators should be fitted at least 150mm above finished floor level (depending on the height of the skirting board) to allow air circulation.

The most common types of radiators are shown below.

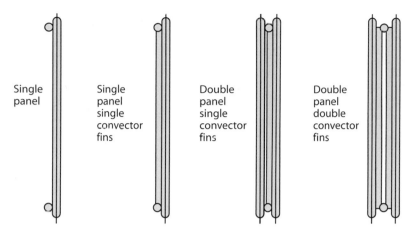

Single panel

Single panel single convector fins

Double panel single convector fins

Double panel double convector fins

The types of panel radiators

Seamed top radiator style

Manufacturers provide a wide range of radiator heights, from 300mm through to 900mm, and lengths from 400mm, increasing by 100/200mm increments through to 3m.

It is important that radiators are fixed according to the manufacturer's instructions if the best output performance is to be achieved. Outputs vary from manufacturer to manufacturer.

There are three different styles of radiator.

- **Seamed top** – this is a very common style of radiator, and was the market leader for many years. Top grilles and side panels are available for this radiator style.

- **Compact** – these have factory-fitted top grilles and side panels, making them a more attractive radiator style. These are currently the most popular radiator style available.

- **Rolled top** – these are the least popular of all radiator styles. They are somewhat old-fashioned-looking with exposed welded seams either side.

Compact radiator style

Domestic panel radiators have ½in BSP female threads at either side and top and bottom, and these will accept a variety of radiator valves. One end of the radiator has an air release valve, with the other end blanked by the use of a plug. These are usually supplied by the radiator manufacturer.

Column radiators

Column radiators (often known as hospital or church radiators) have been available for many years. As the name suggests, they are made up of columns; the more columns the radiator has, the better the heat output. They are increasingly being used with modern heating systems, especially on period refurbishments.

Rolled top radiator style

Column radiators can be made from three different metals, these being traditional cast-iron, steel and aluminium with many modern column radiator designs now being produced by a variety of manufacturers.

Modern column radiators

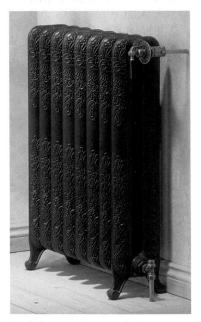

Traditional column radiators

Low-surface-temperature radiators (LSTs)

Low-surface-temperature radiators were specifically designed to conform to the NHS Estates guidance note 'Safe hot water and surface temperature', which states that: 'Heating devices should not exceed 43°C when the system is running at maximum design output.'

This has been adopted not only by the NHS but also local authorities and commercial buildings installations to which the general public may have access, including residential care homes and schools. LSTs are also becoming popular in domestic installations, especially in children's bedrooms and nurseries and where the elderly, infirm or disabled are likely to come into contact with radiators.

Low-surface-temperature radiators

Fan convectors

Fan convectors work on the same basic principle as traditional finned radiators. A finned copper heat exchanger is housed in a casing which also contains a low-volume electrically operated fan. As the heat exchanger becomes hot, a thermostat operates the fan and the warm air is blown into the room. Because the warm air is forced into the room, more heat can be extracted from the hot, circulating water. Once the desired temperature has been reached, the fan is again switched off by the thermostat.

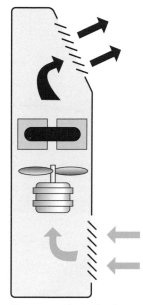

The operation of a fan convector

Fan convectors tend to be larger than traditional radiators and they also require a mains electric connection, usually via a switched fuse spur. There are two separate types of fan convector.

Wall-mounted type

These tend to be quite large in size. The manufacturer's data should be consulted to allow the correct heat output to be selected.

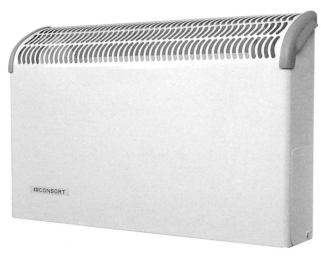

A wall-mounted fan convector

Kick-space type

These are specifically designed for kitchen use where space to mount a radiator is limited. They are installed under a kitchen unit and blow warm air via a grille mounted on the kick plinth.

A kick-space fan convector

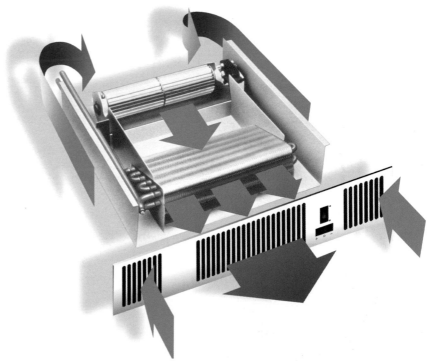

The working principles of a kick-space heater

Tubular towel warmers

These are available in a range of different designs and colours, and are often referred to as designer towel rails. They can be supplied for use with wet central heating systems with an electrical element option, for use during the summer when the heating system is not required. They are usually mounted vertically on the wall and can be installed in bathrooms and kitchens.

Towel warmers with integral panel radiators

Less popular than tubular towel rails, these heat emitters combine a towel rail and radiator in one unit. They allow a towel to be warmed without affecting the convection current from the radiator. They are generally only installed in bathrooms.

Skirting heating

Skirting heating consists of a finned copper tubular heat exchanger in a metal casing that replaces the skirting boards in a room. This system is usually used where unobtrusive heat emitters are required. Skirting heating can be used as perimeter heating below glazing or for background heat in some areas.

The heat output, at 450 watts per metre, is quite low which means that, to be effective, the skirting heating would need to be at floor level on all walls of the room to offset the room heat losses. The heat coverage is very similar to that experienced with specialist underfloor heating.

Another disadvantage is that efficiency is reduced by dust collecting in the fins.

A tubular towel rail

A towel rail with integral panel radiator

Skirting heating

Components required for installation with a range of heat emitters

This part of the unit looks at the most common mechanical controls used on domestic central heating systems:

- thermostatic radiator valves
- wheel-head radiator valves
- lockshield radiator valves
- automatic air vents
- automatic bypass valves
- anti-gravity valves.

Thermostatic radiator valves

These control the temperature of the room by controlling the flow of water through the radiator. They react to air temperature. TRVs have a heat-sensitive head that contains a cartridge, which is filled with a liquid, a gas or a wax, and this expands and contracts with heat. As the room heats up, the wax/gas/liquid cartridge expands and pushes down on a pin on the valve body. The pin closes and opens the valve as the room heats up or cools down. The valve head has a number of temperature settings to allow a range of room temperatures to be selected. Document L1 of the Building Regulations requires that thermostatic radiator valves are installed on new installations to control individual room temperatures, and on all radiators except the radiator where the room thermostat is fitted.

Most TRVs are bi-directional. This means that the can be fitted on either the flow or the return.

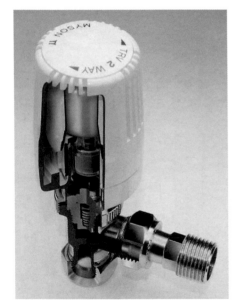

A thermostatic radiator valve

Wheel-head radiator valves

These allow manual control of the radiator and can be turned on or off. The valve is turned on by rotating the wheel head anticlockwise, and turned off by rotating it clockwise.

Lockshield radiator valves

These are designed to be operated only by a plumber – not by the householder. They are adjusted during system balancing to regulate the flow of water through the radiator. The lockshield head covers the valve mechanism, and it can be turned off for radiator removal. Lockshield valves are also available incorporating a drain-off valve.

Automatic air vents

Automatic air vents are fitted where air is expected to collect in the system, usually at high points. They allow the collected air to escape from the system but seal themselves when water arrives at the valve.

A wheel-head radiator valve with lockshield cover

THE CITY & GUILDS TEXTBOOK

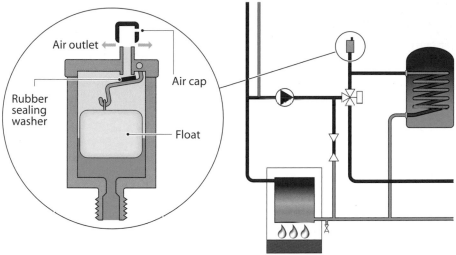

An automatic air valve (AAV)

When water reaches the valve, the float arm rises, closing the valve. As more air reaches the valve the float momentarily drops, allowing the air out of the system. These valves are often used with a check valve that prevents air from being drawn into the system backwards through the valve.

Automatic bypass valves

The automatic bypass valve controls the flow of water across the flow and return circuit of fully pumped heating systems by opening automatically as other paths for the water close, such as circuits with motorised valves and radiator circuits with thermostatic radiator valves. This occurs as the hot water circuit and heating circuit/ thermostatic radiator valves begin to reach their full temperature. As the circuits close, the bypass will gradually open, maintaining circulation through the boiler and reducing noise In the system that is due to water **velocity**. Most boiler manufacturers require a bypass to be fitted to maintain a minimum flow rate through the boiler to prevent overheating.

Velocity

The measurement of the rate at which an object changes its position.

An automatic bypass valve

Automatic bypass valves are much better than fixed bypass valves as, being permanently open, they take the flow of hot water away from the critical parts of the system, which increases the heating time for both hot water and heating circuits. This reduces the efficiency of the system and increases fuel usage.

Anti-gravity valves

Anti-gravity valves prevent unwanted gravity circulation to the upstairs radiators on semi-gravity systems when only the hot water is being heated. They are essential on all semi-gravity systems, and especially in those systems fuelled by solid fuel. Anti-gravity valves should be positioned on the vertical flow to the upstairs heating circuit.

Anti-gravity valves are very similar in design to the single-check valves mentioned in earlier chapters. They only allow water flow in one direction, and when the heating system is off they are in the closed position. In this position, gravity circulation cannot take place. As soon as the central heating circulating pump switches on, the flow of the water opens the valve to allow heating circulation.

Design considerations for underfloor heating

SmartScreen Unit 306

Handout 6 and Worksheet 4

An underfloor heating system provides invisible warmth and creates a uniform heat, eliminating cold spots and hot areas. The temperature of the floor needs to be high enough to warm the room without being uncomfortable underfoot. There is no need for unsightly radiators/convectors because the heat comes from the ground up. Underfloor heating creates a low-temperature heat source that is spread over the entire floor surface area. The key here is the low temperature. Whereas most wet central heating systems containing radiators and convectors operate at around 70–80°C, underfloor heating operates at a much lower temperatures, making this system ideal for air and ground source heat pumps. Typical temperatures are:

- 40–45°C for concrete (screeded) floors
- 50–60°C for timber floor constructions.

Traditional wet central heating systems generate convection currents and radiated heat. Around 20% of the heat is radiated from the hot surface of the radiators, and if furniture is placed in front of the radiator, the radiation emission is reduced. Meanwhile 80% of the heat is in the form of convection currents, or hot air rising. This adds up to a very warm ceiling! Underfloor heating systems, however, rely on both conduction and radiation. The heat from the underfloor heating system conducts through the floor, warming the floor structure and making the floor surface essentially a large storage heater. The heat is then released into the room as radiated heat. Around 50–60% of the heat emission is in the form of radiation, providing a much more comfortable temperature at low room levels when compared with a traditional wet system with radiators. With the whole floor being heated, furniture positioning is no longer a problem because, as the items of furniture gain heat, they too emit warmth.

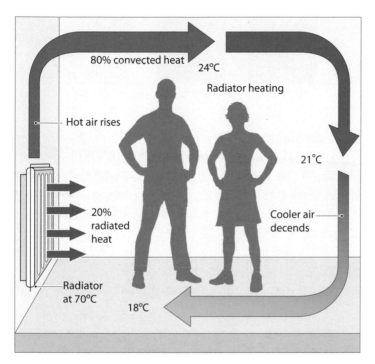

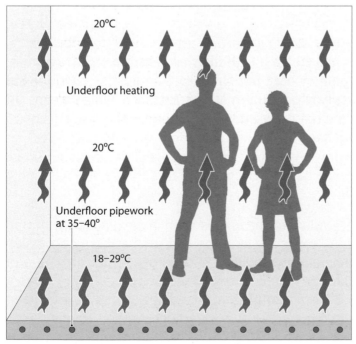

The principle of underfloor heating

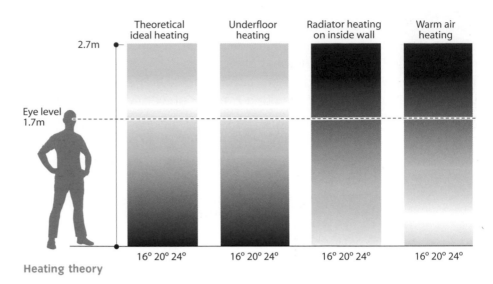

Heating theory

During the design stage, the pipe coils are fixed at specific centres depending on the heat requirement of the room and the heat emission (in watts) per metre of the pipe. The whole floor is then covered with a screed to a specific depth, creating a large thermal storage heat emitter. The water in the pipework circulates from and to a central manifold and the heats the floor. The heat is then released into the room at a steady rate. Once the room has reached the desired temperature, a room thermostat actuates a motorised head on the return manifold and closes the circuit to the room.

Such is the nature of underfloor heating that many fuel types can be used, with some utilising environmentally friendly technology. Gas- and oil-fired boilers are common, as are biomass fuels, solar panels and heat pumps.

Floor coverings are an important aspect for underfloor heating. Some floor coverings create a high thermal resistivity, making it difficult for the heat to permeate through. Carpet underlays and some carpets have particularly poor thermal transmittance, which means the heat is kept in and not released. Thermal resistivity of carpets and floor coverings is described by a TOG rating. The higher the TOG rating, the less heat will get through. Floor coverings used with underfloor heating should have a TOG rating of less than 1 and must never exceed 2.5.

Quite often underfloor heating is used in conjunction with traditional wet radiators, especially in properties such as barn conversions. The higher temperatures required for radiators do not present a problem because the flow water for the underfloor system is blended with the return water via a thermostatic blending valve to maintain a steady temperature required for the underfloor system. Zoning the upstairs and downstairs circuits with two-port motorised zone valves and independent time controls for the heat emitters also helps.

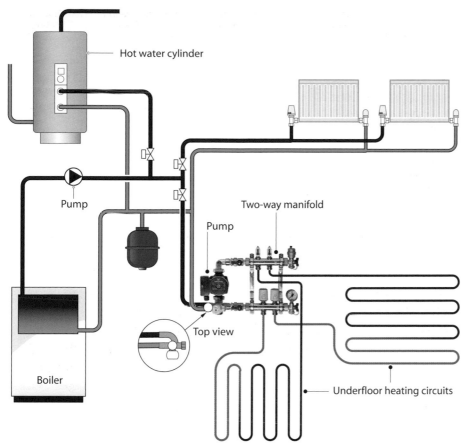

Hot water cylinder

Pump

Two-way manifold

Pump

Top view

Boiler

Underfloor heating circuits

A typical underfloor heating system combined with wet radiators

The advantages of underfloor heating systems

The advantages and disadvantages of underfloor heating systems are described in Table 17.

Table 17: The advantages and disadvantages of underfloor heating

Advantages	Disadvantages
• The pipework is hidden underfloor. This allows better positioning of furniture and interior design • The heat is uniform, giving a much better heat distribution than traditional systems • These systems are very energy efficient, with low running costs	• Not very suitable for existing properties unless a full renovation means the removal of floor surfaces • Can be expensive to install when compared with more traditional systems • Heat-up time is longer, as the floor will need to get to full temperature before releasing heat

Advantages	Disadvantages
<ul><li>Environmentally friendly fuels can be used</li><li>Underfloor heating is almost silent, with low noise levels when compared with other systems</li><li>Cleaner operating with little dust carried on convector currents. This can help people who suffer from allergies, asthma and other breathing problems</li><li>System maintenance is low and decorating becomes easier as there are no radiators to drain and remove</li><li>Individual and accurate room temperatures as every room has its own room thermostat that senses air temperature</li><li>Less possibility of leaks</li><li>Greater safety, as there are no hot surfaces that can burn the elderly and infirm</li></ul>	<ul><li>Slower cool-down temperatures means the floors may still be warm when heat is not required</li><li>Greater installation time</li><li>More electrical installation of controls is required, as each room will need its own room thermostat and associated wiring</li></ul>

The layout features of underfloor heating

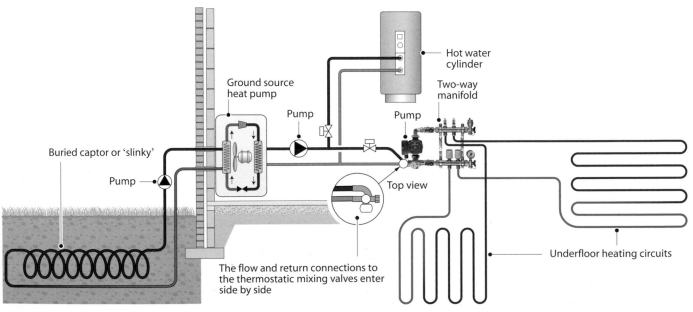

A typical underfloor heating system using a ground source heat pump

Underfloor heating uses a system of continuous pipework, laid under a concrete or timber floor in a particular pattern and at set centre-to-centre pipe distances. Each room served by an underfloor heating system is connected at a central location to a flow and return manifold, which regulates the flow through each circuit. The manifold is connected to flow and return pipework from a central heat source, such as a boiler or a heat pump.

The manifold arrangement also contains a thermostatic mixing valve to control the water to the low temperatures required by the system and an independent pump to circulate the water through every circuit.

Each underfloor heating circuit is individually controlled by a room thermostat, which activates a motorised head on the return manifold to precisely control the heat to the room to suit the needs of the individual.

The operating principles of underfloor heating systems

As you have seen, underfloor heating works by distributing heat via a series of pipes laid under the floor of a room. Certain components are required to do this and ensure that the system warms the room. However, these components must be controlled in such a way that they maintain a steady flow of heat while ensuring that the floor does not become too hot to walk on. This is achieved by the use of:

- manifolds
- a thermostatic mixing (blending) valve
- a circulating pump
- various pipework arrangements to suit the floor and its coverings
- the application of system controls – time and temperature to space heating zones.

Manifolds

In technical terms, the **manifold** is designed to minimise the amount of uncontrolled heat energy from the underfloor pipework. The manifold is at the centre of an underfloor heating system. It is the distribution point where water from the heat source is distributed to all of the individual room circuits and as such, should be positioned as centrally as possible in the property. Room temperature is maintained via thermostatic motorised actuators on the return manifold, while the correct flow rate through each coil is balanced via the flow meters on the flow manifold. Both the flow and return manifolds contain isolation valves for maintenance activities, an automatic air valve to prevent airlocks and a temperature gauge so that the return temperature can be monitored.

Manifold

A manifold, in systems for moving fluids or gases, is a junction of pipes or channels, typically bringing one into many or many into one.

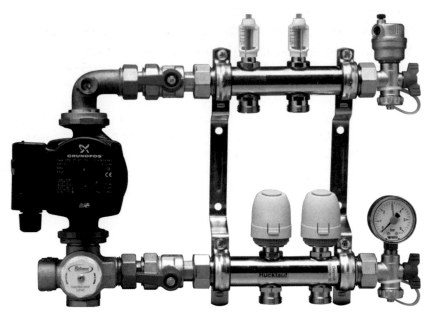

A typical underfloor heating manifold

Most manifolds contain a circulating pump and a thermostatic mixing valve, often called a blending valve.

Thermostatic mixing (blending) valve

The thermostatic mixing or blending valve is designed to mix the flow and return water from the heat source to the required temperature for the underfloor heating circuits. These valves are available in many different formats, with the most common being as part of the circulating pump module shown in the photo to the left. The temperature of the water is variable by the use of an adjustable thermostatic cartridge inside the valve.

Circulating pump

The circulating pump is situated between the thermostatic mixing valve and the flow manifold to circulate the blended water through every circuit. Most models are variable speed.

Underfloor heating pipework arrangements

The success of the underfloor heating system depends on the installation of the underfloor pipework and the floor pattern installed. There are many variations of pipe patterns based on two main pattern types:

- series pattern
- snail pattern.

In general, underfloor heating pipes should not be laid under kitchen or utility room units.

Underfloor heating circulating pump/ blending valve module

Series pattern

The series pattern (also known as the meander pattern) is designed to ensure an even temperature across the floor, especially in systems incorporating long pipework runs. It is often used in areas of high heat loss.

The flow pipe must be directed towards any windows or the coldest part of the room before returning backwards and forwards across the room at the defined pipe spacing centres.

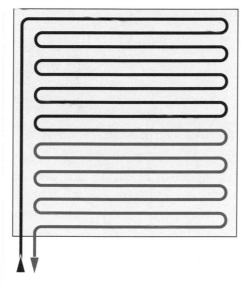

The series pattern

Snail pattern

The snail pattern (also known as the bifilar pattern) is used where an even, uniform temperature is required, such as under hardwood floors or vinyl floor tiles.

The flow pipe runs in ever decreasing circles until the centre of the room is reached. It then reverses direction and returns with parallel runs back to the starting point.

The snail pattern

Application of system controls

The number of homes that require both time and temperature zone control has increased in recent years. In 2006, a survey showed that the average floor area of a domestic property with four bedrooms was around 157m^2 and over 200m^2 for a five-bedroomed domestic property. With properties of this size zoning becomes a necessity, and in 2006 Document L1A/B of the Building Regulations requested that zoning of the heating system must be installed in all properties of 150m^2 or more. This was updated in 2010 to include any property.

In most instances zoning requires the separating of the upstairs circuit from the downstairs or, in the case of single-storey dwellings, separating the living space from the rest of the property. Having separate time and temperature control of the individual circuits is a necessity.

Zoning with separate temperature control

Having separate temperature-controlled zones provides a much better living environment, because different parts of the dwelling can be maintained at different temperatures without relying on a single room to dictate the temperature across the whole system. Lower temperatures can be maintained in rooms within the dwelling that are not occupied, allowing the dwelling to take full advantage of any solar gains, especially in rooms that face south, southeast or southwest. This can be quite pronounced, even in the winter sun. Significant savings on both energy usage and fuel costs can be made by simply taking advantage of the free heat that the Sun can provide. Outside sensors linked to weather compensators and delayed start and optimum start controls help to further reduce energy usage and cost.

Zoning with separate time control

Zoning with separate time control offers another dimension to the concept of zoning by allowing the heating to be controlled at different times of the day in different zones. The heat can be focused in those rooms that are occupied throughout the day, with the heating to other parts of the dwelling timed to come on in the early morning and evening. Separate zones reduce energy usage and costs while maintaining improved comfort levels throughout the property.

Zoning in practice

Zoning is required by Approved Document L1A/B of the Building Regulations 2010, and the installer must make decisions on the best way to arrange those zones to take the best advantage of energy savings while complying with the wishes of the customer/end user and the regulations. The only way that this can be achieved is by talking to the customer and finding out their usage patterns. The main aim of zoning is to avoid overheating areas that require less heat to maintain the desired warmth, or because the set point could be lower than in other areas. The point here is that the number of zones laid down by Document L is the minimum, and there are real benefits to adding additional zones in key areas of the property.

An underfloor system lends itself naturally to zoning, as each room is controlled by a room thermostat, which activates actuators on the individual circuits at the manifold. Further controls can be added where heat emitters and underfloor heating circuits are installed on the same system. In this case, the zones are both individually temperature controlled and timed. The system can also be linked to other advanced controls such as night setback and delayed start.

The choice of controls for the zones should be decided by the predicted activity in those zones. There are many options that can be used individually or collectively to achieve a good system control:

- having individual temperature and timing controls in every zone
- using multi-channel programmers that allow the timing of individual rooms or multiple zones to be set from a single point. This is often more desirable than many individual programmers at different locations within the dwelling
- using thermostatic radiator valves (TRVs) to vary the heat output of individual heat emitters. This can be beneficial where solar gain adds to the room temperature, as they are very fast-reacting in most circumstances. Some TRVs also have electronically timed thermostatic heads which can be linked to a wireless programmer.

Zoning can help make significant energy savings. It allows the optimisation of the heating system, maintaining the dwelling at a comfortable temperature and saving money at the same time.

Positioning components in underfloor central heating systems

For an underfloor heating system to work effectively, the components require careful positioning to ensure that the efficiency of the system is maintained. All too often, systems fail to live up to their potential because of poor positioning of key components.

Manifold position

The longer the circuit, the more energy is needed to push the water around it. Water will always take the path of least resistance, and shorter circuits will always be served first. In many instances, balancing the system will help even out the circulation times so that all circuits receive the heat at the same time, but the system will only be as good as the slowest circuit. If the longest circuit is slow, then once the system is balanced all the circuits will be slow. For this reason the manifold should be positioned centrally within the dwelling. Even if the short circuits become longer, balancing the system is relatively easy.

The manifold installed

A problem that may occur where the manifold is located is that the area may become a potential 'hot spot' on the system because of pipework congestion around the manifold. This can be prevented by insulating the pipework around the manifold until the point at which the pipework enters the room it is serving.

Pipework arrangements (cabling)

There are many variations on the two basic layouts. The pattern should be set out in accordance with the orientation and shape of the room. Window areas may be colder and may require the bulk of the heat in that area. Other considerations include the type of floor construction and the floor coverings. The pipework should be laid in one continuous length without joints. In some instances, the pipe is delivered on a continuous drum of up to 100m to enable large areas to be covered without the need for joints. Large rooms may require more than one zone and the manufacturer's instructions should be checked for maximum floor coverage per zone.

The series pattern laid out

The snail pattern laid out

Pipework installation techniques

There are different installation techniques depending on the type of floor.

- **Solid floors** – there are many types of underfloor heating installation techniques for solid floors. The diagram shows one of the more common types, which uses a plastic grid. The underfloor heating pipe is simply walked into the pre-made castellated grooves for a precise centre-to-centre guide for the pipework using a minimum radius bend.

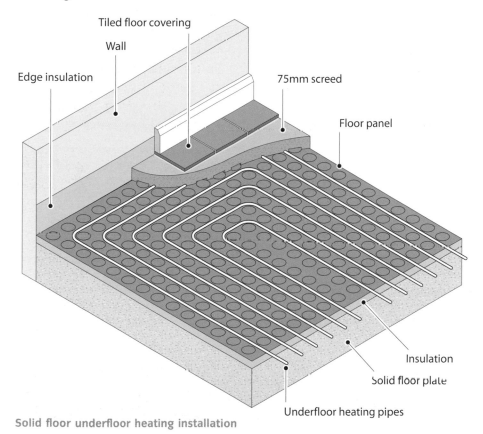

Solid floor underfloor heating installation

Solid floor underfloor heating installation method

The panels are laid onto pre-installed sheets of insulation to ensure good performance and minimal downwards heat loss. Edge insulation is required to allow for expansion of the panels.

Table 18: Key design and installation information for solid floors

Maximum heat output	Approx. 100W/m^2
Recommended design flow temperature	50°C
Maximum circuit length	100m (15mm pipe) 120m (18mm pipe)
Maximum coverage per circuit	12m^2 @ 100mm centres 22m^2 @ 200mm centres 30m^2 @ 300mm centres (18mm pipe only)
Material requirements	
Pipe	8.2m/m^2 @ 100mm centres 4.5m/m^2 @ 200mm centres 3.3m/m^2 @ 300mm centres (18mm pipe only)
Floor-plate usage	1 plate/m^2 allowing for cutting
Edging insulation strip	1.1m/m^2
Conduit pipe	2m/circuit

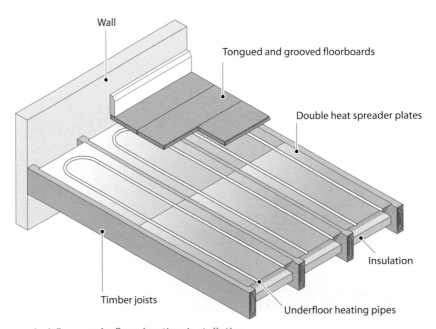

Suspended floor underfloor heating installation

- **Suspended timber floor** – this system is designed for use under timber suspended floors. It uses aluminium double heat spreader plates to transmit heat evenly across the finished floor surface.

This system is suitable for any timber suspended floor with joist widths up to 450mm. The heat plates are simply fixed to the joists using small flat-headed nails or staples. A layer of insulation must be placed below the plates to prevent the heat penetrating downwards.

Where the pipework must cross the joists, the joists must be drilled in accordance with the Building Regulations.

Suspended floor underfloor heating installation method

Table 19: Key design and installation information for suspended floors

Maximum heat output	Approx. 70W/m²
Recommended design flow temperature	60°C
Maximum circuit length	80m (15mm pipe)
Maximum coverage per circuit using a double spreader plate	17m² @ 225mm centres
Material requirements	
Pipe	4.5m/m² @ 100mm centres
Heat spreader plates	2 plates/m²

■ **Floating floor** – this system is designed for use where a solid floor installation is not suitable due to structural limitations. It can be installed directly onto finished concrete or timber floors.

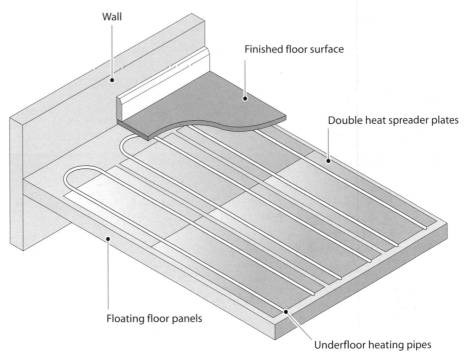

Floating floor underfloor heating installation

■ The pipework is laid on top of 50mm-thick polystyrene panels, each having a thermal transmittance of 0.036Wm^{2}K. The insulation has pre-formed grooves that the pipe clips into after the heat spreader plates have been fitted. The insulation is not fixed and 'floats' on the top of the sub-floor. The finished flooring can then be laid directly onto the top of the pipework, completing the 'floating' structure.

Table 20: Key design and installation information for floating floors

Maximum heat output	Approx. 70W/m^2
Recommended design flow temperature	60°C
Maximum circuit length	80m (15mm pipe) 100mm (18mm pipe)
Maximum coverage per circuit using a double spreader plate	28.5m^2 @ 300mm centres (15mm pipe) 30m^2 @ 300mm centres (18mm pipe)

Material requirements

Pipe	3.1m/m² @ 300mm centres
Single heat spreader plates	3 plates/m²
Floating floor panel	1 panel/1.4m²

Floating floor underfloor heating installation method

Know how to de-commission, commission, and fault-find on central heating systems (LO5)

There are nine assessment criteria to this Outcome:

1 Interpret information required when testing, commissioning and fault-finding on central heating systems.

2 Describe means of safeguarding customer property.

3 Explain the procedure for de-commissioning a central heating system.

4 Explain the use of bespoke tools in commissioning a central heating system.

5 Describe the procedure for commissioning a central heating system.

6 Describe the procedure for commissioning an underfloor heating system.

7 Identify additives used in central heating systems.

8 Describe the procedure for power flushing a central heating system.

9 Describe how to rectify faults in central heating systems.

Range	
Information	Manufacturers' instructions, component instructions, central heating design guide, component instructions, fault flow charts, helpline, internet
Safeguarding	Dust sheets, warning notices, appropriate clothing, move delicate items
Procedure for de-commissioning	Isolate power, isolate incoming water supply, cap off, label, inform relevant people, attach hose, drain, vent, ensure system is fully drained
Bespoke tools	Strap-on thermostats, infrared thermometer, vent key
Procedure for commissioning a central heating system	Remove thermostatic heads, cold flush, fill, vent, heat up system, hot flush, re-fill and add inhibitor, balance, adjust (as required) bypass, ensure operation of boiler interlock, hand over to customer, complete documentation
Procedure for commissioning an underfloor heating system	Connect to mains water supply, connect second hose to drain, fill, vent, balance each UFH zone
Additives	Inhibitor, cleansing agent, restoring agent, anti-freeze, leak sealers
Procedure for power flushing	Safeguard property, connect flushing machine between pump, boiler or heat emitter, flush system clear, reconnect, refill with additives
Common faults	Pumping over, persistent venting, micro air leaks, radiators not getting warm at the top, cold spots, stuck TRVs, motorised valves not operating, heat when no demand, component failure, leaks

Interpret information when testing, commissioning and fault-finding on central heating systems

Inadequate commissioning, system flushing and maintenance operations can inflict damage on even the best-designed and well-installed central heating system. Building debris and swarf (pipe filings) can easily block pipes, and these can also promote bacteriological growth. In addition, excess flux used during the installation can cause corrosion and may lead to system leakage and component failure.

It is obvious, then, that correct commissioning procedures must be followed if the problems stated are to be avoided. The following documents must be consulted:

- Water Supply (Water Fittings) Regulations 1999
- British Standards BS EN 12828, BS EN 12831, BS EN 14336 and BS 7593
- Building Regulations Approved Document L1A/B
- Domestic Building Services Compliance Guide
- manufacturers' instructions for any equipment and appliances.

Water Supply (Water Fittings) Regulations 1999

These are the national requirements for the design, installation, testing and maintenance of cold and hot water systems in England and Wales (Scotland has its own, almost identical, Scottish Water Byelaws 2004). Their purpose is to prevent contamination, wastage, misuse, undue consumption and erroneous metering of the water supply used for domestic purposes. Schedule 2 of the regulations states that:

'The whole installation should be appropriately pressure tested, details of which can be found in the Water Regulations Guide (Section 4: Guidance clauses G12.1 to G12.3). This requires that a pressure test of 1.5 times the maximum operating pressure for the installation or any relevant part.'

This document should be read in conjunction with the British Standards.

British Standards

The main British Standards for the design, installation, commissioning and testing of central heating systems are:

- **BS EN 12828:2003** – Heating systems in buildings – Design for water-based heating systems
- **BS EN 1283:2003** – Heating systems in buildings – Method for calculation of the design heat load

- **BS EN 14336:2004** – Heating systems in buildings – Installation and commissioning of water-based heating systems
- **BS 7593:2006** – Code of practice for treatment of water in domestic hot water central heating systems. BS 7593 gives guidance on the potential problems and the remedies required to maintain system efficiency in line with Approved Document L1, thereby increasing the lifespan of hot water central heating systems.

For commissioning and testing of systems, the key British Standards are BS EN 14336 and BS 7593. The following sections are of relevance:

- **Section 5 Pre-commissioning checks**
  - 5.1 Objective
  - 5.2 State of the system
  - 5.3 Water tightness test
  - 5.4 Pressure test
  - 5.5 System flushing and cleaning
  - 5.6 System filling and venting
  - 5.7 Frost precautions
  - 5.8 Operational checks
  - 5.9 Static completion records
- **Section 6 Setting to work**
- **Section 7 Balancing water flow rates**
- **Section 8 Adjusting of controls**
- **Section 9 Handover**
  - 9.1 Objective
  - 9.2 Documents for operation, maintenance and use
  - 9.3 Instructions on operation and use
  - 9.4 Handover documentation
- **Annex A: Guide to good practice for water tightness test**
- **Annex B: Guide to good practice for pressure testing**
- **Annex C: Guide to good practice for system flushing and cleaning**
- **Annex D: Guide to good practice for operational tests**
- **Annex E: Guide to good practice for static completion**
- **Annex F: Guide to good practice for setting to work**
- **Annex G: Guide to good practice for balancing water flow rates**
- **Annex H: Guide to good practice for setting of control systems**

These will be looked at a later in the unit.

Building Regulations

The Building Regulations make reference to commissioning of central heating systems. These are mentioned briefly in Approved Document L1A/B. Here it states that:

(2) Where this regulation applies the person carrying out the work shall, for the purpose of ensuring compliance with paragraph F1(2) or L1(b) of Schedule 1, give to the local authority a notice confirming that the fixed building services have been commissioned in accordance with a procedure approved by the Secretary of State. L1(b) of Schedule 1, give to the local authority a notice confirming that the fixed building services have been commissioned in accordance with a procedure approved by the Secretary of State.

(3) The notice shall be given to the local authority:
- Not later than the date on which the notice required by regulation 15(4) is required to be given; or
- Where that regulation does not apply, not more than 30 days after completion of the work.

Domestic Building Services Compliance Guide

The Domestic Building Services Compliance Guide gives specific instructions regarding the commissioning of central heating systems. In Section 2 of the guide, it states that:

On completion of the installation of a boiler or hot water storage system, together with associated equipment such as pipework, pumps and controls, the equipment should be commissioned in accordance with the manufacturer's instructions. These instructions will be specific to the particular boiler or hot water storage system.

The installer should give a full explanation of the system and its operation to the user, including the manufacturer's user manual where provided.

Manufacturers' instructions

Where appliances and equipment are installed on a system, the manufacturer's instructions are a key document when undertaking testing and commissioning procedures, and it is important that these are used correctly for both installation and commissioning operations. Only the manufacturer will know the correct procedures that should be used to safely put their equipment into operation so that it performs to its maximum specification. Remember the following.

- Always read the instructions before operations begin.
- Always follow the procedures in the correct order.

- Always hand the instructions over to the customer on completion.
- Failure to follow the instructions may invalidate the manufacturer's warranty.

Safeguarding customer property

A new-build house and an occupied dwelling require two completely different styles of working when on site. While many of the working practices we use on site can be used in an occupied dwelling, the overriding emphasis in this case is care and attention to detail.

There are three concerns when working in an occupied dwelling:

- protecting the customer's property
- protecting the building fabric
- installing equipment in accordance with the customer's wishes while maintaining the quality of the installation according to the regulations in place.

There have been many instances where a good installation has been marred by carelessness by the plumber, and failing to liaise with the customer. This can inevitably result in disputes, withholding of money owed and, occasionally, court action.

While many companies have their own style of working, many employ plumbers for specific tasks – consider for example, operatives who work on site and those who work in private houses.

Working in private houses

Many customers complain about the lack of information given to them. In many cases, this is down to poor customer liaison. Before an installation takes place, ensure the following.

- Make sure the customer knows what day and time you will be arriving; agree a start time with the customer and stick to it. Early-morning arrivals are not always welcome.
- Walk around the house with the customer. Point out any existing damage to furniture, fixtures, carpets and wall coverings. This will prevent any misunderstandings over damage and marks already in place.
- Point out which carpets and pieces of furniture or delicate items will need to be removed before you begin work, and ask the customer to remove them. If you are going to be working outside, politely ask the customer to move their car before you begin work so that it does not get damaged.
- Cover all furniture, carpets and fixtures that cannot be removed in the area where you are going to work with dust sheets.

- Agree with the customer the position of radiators, boilers and all visible pipework before work begins.

- Keep the customer informed about any problems that arise that may require a decision from them.

- Let the customer know when any of the services (water, gas or electricity) or appliances (such as the boiler) are going to be turned off or taken out of service, and ensure that they have collected enough water for the period of temporary de-commission. If working on a central heating system, ensure they have access to other forms of heat, especially during cold weather.

- Position warning notices at points of isolation, ie stop taps and isolation valves, so that the customer knows not to turn these on.

- Always wear appropriate clothing and footwear when working in a customer's dwelling.

Protection of the building fabric and its surroundings

You have seen how a customer's personal belongings should be protected from dust and damage from the installation process, but there are other ways that we can protect the building and its surroundings.

- When soldering is taking place in the building, the risk of fire is ever present. To protect the building fabric, a heatproof soldering mat should be used. Remember, though, that this will not provide protection if the flame is directly on the mat. A certain amount of angle should be applied to the blowtorch, if possible, to deflect the heat away from the wall/floor/ceiling/skirting board. There are three different types of mat available that will resist temperatures of 600°C, 1,000°C and 1,300°C.

  One other way you can protect against heat is to use heat-dissipating spray gel. It offers protection against the scorching of wallpaper and paint and loosening of existing joints, and it also reduces the risk of fire by protecting surfaces and dissipating heat.

- When working outside the building, protect the customer's garden by the use of walk boards across flower beds and protective sheeting across grass lawns. Do not dig ladders into lawns.

- When drilling walls, to prevent blowing the surface of the backside of the wall you are drilling, first drill a small pilot hole and drill from both sides. This will ensure that the wall surface around the hole is not damaged.

- Before drilling a wall, check it first with a cable/pipe detector to ensure that there are no services already in the wall.

■ When removing old equipment, where there is a risk of spillage of dirty water, for example from removing old radiators, to prevent the spillage, turn the radiator upside down so that the valve tails are at the top. Always remember to protect carpets that cannot be removed during simple maintenance operations such as these.

De-commissioning a central heating system

Occasionally, systems will require isolation for repairs, renewal of appliances and extensions to systems or when systems or appliances are being permanently removed. This is known as de-commissioning. De-commissioning takes two forms:

■ **Temporary de-commissioning** – this is where systems are isolated for a period of time so that work can be performed. Eventually the system will be re-commissioned and put back into normal operation.

■ **Permanent de-commissioning** – when a system or an appliance is taken out of use, it has to be permanently de-commissioned. This will require that the system is isolated and drained, the appliance(s) removed and the pipework cut back, removed and capped off.

De-commissioning central heating systems follows much the same process as de-commissioning other systems you have looked at. There are a number of scenarios in which systems would need to be de-commissioned:

■ where the system is being completely stripped out prior to a new system installation or where the building is being demolished

■ where the boiler is being replaced and the feed and expansion cistern is being taken out

■ where the system is being added to or altered

■ where system components such as radiators are being permanently taken out

■ for general maintenance activities such as:
  ● pump replacement
  ● radiator replacement
  ● replacement of valves and other controls.

Always remember to:

■ keep the customer and/or other trades informed of the work being carried out, ie when the system is being isolated and the expected length of time for which it will be out of service

■ ensure that any services, such as electricity, gas, etc are safely isolated and pipework capped

- use warning notices, such as 'Do not use' or 'System drained' on any taps, valves, appliances, electrical components and so on.

Bespoke commissioning tools

Bespoke tools used in commissioning a central heating system include the following:

- **strap-on thermostats** – used to check the flow and return temperatures on radiators and boilers during the commissioning and benchmarking processes
- **infrared thermometers** – used to check the temperatures of hot water and heat emitters
- **vent keys** – small keys usually made of brass or aluminium, with a square key slot, used to open the air valves on radiators during the system filling process.

Commissioning a central heating system

Before soundness testing a central heating system, visual inspections of the installation should take place. BS EN 14336 states that:

- all plant items must be in accordance with the design, drawings, specifications and, where applicable, the manufacturers' instructions
- correct installation procedures must be followed
- the standards of installation must be met
- there should be availability of a fuel supply and the flue gas removal system should be correctly installed.

The visual inspection should include:

- Walking around the installation. Check that you are happy that the installation is correct and meets installation standards.
- Check that the power supply and fuel supply (gas, oil, etc) is off and cannot be inadvertently switched on. If necessary, place warning notices at key isolation points.
- Check that all radiator valves and air release valves are closed.
- Check that the drain-off valves for the system are closed.
- Check that all motorised valves have been manually opened.
- Check that the pump has been removed and a temporary section of pipe installed. This is to prevent the pump being damaged by any debris that has found its way into the system.
- Ensure that all room and cylinder thermostats are in the 'off' position.

- Check that all capillary joints are soldered and that all compression joints are fully tightened.
- Check that enough pipe clips, supports and brackets are installed and that all pipework is secure.
- Check that the equipment (boiler, pump, motorised valves, expansion vessels, etc) is installed correctly and that all joints and unions on and around the equipment are tight.
- Check that cisterns and tanks are supported correctly and that float-operated valves (FOVs) are provisionally set to the correct water level.

Initial system fill

The initial system fill is always conducted at the normal operating pressure of the system. BS EN 14336 is very clear:

> 'The heating system shall be water tight and tested for leakage…This test may be an independent test or a combined test for water tightness and pressure verification.'

The system must be filled with fluid category 1 water direct from the water undertaker's mains cold water supply. It is usual to conduct the fill-in stages so that the filling process can be managed comfortably. There are several reasons for this.

- Filling the system in a series of stages allows the operatives time to check for leaks stage by stage. Only when the stage being filled is leak free should the next stage be filled.
- Airlocks from cistern-fed open-vented systems are less likely to occur, as each stage is filled slowly and methodically. Any problems can be assessed and rectified as the filling progresses without the need to isolate the whole system and initiate a full drain-down.
- Where a sealed system has been installed, the system should be filled until the pressure gauge reaches the normal fill pressure, usually 1 bar.

Once the system pressure has stabilised and the cistern (if an open-vent system) is full, the furthest radiator on the index circuit can be opened and the radiator bled of air. Where sealed systems are concerned, this will cause the pressure to drop. The system should then be recharged with water to normal operating pressure. Do not be tempted to overfill the system just for the sake of filling the system with water, as this can often cause the pressure relief valve to open and discharge water.

Working back towards the boiler, open and fill the downstairs radiators first. This will ensure that airlocks do not occur on any pipework drops to the lower radiators. Once the downstairs circuit is full, the upstairs circuit can be filled working from the furthest radiator back towards the boiler.

If the system has been connected to the heat exchanger coil of a hot water storage vessel, open and vent the air from any automatic air valves.

When the system has been filled with water, it should be allowed to stabilise and any FOVs should be allowed to shut off and a check made to see where the water line is to ensure there is room for the expansion of water. Where a sealed system is installed, the water pressure should be topped up and the system allowed to stabilise. The system will then be deemed to be at normal operating pressure.

Once the filling process is complete, another thorough visual inspection should take place to check for any possible leakage. The system is then ready for pressure testing.

Pressure testing procedures

Pressure testing can commence when the initial fill to test the pipework integrity has been completed. Again, on large systems, this is best done in stages to avoid any possible problems.

British Standards requirements

BS EN 14336:2004 – Heating systems in buildings – Installation and commissioning of water-based heating systems is very specific with regard to the testing of central heating systems. It states that:

'The heating system shall be pressure tested to a pressure at least 30% greater than the working pressure for an adequate period, as a minimum of 2 hours duration. A suggested method is given in Annex B.'

Annex B of BS EN 14336

B.2.2 Hydraulic pressure testing
B.2.2.1 Preparations

When preparing a hydraulic pressure testing, the following procedure should be applied:

- Blank, plug or seal off all open ends.
- Remove and/or blank off vulnerable items, fittings and plant pressure switches and expansion bellows.
- Close all valves at the limits of the test section of the pipework. Plug the valves if they are not tight, or could be subjected to vibration or tampering.
- Open all valves in the enclosed test section.
- Check that all high points have vents, and that these vents are closed.
- Check that the testing pressure gauge or manometer is functioning, has the correct range and has been recently calibrated.
- Check that there are adequate drain cocks, a hose is available and that it will reach from the cocks to the drain.

- Assess the best time to start the test in view of the duration required after completion of all the preliminaries.

B.2.2.2 During tests

For a hydraulic pressure testing, the following procedure should be applied.

- When filling the system with water or other liquid, 'walk' the system continuously checking for leaks by the noise of escaping air or signs of liquid leakage.
- Release air from high points systematically up through the system.
- When the system is full of water, raise the pressure to test pressure and seal.
- Should the pressure fall, check that stop valves are not leaking and then 'walk' the system again checking for leaks.
- When satisfied that the system is sound, have the test witnessed, eg by the clerk of works or the client's representative, and obtain relevant signatures.

Water Regulations requirements

In most domestic situations, pressure testing of central heating systems follows closely the requirements of the Water Supply (Water Fittings) Regulations 1999. The procedure for this is given in BS 6700 and BS EN 806. These documents suggest that the method of testing should relate to the pipework materials installed and give specific tests both for systems with metallic pipes and those with elastomeric (plastic) pipes. This subject was covered in detail in Unit 303.

Planning the test

Before the test is conducted, a risk assessment should be carried out. Pressure testing involves stored energy, the possibility of blast and the potential hazards of high-velocity missile formation due to pipe fracture and fitting failure. A safe system of work should be adopted and a permit to work sought where necessary. Personal protective equipment (PPE) should also be used.

The following factors should be carefully considered:

- Is the test being used appropriate for the service and the building environment?
- Will it be necessary to divide the vertical pipework into sections to limit the pressures in multi-storey buildings?
- Will the test leave pockets of water that might cause frost damage or corrosion later?
- Can all valves and equipment withstand the test pressure? If not, these will need to be removed and temporary pipework installed.

- Are there enough operatives available to conduct the test safely?
- Can different services be inter-connected as a temporary measure to enable simultaneous testing?
- How long will it take to fill the system using the available water supply?
- When should the test be started when the size of the system is considered?
- Preparation should also be taken into account.

Preparing for the test

- Check that the high points of the system have an air vent to help with the removal of air from the system during the test. These should be closed to prevent accidental leakage.
- Blank or plug any open ends, and isolate any valves at the limit of the test when the test is being conducted in stages.
- Remove any vulnerable equipment and components and install temporary piping.
- Open the valves within the section to be tested.
- Important! Check that the test pump is working correctly and that the pressure gauge is calibrated and functioning correctly.
- Attach the test pump to the pipework and install extra pressure gauges if necessary.
- Check that a suitable hose is available for draining-down purposes.

The hydraulic test procedure

1 Using the test pump, begin to fill the system. When the pressure shows signs of rising, stop and walk the route of the section under test. Listen for any sounds of escaping air and visually check for any signs of leakage.
2 Release air from the high points of the system or section and completely fill the system with water.
3 When the system is full and free of air, pump the system up to the required test pressure.
4 If the pressure falls, check that any isolated valves are not passing water and visually check for leaks.
5 Once the test has been proven sound, the test should be witnessed and a signature obtained on the test certificate.
6 When the pressure is released, open any air vents and taps to atmosphere before draining down the system.
7 Refit any vulnerable pieces of equipment, components and appliances.

Flushing requirements of central heating systems

BS EN 14336 is very specific with regard to the flushing requirements of central heating systems. It states that:

> 'Systems shall, if necessary, be cleaned and/or flushed. If the system is not to be used immediately, consideration shall be given to whether the system is to be left full or empty.'

This section of the unit looks at the requirements of BS EN 14366 system flushing, paying particular attention to:

- cold and hot system flushing
- system additives:
  - cleansers
  - neutralisers
  - corrosion inhibitors
- power flushing.

Cold and hot system flushing

- **Cold flushing** – after the system has been filled with water and checked for leakage, it should be drained and completely emptied. This will remove much of the debris left over from the installation process, such as copper filings, which may otherwise foul the moving parts of the system or become lodged in crucial sections such as the boiler heat exchanger. Once the system has drained, those components of the system removed prior to testing, such as the circulating pump, can be re-fitted and the system re-filled with fresh water.

- **Hot flushing** – a hot flush is conducted to remove any excess jointing compounds, flux residues and oils that may be present after the installation has been completed. The hot water separates these from the pipework and components. With the system now full, it should be run to its maximum operating temperature. Before this can take place, it is important to remember the following.

  - The electricity supply to the system must be tested and switched on. Always check to ensure that the correct size of fuse (3 amps) has been used in the switched fused spur.
  - The fuel supply to the boiler must have been tested in line with the regulations in force and been turned on.
  - All radiator valves should be open to allow water circulation to take place.
  - All thermostats should be calling for heat.

When the system has reached temperature, it should be completely isolated from the energy and fuel sources to prevent accidental operation of the system, and drained down while the water is still hot.

Balancing a central heating system

Balancing a central heating system is one of the most important parts of the commissioning process. This means adjusting all of the various thermostats, thermostatic valves and the circulating pump speed to give the desired temperatures in every room in the dwelling, while maximising the efficiency of the system, thus saving money and energy.

All radiators (except one) will have a thermostatic valve, usually on the flow, and a lockshield balancing valve on the other. The idea is to achieve a similar temperature drop of 10°C across all radiators on the system. If the pump rate is set too high, then the temperature drop will be less than this on a correctly balanced system, so it is important that the circulating pump speed is set correctly.

Before beginning the balancing process, it is a good idea to record the temperatures and valve settings of each radiator. If anything should go wrong, this reference point will help when starting the process again. If zone valves are fitted, it is especially important that these valves are open during the balancing operation.

Balancing a central heating system can be quite a lengthy process. A simple chart or table helps to record all the necessary information. Each radiator in turn can be adjusted, and the thermostatic radiator valve and lockshield valve settings, flow and return temperatures, and the vital temperature difference can all be recorded. An example of the record sheet is shown below.

Room	Radiator	Reading	TRV setting	Lockshield valve setting	Flow temperature	Return Temperature	Temperature difference
		Original					
		1					
		2					
		3					
		4					

The sample record sheet shows the record for just one room. Each room would require the same information to be recorded.

The balancing process

1 Record the initial TRV and lockshield valve settings of all radiators. Also record the room thermostat temperature, the boiler temperature setting and the pump speed.

2 Open all zone valves.

3 Open all TRVs fully.

4 Open all lockshield valves fully. It is important that you record how many turns it takes to open the lockshield valves from fully closed to fully open.

5 Set the room thermostats to maximum.

6 While the system is off and cold, bleed all radiators to remove any air.

7 Check the system pressure and reset if necessary.

8 Turn on the boiler.

9 As the system begins to warm up, visit each radiator in turn and check which is the flow and return of each radiator. Make a note of which radiators get hot first.

10 While the system is warming, turn down the lockshield valves of the radiators that are heating up the quickest halfway so that the cooler radiators catch up. This gives an approximate balance.

11 Now let the system stabilise for about 1 hour.

12 Using an infrared thermometer, record the first set of temperatures on the chart. Don't forget to calculate the temperature difference. This should be about 10°C, but as this is only the first set of figures it is unlikely to be at this point.

13 On the radiators with the smallest temperature difference, close the lockshield valves a little, recording how many turns were made. The radiators with the highest temperature difference should be left fully open.

14 Repeat steps 11, 12 and 13 until all radiators are fully hot and the temperature differences are as near to 10°C as possible.

15 Adjust the boiler temperature to give a flow temperature of 80°C. Then, adjust the pump speed to give a temperature difference of 10°C across the radiator flows and returns. Take care with this, as altering the pump speed may mean that steps 12 and 13 may need to be repeated to give the correct temperature difference.

16 Adjust all TRVs to give the desired temperatures in all rooms. This is best done over several days, as rooms often take time to warm up and cool down.

17 Now set the room thermostats to the desired temperatures.

18 Turn the boiler down to the lowest setting needed to maintain the temperatures required. It can always be increased later if needed.

19 Measure and record the flow and return temperatures at the boiler. This information should be kept with all other system records to help with any fault-finding procedures in the future.

   a Unbalanced systems – poor circulation.

   b Poor boiler connection into a low loss header.

   c Remedial work associated with defective components.

Handover to the customer or end user

When the system has been tested and commissioned, it can then be handed over to the customer. The customer will require all the documentation relating to the installation, which should be presented to them in a file containing the following:

- all manufacturers' installation, operation and servicing manuals for the boilers, heat emitters and any other external controls such as motorised zone valves, pumps and temperature/timing controls fitted to the installation
- commissioning records and certificates
- Building Regulations compliance certificate
- an 'as fitted' drawing showing the position of all isolation valves, drain-off valves, strainers etc, and all electrical controls.

The customer must be shown around the system so they understand the operating principles of any controls, time clocks and thermostats. Emergency isolation points on the system should be pointed out, and a demonstration given of the correct isolation procedure in the event of an emergency. Explain to the customer how the systems work, and ask if they have any questions. Finally, highlight the need for regular servicing of the appliances and leave emergency contact numbers.

Commissioning an underfloor heating system

Before commissioning takes place, it is recommended that the mixing valve and all other valves and pipework in the heating circuits are thoroughly flushed with water to remove flux and debris before the final filling and venting takes place.

Initial system fill

1 Close the isolating valves on the flow and return manifolds.
2 Connect a hosepipe to the flow manifold drain-off point, with the other end connected to the mains cold water supply.
3 Connect a second hose to the return manifold drain-off point. The other end must discharge over a drain gully.
4 Turn on the mains cold water and open each circuit of the underfloor system in turn until the water is running smoothly through the system and any air has been removed.
5 Turn off the water, close the drain points and remove the hoses.
6 Connect a hydraulic test pump and pressurise the system to 6 bar. Leave for a period of 1 hour.
7 Once the pressure test is complete, reduce the pressure to 3 bar. Now the over screed can be laid.

Note: The screed should be allowed to dry thoroughly before the heat is turned on.

System balancing

1 Set the correct temperature at the manifold mixing/blending valve. Ensure that the boiler is operating and the correct temperature is being supplied at the mixing valve. Open the manifold isolating valves. Adjust the mixing valve as required and check the temperature of the water being supplied to the circuits using the dial thermometers fitted to the manifold.

2 Ensure that the manifold pump is set to a suitable speed.

3 Open the actuators to each circuit and adjust the flow rate through each circuit to the manufacturer's flow rate recommendations, using the flow meters on the return manifold. Repeat this process for each circuit.

4 Check the operation of each circuit actuator by operating the individual room thermostats.

Additives in central heating systems

BS EN 14336 makes recommendations about chemical cleaning. It states that:

'For chemical cleaning, the following procedure should be applied:

- Chemical cleaning should be preceded by flushing with frequent sample testing as necessary;

- The system shall be completely flushed and water filled with or without inhibitor, in accordance with the specification;

- Where the whole system is not being chemically cleaned at the same time, it is recommended that the isolating valves be locked in order to avoid pollution from untreated sections.'

Chemical cleanser

Central heating systems can become contaminated with mineral oils that are used to protect steel components, such as radiators, from corrosion during the manufacturing process and excess flux residues as a result of the installation process. If these oils are not removed, component failure becomes a greater risk because the oil attacks the rubber parts that are present in components, such as motorised valves and thermostatic radiator valves. Mineral oils can also lead to eventual pump failure. Flux residues lead to corrosion of the pipes and fittings, especially in those systems using copper tubes. The risk of failures of this kind is eliminated by the use of a chemical cleanser which is administered in accordance with the recommendations in BS 7596:2006. The cleanser should then be circulated around the system for a period of 1 hour with the boiler on, after which the

System cleaner

system should be drained down and the system flushed to remove the cleanser until the water runs clear.

Central heating inhibitor

Untreated central heating systems are prone to electrolytic corrosion. The results of this are metal deposits that form a thick black oxide sludge which blocks boiler heat exchangers and radiators, and fouls circulating pumps, causing them to seize. In severe cases, sludging can result in component and even boiler failure.

In some areas of the UK, where temporary hard water exists, boilers and the associated system pipework are susceptible to scaling by the calcium carbonate deposits (limescale) present in the water. Again, this can cause major problems with loss of boiler efficiency, boiler noise and component failure.

System corrosion and scaling can be prevented by adding a central heating inhibitor to the system water via the feed and expansion cistern in open-vented systems, or via a radiator in sealed systems. Approved Document L of the Building Regulations stipulates that where the water undertaker's cold water main exceeds 200ppm of calcium carbonate, the feed water must be chemically treated to reduce the rate of limescale accumulation.

To be completely effective, the inhibitor must be administered at the correct dosage. If not enough inhibitor enters the system, its protective effect will be reduced. A 1-litre bottle of inhibitor is enough for a 10-radiator system containing 100 litres of water.

One radiator = 10 litres. Double panel radiators count as two radiators.

Central heating neutralisers

Older central heating systems may require de-sludging and de-scaling using an acid-based cleanser to remove any hard, encrusted deposits that have formed on the inside of the pipework, radiators and components. Any water containing an acid is harmful to the environment and causes major problems for water undertakers when discharged down a drain or sewer. A neutraliser, administered through the feed and expansion cistern, injected through a radiator or via a power flushing unit, will calm the action of the cleanser on the inside of the system, neutralising its effects. The system should be thoroughly flushed and the water tested to ensure that the system is free of both acid cleanser and neutraliser. An inhibitor can then be put into the system once the system has been flushed through.

Central heating anti-freeze

When added to a central heating system, antifreeze offers protection against freezing at temperatures as low as −15°C. It can be added to any home central heating system.

Black oxide sludge

Scaled pipework

System inhibitor

System neutraliser

Leak sealer

Leak sealer can be used when small undetectable leaks, such as micro-leaks, cause system pressure loss or continual air infiltration leading to corrosion. A leak sealer will:

- reduce pressure loss from sealed systems
- seal most weeps and small inaccessible leaks
- not cause blockages of the pump, air vents, etc
- be compatible with all metals and materials commonly used in central heating systems.

Power flushing a central heating system

When replacing boilers, or dealing with blocked pipework or radiators, a power flush may be required to remove any sludge within the system. In most cases, where a new boiler is being installed, a power flush is required as part of the warranty. The manufacturer's warranty will be void if this is not carried out.

Power flushing involves using a high-powered pump to circulate strong, often acid-based cleaning chemicals and de-sludging agents through the system. These powerful chemicals strip the old corrosion residue from the system, ensuring that the system does not contain sediment which may be harmful to new boilers, controls and valves.

The power flushing unit is connected to the heating system, often by removing the pump or a radiator.

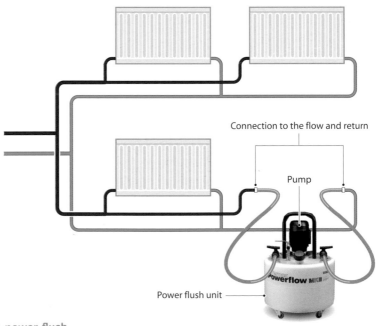

Connection to the flow and return

Pump

Power flush unit

System power flush

Power flush pump

After power flushing is complete, an inhibitor may be added to the system water to keep the system free from corrosion.

Rectifying faults in central heating systems

Table 21 describes some of the more common faults that occur in central heating systems.

Table 21: Faults in central heating components

Component	Known faults	Symptoms	Repairs
Filling and venting – open-vent systems	• Blocked cold-feed pipe to the system or blocked air separator	• System not filling after drain-down due to a blockage of sediment	• The affected section of pipe or the air separator must be removed and replaced
	• System discharging water from the feed and expansion cistern overflow when heating up	• Water level in the feed and expansion cistern too high and will not accommodate the expansion of water	• Lower the water level and reset the FOV to shut off at a lower level
	• Airlocks in the system	• Usually occur at high spots on the installation pipework	• Fit air release valves at all high spots in the pipework
		• Often occur with older systems where the feed and vent pipes are combined	• The feed and vent pipes must be separated or the problem will recur every time the system is drained down

Component	Known faults	Symptoms	Repairs
Filling and venting – sealed systems	• Service valve to the filling loop passing water	• Service valve worn	• Replace service valve
	• Rising pressure causing the pressure relief valve to discharge water	• Service valve open slightly or worn (see fault above)	• Check the service valve, replace as necessary and remove the filling loop
		• Pin hole in the hot water plate heat exchanger causing the cold water to pass through to the heating system (combination boilers only)	• Replace the plate heat exchanger
	• Radiators will not vent	• No pressure	• Top up the pressure at the filling loop and re-try venting procedure
Pumps	• Worn/broken impeller	• Motor working but water not being pumped	• No repair possible • Replace the pump
	• Burnt-out motor	• Voltage detected at the pump terminals but pump not working	• No repair possible • Replace the pump
	• Cracked casing	• Water leaking from the pump body	• No repair possible • Replace the pump
	• Faulty capacitor	• Slow starting pump	• Replace the capacitor if possible • Check manufacturer's instructions

Component	Known faults	Symptoms	Repairs
Expansion vessels	• Pressure loss due to faulty Schrader valve	• No pressure in the expansion vessel • Water discharging from the pressure relief valve during water heat-up	• Pump air into the expansion vessel using a foot pump and check the Schrader valve with leak detector fluid • Check for bubbles • Replace Schrader valve as necessary
	• Ruptured bladder/diaphragm	• Water discharging from the Schrader valve. Water discharging from the pressure relief valve on water heat up	• It is possible to replace the bladder/diaphragm of some accumulators. Check the manufacturer's instructions
Expansion (pressure) relief valves	• Water dripping intermittently when the water is being heated	• Usually an indication that the expansion vessel has lost its air charge or internal expansion bubble has disappeared	• Check and recharge the expansion vessel or internal air bubble
Motorised valves	• Motorised valve not activating when thermostats calling for heat	• Faulty actuator or faulty motor	• Replace valve actuator head • Replace valve motor
	• Valve not shutting off	• As most motorised valves close when de-energised, the problem is most likely to be a seized valve or broken valve spring	• Check the operation of the valve with the manual lever • If the valve moves freely, then the valve spring is broken, so replace the valve head • If the valve will not move, the valve is seized and must be replaced
	• Valve leaking from below the actuator head	• Valve spindle seal has worn	• Replace the valve

Component	Known faults	Symptoms	Repairs
Cylinder thermostats	• Hot water too hot	• System thermostat is not operating at the correct temperature	• Check the temperature of the hot water with a thermometer against the setting on the thermostat • Replace the thermostat as necessary
	• No hot water	• System thermostat not operating	• Check the thermostat with a GS38 electrical voltage indicator for correct on/off functions • Replace as necessary
Room thermostats – hard-wired	• Thermostat not activating the heating – radiators cold	• Dirt/dust on the sensors	• Follow safe isolation procedure and clean the dust from the thermostat
		• Room thermostat not working • Faulty thermostat	• Check the thermostat with a GS38 electrical voltage indicator for correct on/off functions • Replace as necessary
Room thermostats – wireless	• Thermostat not activating the heating – radiators cold	• Loss of radio connection between the thermostat and the boiler	• Follow manufacturer's instructions to re-establish the signal • Check and replace thermostat batteries
		• Room thermostat not operating correctly	• Replace thermostat

Component	Known faults	Symptoms	Repairs
High limit thermostats	• No hot water	• Usually an indication that the system thermostat has malfunctioned and the high limit thermostat has activated to isolate the heat source	• Check the main system thermostat and reset the high limit thermostat
Pressure (bourdon) gauges	• Sticking pressure indicator needle	• Gauge not reading the correct pressure and does not move when the pressure is raised or lowered	• No repair possible. Replace the gauge
Weather compensation/delayed and optimum start controls		• As these controls often contain multiple sensors, it is recommended that the manufacturer's installation instructions are referred to whenever possible. Alternatively, contact the manufacturer's technical help support line	
Radiator/heat emitter	• Radiator/emitter is blocked with black oxide sludge	• Emitter heating at the top and sides only. Middle of the emitter is cold	• Remove radiator and flush through with clean water • Undertake a full, chemical system power flush to remove the system sludge build-up

Safe isolation of central heating systems or components

Repair and maintenance tasks on central heating systems, appliances and valves are essential to ensure the continuing correct operation of the system. The term used to describe isolating a water supply during maintenance operations is 'temporary de-commissioning'. There are basically two types:

- planned preventative maintenance
- unplanned/emergency maintenance.

When a maintenance task involves isolating the central heating system, a notice should be placed at the point of isolation stating 'System off – do not turn on' to prevent accidental turn-on of the system. Where key components such as the expansion vessel, pressure relief valve, motorised valves and thermostats are found to be faulty, the system should be isolated and temporarily de-commissioned until replacement parts are obtained and fitted. If a component requires removal and replacement, it is always a good idea to cap off any open ends until the new component is installed. When removing radiators fitted with thermostatic radiator valves, the TRVs should be shut off using the proprietary cap that came with them.

Where the equipment also uses an electrical supply, safe isolation of the electricity supply is vital. The Safe Isolation Procedure should be followed and the fuse/supply locked off for safety (see Unit 201 for the Safe Isolation Procedure).

A record of all repairs and maintenance tasks completed will need to be made on the maintenance schedule at the time of completion, including their location, the date when they were carried out and the type of tests performed. This is for future reference.

Where appliance servicing is carried out, the manufacturers' installation and servicing instructions should be consulted. Any replacement parts may be obtained from the manufacturers.

Do not forget to keep the householder/responsible person informed of the areas that are going to be isolated during maintenance tasks and operations.

Procedures for carrying out diagnostic tests to locate faults

With central heating components, there are simple diagnostic tests that can be performed on some of the common components and controls to identify exactly what the fault is:

- circulating pumps
- sealed heating system components
- control components.

Circulating pumps

The first time that the customer notices that the pump isn't working is when the radiators fail to get hot or the boiler makes unusual noises. In most cases, the boiler will shut down on the high limit energy cut-out because the hot water is not being circulated away.

If pump failure is suspected carry out the following steps.

1 Check that the rest of the system is operating, eg that the room and cylinder thermostats operate the motorised valves.
2 Turn off the switched fuse spur.
3 Remove the centre pump bleed screw. This will expose the pump impeller shaft. Check that it rotates freely with a small screwdriver.
   a If it does, the pump is not seized up – go to step 4.
   b If it does not, free the shaft by rotating it several times and try

the system again. In most cases the pump will now operate satisfactorily.

4   With the electricity on, check that there is electricity at the pump live and neutral terminals with a suitable test lamp or GS38 proving unit. Do not use a multi-meter to check for voltage. Exercise caution, as the system will be live!

5   If 230V is detected, it can be assumed that the pump is faulty and will need to be replaced.

To replace the circulating pump carry out the following steps.

1   Ensure that the new pump will fit. Some older pumps were larger and a pump extension might need to be used.

2   Isolate the heating system from the electricity supply using the Safe Isolation Procedure, as detailed in Unit 201.

3   Remove the electrical cover and disconnect the live, neutral and earth wires.

4   Isolate the pump using the gate valves on either side.

5   Placing a bowl under the pump to catch the water, break open the pump unions. If the pump has been in for some years, this may prove difficult. If the unions will not move, try tapping them all the way around with a small hammer to loosen them.

6   Carefully remove the pump, noting the direction of flow.

7   Clean the pump valves where the washer sits to ensure a good water-tight joint when the new pump is installed.

8   Install the new pump, ensuring that the direction of flow is correct. If possible, install the pump with a slight upward angle towards the bleed screw, as this helps with venting the air and prevents excessive wear on the pump bearings. Do not forget the rubber pump washers between the pump and the valves. Do not use any jointing compound unless this is necessary.

9   Re-connect the live, neutral and earth wires to the new pump. Ensure that this is done correctly. Replace the terminal box cover on the pump.

10 Turn on the pump valves and bleed the pump of air. Check for leaks.

11 Re-instate the electricity supply and test.

Sealed heating system components

Sealed heating system components occasionally malfunction. Fortunately, these malfunctions are very easy to diagnose. Sealed heating system components are:

- expansion vessel
- filling loop
- expansion relief valve
- pressure gauge.

Table 22: Expansion vessel fault-finding table

Fault	Probable cause	Recommended solution
No air charge in the expansion vessel	Faulty Schrader valve	• Recharge the vessel with air and check the Schrader valve with leak detection fluid • If the valve is leaking, replace the expansion vessel
Water detected at the Schrader valve	Expansion vessel full of water due to a ruptured membrane/diaphragm in the expansion vessel	• Replace the membrane if possible (check the manufacturer's instructions) • If not, replace the expansion vessel with one of similar capacity

Note: If a faulty expansion vessel is diagnosed, the system should be isolated and temporarily de-commissioned until a replacement vessel is obtained and fitted.

Filling loops must be disconnected from the system in line with the Water Supply (Water Fittings) Regulations 1999. However, some installers choose to ignore this fact and the filling loop is left connected to the system. This constitutes a cross-connection between a category 1 fluid and a category 3 fluid and is a breach of the Water Regulations. After any fault-finding process, the filling loop must be disconnected.

Table 23: Filling loop fault-finding table (assumes the filling loop has been left connected)

Fault	Probable cause	Recommended solution
System pressure rising constantly	Faulty service valve	• Replace the service valve and disconnect the filling loop
Water detected when the filling loop is removed	See above. Check the double-check valve as this may be passing water under back pressure	• See above • Replace double-check valve and leave the filling loop disconnected after refilling

Table 24: Expansion relief valve fault-finding table

Fault	Probable cause	Recommended solution
Water discharges from the pressure relief valve	**Intermittently** • Expansion vessel has lost its air charge	• Recharge the air bubble by draining down • Check and recharge expansion vessel as necessary. See expansion vessel fault-finding chart
	Continually • Pressure relief valve has dirt under the valve seat • Pressure relief valve seating damaged	• Replace pressure relief valve as required

If a fault with the pressure gauge is suspected, simply tap the front of the gauge to see if the needle moves. If the gauge continues to give a false reading, replace the pressure gauge.

Control components

Control components include motorised valves, thermostats and time clocks. These are best diagnosed through the system type. In other words, the type of system may dictate the diagnostics that are performed. Remember the fault-finding rule with controls – first, check that you have wired the system correctly. Only start suspecting component faults when you are sure that the system is correctly wired.

Make sure that the terminals have been wired correctly.

- Terminal C (common) is the left-hand terminal
- Terminal 1 is the middle terminal
- Terminal 2 is the right-hand terminal.

Disconnect Terminals 1 and 2 while the checks are taking place. This will prevent false readings due to backfeed.

The cylinder thermostat is faulty if:

- Terminal 1 does not become live when calling for hot water, or
- Terminal 2 does not become live when satisfied (Terminal C must be live in both cases).

Make sure the room thermostat terminals have been wired correctly.

- Disconnect Terminal 3 while the checks are taking place – this will prevent false readings due to backfeed.

- Remove wire from Terminal 3.
- Make sure Terminal 1 is live.
- Turn the room thermostat to call for heat – if live is not detected on Terminal 3, the thermostat is faulty.

The mid-position valve on a Y-plan system is faulty if the valve does not operate as described in the following checks. The checks should be made in the correct order from 1 to 6.

The valve is open for heating only:

Check 1:

- Switch off the electricity supply.
- Disconnect grey and white wires from appropriate junction box terminals.
- Reconnect both grey and white wires to permanent live terminal in junction box.

Check 2:

- Switch on the electricity supply. The valve should move to the fully open heating position at Port A.
- The motor should stop automatically when Port A is open. The valve should remain in this position as long as electricity is applied to white and grey wires.
- With Port A fully open, the orange wire becomes live to start pump and boiler. This can be checked by feeling that the Port A heating outlet is getting progressively hotter.

The valve is open for domestic hot water only:

Check 3:

- Switch off the electricity supply.
- The valve should now automatically return to open domestic hot water Port B.
- Heating Port A should close.

Check 4:

- Isolate the grey and white wires and tape over them to make them safe.
- Remove the cylinder stat wire from Terminal 6 in the wiring centre box and connect to the permanent live terminal.
- Switch on the fused spur.
- The cylinder thermostat must be set to call for heat, and the pump and boiler should start.

The valve is open for both domestic hot water and heating:

Check 5:

- Switch off the electricity supply.
- Replace the cylinder thermostat wire to Terminal 6.
- Isolate and make safe the grey wire by taping it over, and connect the white wire to the permanent live.
- Switch on the electricity supply. The motor should now move to the mid position and stop automatically.
- The cylinder thermostat must be set to call for heat.
- Both Ports A and B should now be open for hot water and heating, and the boiler and pump should start. This can be checked by feeling the Port A heating outlet and the Port B hot water outlet to see if they are getting progressively hotter.

Check 6:

- Switch off the electricity supply.
- Reconnect the white and grey wires to their junction box terminals.
- If check 5 completes satisfactorily, the problem is not the mid-position valve. The fault is elsewhere in the circuit.

Two-port motorised zone valves on S-plan and S-plan plus systems are faulty if you observe the following:

- The motor fails to operate when live is applied to the brown wire and neutral to the blue wire. (The motor can be viewed with the valve cover removed.) The motor should stop automatically when the valve is fully open and will stay in this condition as long as live is applied to the brown wire. The valve automatically closes under the spring return when live is removed from the brown wire.
- The orange wire only becomes live after the valve has fully opened (make sure the grey wire is live).
- The boiler continues to run when the cylinder thermostat and/or room thermostat is satisfied and/or the clock is in the 'off' position.

The programmer should only be suspected as faulty:

- after a check that any links required are in place
- after a check that the programmer has power to the correct terminals
- after a check that the programmer timing is set up correctly
- if a 230V reading does not appear at the heating 'on' terminal when 'heating only' is selected either on continuous or timed

- if a 230V reading does not appear at the hot water 'on' terminal when 'hot water only' is selected either on continuous or timed
- if a 230V reading does not appear at the hot water 'off' terminal with 'hot water off' on the programmer.

Be able to de-commission, install, commission and fault-find on sophisticated central heating systems and their components (LO6)

There are nine assessment criteria to this Outcome:

1 Confirm safe isolation of all electrical and water supplies.
2 Install pipework to S-plan heating systems and underfloor heating manifolds.
3 Install components required for a boiler interlock.
4 Demonstrate 'dead' testing of boiler interlock systems.
5 Carry out visual inspections of pipework and components.
6 Commission central heating systems and components.
7 Commission underfloor heating systems.
8 Demonstrate procedures for de-commissioning.
9 Resolve faults in central heating systems.

Range	
Components	Motorised valves, auto bypass, room stat, programmer, cylinder stat
'Dead' testing	Earth continuity, resistance to earth, continuity, short circuit
Faults	Pumping over, dragging air in, motorised valves not operating, heat when no demand, component failure

This Outcome is part of the practical assessment that will be conducted within the workshop environment.

There are five practical tasks:

- **Task CH1** – safely isolate all electrical and water supplies. This task involves using the correct isolation procedure to isolate the electricity supply of a given circuit. The recommended time for the task is 1 hour. You must:
  - confirm that all documentation required to complete the task is available

- confirm that tools and equipment required to complete the task are available and suitable
- liaise with the customer regarding isolation of supplies
- identify a suitable electrical isolation point and turn off the supply
- check the voltage indicator on a known supply or proving device
- confirm that electrical supply was isolated using a voltage indicator
- re-check the voltage indicator on a known supply or proving device
- lock off and label the electrical isolation point
- isolate the water supply and confirm that there was no flow
- label the water supply valve.

- **Task CH2** – install and commission a central heating system, including underfloor heating pipework. This task involves the installation of central heating pipework and connection to an underfloor heating manifold. It also involves the testing and commissioning of the installation in line with current industry standards. The recommended time for the task is 9 hours. You must:
  - confirm that all documentation required to complete the task is available
  - confirm the suitability of materials and components
  - plan the installation and prepared work area for installation
  - install connections to the pre-installed boiler, radiators and underfloor pipework circuit to industry standards
  - install flow and return connections to the hot water cylinder to industry standards
  - correctly position all system components
  - correctly install all sealed system components
  - prepare the work area for the commissioning process
  - ensure that the customer is aware that the commissioning procedure is about to take place
  - visually inspect all pipework connections
  - visually check that all valves and taps are set in the correct position for the commissioning process
  - visually check that the system is ready for a soundness test
  - test the central heating system for soundness
  - carry out recommended flushing procedures on the system
  - fill and bleed the air from the system and underfloor pipework
  - test the boiler, system and controls for correct operation

- set flow rates and adjust the blending valve on the underfloor pipework manifold
- balance the system
- complete a benchmark/commissioning certificate.

- **Task CH3** – install and test electrical control components and wiring. The task involves the installation and testing of central heating electrical components and controls from a pre-installed switch fused spur. The recommended time for the task is 6 hours. You must:
  - confirm that all documentation required to complete the task is available
  - confirm suitability of materials and components for task to be carried out
  - plan the installation and prepare the work area
  - confirm that the electrical supply is safely isolated
  - confirm that electrical components are correctly mounted and installed
  - correctly install system wiring from a 13 amp pre-installed 3 amp fuse spur
  - visually check that the system is ready for testing
  - notify relevant persons that testing of the installation is about to commence
  - check that the correct size of fuse is installed
  - carry out an earth continuity check
  - carry out a short-circuit check
  - carry out a resistance to earth check
  - carry out a continuity check.

- **Task CH4** – de-commissioning a central heating system. The task involves permanently de-commissioning the central heating installation and terminating gas, electrical and water supplies to industry standards. The recommended time for the task is 2 hours. You must:
  - inform the customer that de-commissioning is about to take place
  - confirm that the appliances are turned off
  - confirm isolation of electrical, gas and water supplies
  - lock off/cap and label supplies
  - drain the system from appropriate drain-off points to the correct drainage point
  - remove the pipework, components and electrical cables and dispose of them, using appropriate methods
  - make good any damaged surfaces as required.

Supporting evidence for this task includes:

- pipework diagram of the heating system produced by the candidate
- wiring diagram of the electrical control system produced by the candidate
- photograph(s) of the installation in progress
- photograph(s) of the completed installation
- completed benchmark/commissioning certificate.

■ **Task CH5** – identify and repair faults on central heating systems. The task involves identifying and successfully rectifying faults that have been set by your assessor prior to the task taking place. The recommended time for the task is 4 hours. You must:

- check the installation to confirm compliance with industry requirements and manufacturers' instructions
- successfully diagnose a fault set by your assessor
- rectify and replace any components in a safe manner, eg using electrical isolation
- re-instate the services and re-commission the system including heat testing (where appropriate)
- complete fault identification records for the work activity.

Before undertaking the assessments, discuss with your tutor any concerns you may have. Be clear about what is required of you and ask the appropriate questions. There are time limits for each of the five tasks, and photographic and written evidence is required.

Good luck with your assessment!

Conclusion

You will have seen, as you worked through this unit, that central heating systems are often very complex in nature and require much planning, both in their design and their installation. You have investigated the new technology that is emerging and the problems that arise when things go wrong, either as a result of a fault developing or through poor installation techniques.

The way we live our lives means that a modern, efficient and economical central heating system is now considered a 'must have' for any modern home. You, as an installer, need to keep your skills up to date so that you are able to cope with the changes happening within the industry. You need to be able to respond to the new methods of installation, to innovative ideas and, most importantly, to the wishes of the customer.

Test your knowledge questions

1 Every radiator must be fitted with a thermostatic radiator valve. True or false?

2 Name two types of heat pump.

3 Where would a low loss header typically be used?

4 What is the basic difference between an S-plan system and an S-plan plus system?

5 On which kind of system would you find a three-port diverter valve?

6 List the sealed system components.

7 Where could you install a type CA backflow prevention device?

8 What is the typical temperature of an underfloor heating system laid under a concrete screed?

9 Identify the following component.

10 Identify the following pattern of underfloor heating.

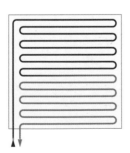

11 Briefly describe the term 'zoning'.

12 Using the formula below, calculate the size of the expansion vessel required for a large domestic central heating system containg 200 litres of water.

$$V = \frac{eC}{1 - \frac{p_1}{p_2}}$$

V = the total volume of the expansion vessel

C = 200 litres

P_1 = 1.5 + 1

P_2 = 6 + 1

e = 0.0324

13 Which type of zone valve is used on an S-plan system?

14 What function does a weather compensator provide?

15 What are the three benefits of choosing the right system of central heating control?

16 What does the acronym CHeSS stand for?

17 In what situation would you use a power flush?

18 What must you do before removing a central heating circulating pump?

19 Why might you use a central heating inhibitor?

20 What is a neutraliser used for?

Assessment checklist

What you now know (Learning Outcome)	What you can now do (Assessment criteria)	Where this is found (Page number)
1. Understand complex domestic heating systems layouts and controls	1.1 Identify documents relating to central heating design and installation.	361–363
	1.2 Describe pipework layouts for complex central heating systems.	364
	1.3 Describe the working principles of key components in a complex central heating system.	367–379
	1.4 Explain boiler interlock.	379
	1.5 Compare the relationship of positive and negative pressures in relation to feed, vent and pump positions.	379–381
	1.6 Describe the wiring arrangements required for S- and Y-plan heating systems and components.	381–387
	1.7 Identify alternative methods of wiring arrangements.	387
	1.8 Describe the procedures for safely isolating supplies.	387–389
	1.9 Describe testing of wiring in domestic heating systems.	390–394
	1.10 Describe the working principles of low loss headers.	394–396
	1.11 Explain the effect bore diameter of tube has on heat loads.	396–397
2. Understand the layouts and operating principles of sealed systems	2.1 Identify components required for sealed central heating systems.	398–401
	2.2 Describe the safety hazards associated with sealed central heating systems.	401–402
	2.3 Describe the advantages of sealed central heating systems.	402–403
	2.4 Explain layout requirements for sealed system components.	403–406
	2.5 Calculate the size of pressure vessels for sealed central heating systems.	406–408
3. Understand the types of boiler in domestic central heating systems	3.1 Identify fuel sources for central heating.	409–413
	3.2 Identify components of a gas central heating boiler.	413–416
	3.3 Describe the operating principles of boilers.	416–426
	3.4 Describe different flueing arrangements for boilers.	426–429
	3.5 Explain the reason for pump overrun on boilers.	429

What you now know (Learning Outcome)	What you can now do (Assessment criteria)	Where this is found (Page number)
4. Understand the types of heat emitters used in underfloor heating systems	4.1 Justify selection of heat emitters used in plumbing systems.	430–435
	4.2 Describe components required for installation with a range of heat emitters.	436–438
	4.3 Describe the design considerations for underfloor heating.	438–441
	4.4 Describe the advantages of underfloor heating systems.	441–443
	4.5 Describe the operating principles of underfloor heating systems.	443–453
5. Know how to de-commission, commission, and fault-find on central heating systems	5.1 Interpret information required when testing, commissioning and fault-finding on central heating systems.	455–458
	5.2 Describe means of safeguarding customer property.	458–460
	5.3 Explain the procedure for de-commissioning a central heating system.	460–461
	5.4 Explain the use of bespoke tools in commissioning a central heating system.	461
	5.5 Describe the procedure for commissioning a central heating system.	461–469
	5.6 Describe the procedure for commissioning an underfloor heating system.	469–470
	5.7 Identify additives used in central heating systems.	470–472
	5.8 Describe the procedure for power flushing a central heating system.	472
	5.9 Describe how to rectify faults in central heating systems.	473
6. Be able to de-commission, install, commission and fault-find on sophisticated central heating systems and their components	6.1 Confirm safe isolation of all electrical and water supplies.	484–487
	6.2 Install pipework to S-plan heating systems and underfloor heating manifolds.	484–487
	6.3 Install components required for a boiler interlock.	484–487
	6.4 Demonstrate 'dead' testing of boiler interlock systems.	484–487
	6.5 Carry out visual inspections of pipework and components.	484–487
	6.6 Commission central heating systems and components.	484–487
	6.7 Commission underfloor heating systems.	484–487
	6.8 Demonstrate procedures for de-commissioning.	484–487
	6.9 Resolve faults in central heating systems.	484–487

TEST YOUR KNOWLEDGE ANSWERS

Unit 201

1 **d)** A safe working environment with adequate welfare facilities.

2 **b)** Continuously reviewed as work progresses.

3 **d)** BS 7671: IET Wiring Regulations.

4 **c)** Where there is a risk of a fall liable to cause personal injury.

5 **b)** Electricity at Work Regulations 1989.

6 **a)** Stop an activity immediately.

7 **c)** Housekeeping.

8 **c)** Placed in an appropriate container.

9 **d)** Safety footwear.

10 **b)** A prohibition.

11 **d)** Advisory.

12 **c)** Stop work and inform duty holder.

13 **d)** Eyes.

14 **c)** Bump cap.

15 **c)** 50mA.

16 **b)** Metallic parts of the casing.

17 **d)** GS38.

18 **c)** Shout for help.

19 **a)** Fuel, oxygen, heat.

20 **a)** Powder, carbon dioxide.

Unit 301

1 **c)** Solar thermal.

2 **d)** Hot water.

3 **a)** Photovoltaic.

4 **b)** A feed-in tariff.

5 **b)** The need for large ground area.

6 **c)** The refrigeration cycle.

7 **d)** Wood ships.

8 **b)** Drinking water.

9 **b)** Reliable water flow.

10 **b)** An inverter.

11 **a)** North.

12 **c)** 200mm.

13 **a)** Deep boreholes.

14 **b)** 3.

15 **b)** At the side or rear of a building.

16 **d)** Is connected to the national grid.

17 **b)** Adequate greywater availability.

18 **c)** Need for large storage area.

19 **b)** There is little or no input at night.

20 **d)** 5m.

Unit 303

1 The Water Industry Act 1999.

2 BS EN 806 and BS 6700.

3 (1) Waste, (2) misuse, (3) undue consumption, (4) contamination, (5) erroneous metering.

4 The Private Water Supply Regulations 2009.

5 RPZ valve.

6 Approved Document G.

7 Indirect boosting with pressure vessel.

8 An electro-mechanical valve.

9 A cistern that provides a break in supply between the water undertaker's supply and the installation.

10 The detection of the water level to activate various pieces of remote equipment.

11 An electronic pressure sensor.

12 Up to 1,000 litres per day.

13 A centrifugal pump designed to be fully immersed in water.

14 Rivers, streams and springs.

15 The multiple barrier principle.

16 Forcing water under pressure through a semi-permeable membrane.

17 Disinfection of private water supplies.

18 (1) Pumped supply direct from a borehole, and (2) gravity supply from a catchment tank.

19 Multi-stage surface pump set.

20 Bladder-type accumulator.

21 $2.5 \times 0.9 = 2.25$ bar.

22 A reduced pressure zone (RPZ) valve.

23 Verifiable check valves have a test point to check for correct operation.

24 Fluid category 1.

25 Fluid category 5.

26 (1) Back pressure, (2) back siphonage.

27 Non-concussive.

28 Type HC diverter.

29 To limit the flow to a predetermined flow rate.

30 Inlet type.

31 Water hammer.

32 Using a weir gauge.

33 30ppm.

34 With water taken from the water undertaker's main.

Unit 304

1 A centralised system is one in which hot water is delivered from a central point to all hot water outlets.

2 The open vent is open to the atmosphere, ensuring that the water does not exceed 100°C. The vent pipe acts as a safety relief should the system become overheated.

3 Unvented hot water storage systems are fed directly from the mains cold water supply and utilise an expansion vessel or internal air bubble to allow for expansion.

4 (1) Unvented hot water storage vessel using an internal expansion air gap, and (2) an unvented

hot water storage vessel using an external expansion vessel.

5 (1) Storage hot water systems, and (2) instantaneous hot water heaters.

6 Building Regulations Approved Document G3.

7

Maximum recommended lengths of uninsulated hot water pipes	
Outside diameter (mm)	Max. length (m)
12	20
Over 12 and up to 22	12
Over 22 and up to 28	8
Over 28	3

8 An indirect unvented hot water storage cylinder with internal expansion.

9 (1) Isolation valve, (2) in-line strainer, (3) pressure-reducing valve, (4) single-check valve, (5) expansion relief valve, (6) expansion vessel, (7) temperature relief valve.

10 Boyle's law – the volume of a gas is inversely proportional to its pressure provided the temperature remains constant.

11 65°C.

12 The non-return valve safeguards against reverse circulation of water.

13 (1) The control thermostat, (2) the overheat thermostat, (3) the temperature relief valve.

14 (1) Isolation valve, (2) in-line strainer, (3) pressure-reducing valve, (4) single-check valve, (5) expansion vessel, (6) pressure relief/expansion relief valve.

15 Trace heating is used as an alternative to secondary circulation. It uses an electric heating element to heat the water.

16 A weir gauge.

17 A visual inspection of the installation.

18 (1) D1 pipework, (2) tundish, (3) D2 pipework.

Unit 306

1 False. A TRV should not be fitted in a room that contains a room thermostat.

2 (1) Ground source heat pump, (2) air source heat pump.

3 On systems that contain multiple zones and different heat emitters.

4 An S-plan plus system allows for zoning of the upstairs and downstairs heating circuits in properties over 150m^2.

5 A three-port diverter valve would be used on a W-plan system.

6 Expansion vessel, pressure gauge, expansion relief valve, filling loop or type CA backflow prevention device.

7 A type CA backflow prevention device would be used as an alternative method of filling a sealed heating system rather than a filling loop.

8 40°C to 45°C.

9 An underfloor heating manifold arrangement.

10 Series pattern.

11 This is where a heating system has separate time and temperature control for different parts of a dwelling.

12 10.07 litres.

13 A two-port motorised zone valve.

14 As the external temperature rises, the weather compensator reduces the flow and return temperatures to compensate for the warmer external temperature.

15 (1) Improved energy efficiency, (2) reduced CO_2 levels, (3) reduced fuel bills.

16 Central Heating System Specification.

17 Power flushing is used to remove sludge from a system prior to the installation of a new boiler.

18 Isolate the pump from the electrical supply using the Safe Isolation Procedure.

19 Inhibitors are used to prevent scaling and corrosion within central heating systems.

20 Neutralisers are used to calm the acidic effect of de-scalers and de-sludgers in central heating systems.

GLOSSARY

abrade To scrape or wear away.

abutment The junction between a pitched roof and a vertical wall.

acceleration A measure of the rate at which a body increases its velocity.

acceleration due to gravity The rate of change of velocity of an object due to the gravitational pull of the Earth.

acetylene (C_2H_2) A flammable gas used in conjunction with oxygen for welding.

acrylonitrile butadiene styrene (ABS) A type of thermoplastic used for waste pipes, soil pipes, underground drainage, gutters and rainwater pipes. Can be solvent welded.

Acts of Parliament These create new laws or change an existing one.

adhesion The way that water tends to stick to whatever it comes into contact with.

air admittance valve Allows air into a stub stack to prevent the loss of trap seals.

air changes The amount of air movement within the building.

air gap A physical unrestricted open space between the wholesome water and the possible contamination.

air separator A fitting designed to correctly position the feed and vent pipes on a central heating system to ensure that the neutral point is automatically built in the system.

air temperature The temperature of the air within a building.

air velocity The speed at which air travels through a building.

alloy A mixture of two or more metals.

alternating current (AC) An electrical current that reverses its direction of travel constantly and uniformly throughout the circuit.

ambient This relates to the feeling of surroundings or atmosphere. For example, the ambient temperature of the liquid inside a cylinder, or even the ambience of a nightclub.

ampere The unit of electrical current.

annealing A process that involves heating copper to a cherry-red colour and then quenching it in water. This softens the copper tube so that the copper can be worked without fracturing, rippling or deforming.

anodic corrosion protection A form of corrosion protection that uses a sacrificial anode to distract the corrosion away from vulnerable parts of the system.

anodising Coating one metal with another by electrolysis to form a protective barrier from corrosion.

anti-gravity valves A valve used in older central heating systems to stop unwanted gravity hot water circulation. Often called a dumb ball valve.

Approved Codes of Practice (ACoP) Documents giving practical guidance on complying with the Regulations.

aquifers Water-bearing rocks below the Earth's surface.

arcing Electricity flowing through the air from one conductor to another – it can produce visible flashes and flames.

architect The designer of a building or structure.

artesian wells and springs Water that rises from underground water-bearing rock layers under its own pressure.

asbestos A naturally occurring fibrous material that has been a popular building material since the 1950s, now known to cause serious and fatal illness.

atmospheric pressure The amount of force or pressure exerted by the atmosphere on the Earth and the objects located on it.

atom A fundamental piece of matter made up of three kinds of particles called subatomic particles: protons, neutrons and electrons.

audit To conduct a systematic review to make sure standards and management systems are being followed.

automatic bypass valve A spring-loaded valve used on fully pumped heating systems; it is designed to automatically open when other paths for water flow begin to close.

automatic urinal flushing cistern Used to flush urinals.

back boiler A boiler made from a non-ferrous metal that is situated behind a real fire. Used with a direct cylinder.

back siphonage A vacuum that can suck water backwards causing contamination of the water supply.

backflow The flowing of water in the wrong direction due to loss of system pressure.

backflow prevention device A mechanical device, usually a fitting, designed to prevent contamination of water through backflow or back siphonage.

backflow protection Protection of contamination of water through backflow or back siphonage.

banjo-type bath waste fitting A type of waste fitting fitted to a bath that connects an overflow to the waste trap.

barbed shanked nail A nail with grooves cut into the shank. This makes the nail difficult to pull out once it has been driven into the wood.

batch feed boiler A solid fuel boiler where the fuel is fed by hand.

batter or slope The angle in relation to the horizontal surface, of the trench walls of an excavation, to prevent the walls collapsing.

Bernoulli's principle Bernoulli's principle states that when a pipe is suddenly reduced in size, the velocity of the water increases but the pressure decreases. The principle can also work in reverse. If a pipe suddenly increases in size, the velocity will decrease but the pressure will increase slightly.

Bill of Quantities (BOQ) A document used in tendering in the construction industry in which materials, parts and labour (and their costs) are itemised. It also (ideally) details the terms and conditions of the construction or repair contract and itemises all work to enable a contractor to price the work for which he or she is bidding.

biomass Any plant or animal matter used directly as a fuel or that has been converted into other fuel types before combustion.

black water Water and effluent from WCs and kitchen sinks that can only be treated by a water undertaker at a sewage works.

blackheart fittings A type of fitting for low carbon steel pipe with a tapered female thread.

boiler cycling This happens when a heating system has reached temperature, and the boiler shuts down. A few minutes later the boiler will fire up again to top the temperature up as the system loses heat, and after a few seconds shuts down again. This constant firing up and shutting down as the system water cools slightly wastes a lot of fuel energy.

boreholes Man-made wells that are drilled directly to a below-ground water source.

Boyle's law A gas law that states that the volume of a sample of gas at a given temperature varies inversely with the applied pressure.

branch ventilating pipe Used on the ventilated branch discharge system of sanitary pipework to ventilate excessively long waste pipe runs.

BS 1566-1:2002 The British Standard for copper indirect hot water storage cylinders.

BS 6700:2006+A1:2009 The main British standard for the installation of hot and cold water installations in dwellings.

BS 7671 The national standard to which all wiring, industrial or domestic, should conform.

BS 8000-13:1989 The Code of Practice for the workmanship on site relating to the installation of sanitation systems.

BS EN 12056-3:2000 The British and European Standard for the installation of rainwater and guttering systems.

BS EN 12056-5:2000 The British and European Standard for the installation of sanitary pipework.

BS EN 12588:2006 The British and European Standard for rolled (milled) sheet lead.

BSP or BSPT Stands for British Standard pipes and British Standard pipe threads, and relates to the type of thread we use on screwed low carbon steel pipes and fittings. Although the pipe is measured in mm, it is universally referred to in imperial measurements, eg ½-inch BSPT (½-inch British Standard pipe thread).

building control officer Responsible for ensuring that regulations on public health, safety, energy conservation and disabled access are met.

Building Regulations Approved Document F: Ventilation Document dealing with indoor air quality to ensure buildings are properly ventilated.

Building Regulations Approved Document H3 The main document concerning the installation of rainwater discharge systems.

Building Regulations Approved Document L: Conservation of fuel and power: 2010 (Part J in Scotland and Part F in Northern Ireland) Document controlling the insulation values of building elements, the heating efficiency of boilers, the insulation and controls for heating appliances and systems together with hot water storage, lighting efficiency and air permeability of the structure.

building services engineer Designer of the internal services within the building such as heating and ventilation, hot and cold water supplies, air conditioning and drainage. Many Building Services Engineers are members of the Chartered Institution of Building Services Engineers (www.cibse.org).

business opportunity In this context, the opportunity to make profit from the work or contract.

C

calorific value The amount of energy released when a known volume of gas, oil or coal is completely combusted under specified conditions. Solid and liquid fuels are measured in megajoules per kilogram (MJ/kg) and gases are measured in megajoules per cubic metre (MJ/m³).

capillary attraction The process where water (or any fluid) can be drawn upwards through small gaps against the action of gravity.

capillary fitting A fitting for copper tubes that uses the principle of capillary attraction to draw solder into the joint when heated.

carbon footprint The amount of carbon dioxide released into the atmosphere as a result of the activities of a particular individual, organisation or community.

Carbon Trust An independent, non-profit organisation set up by the UK Government with support from businesses to encourage and promote the development of low-carbon technologies.

carburising flame A sooty flame containing too much acetylene.

carcinogenic A substance that causes cancer.

celsius (°C) A common unit of temperature that has a zero point (0°C), which corresponds to the temperature at which water will freeze.

Central Heating System Specifications (CHeSS) 2008 CE51 Produced by the Building Research Energy Conservation Support Unit (BRECSU) to create a set of common standards for energy efficiency which domestic heating installers and manufacturers should work towards.

centralised hot water systems Those systems where the source of hot water is sited centrally in the property for distribution to all of the hot water outlets.

centre to centre Measuring from the centre line of one pipe to the centre line of another so that all the tube centres are uniform. This ensures that the pipework will look perfectly parallel because all of the tubes will be at equal distance from each other.

ceramic discs Two thin close-fitting, slotted ceramic plates that control the flow of water from a tap.

chamfer To take off a sharp edge at an angle. If we chamfer a pipe end, we are taking the sharp, square edge off the pipe.

Charles's law A gas law discovered by Jacques Charles which states that the volume of a quantity of gas, held at constant pressure, varies directly with the Kelvin temperature.

chased In the case of pipework or cables, this means they are fitted inside a cut made in a wall.

chlorine A chemical added to water for sterilisation purposes.

cistern A vessel for storing cold water that is only subjected to atmospheric pressure.

civil engineer Designer of roads into and out of the building along with any bridges, tunnels, etc that may be required.

clerk of works (CoW) The architect's representative on site. He or she ensures that the building is constructed in accordance with the drawings while maintaining quality at all times.

Climate Change Act 2008 Sets a target for the UK to reduce carbon emissions to 80% below 1990 levels by 2050.

coal A heavy hydrocarbon that releases high content of sulphur dioxide and carbon dioxide when burnt.

cohesion The way in which the water molecules 'stick' together to form a mass rather than staying as individuals.

coke Produced by heating coal in an oven which reduces both sulphur and carbon dioxide content. Known as a smokeless fuel.

combination boiler A boiler that supplies both instantaneous hot water and central heating from the same appliance.

combined cooling, heat and power (CCHP) Uses the excess heat from electricity generation to achieve additional building heating or cooling.

combined heat and power (CHP) A plant where electricity is generated and the excess heat generated is used for heating.

combined storage and feed cistern Stores water for the domestic hot water system and the indirect system of cold water to the appliances, wash hand basin, bath, WC, washing machine, etc.

combined system A system of below-ground drainage where both rainwater and foul water discharge into the same drain.

combustible Able to catch fire and burn easily.

combustion A chemical reaction in which a substance (the fuel) reacts violently with oxygen to produce heat and light.

commissioning The process of bringing a system or appliance into full working operation through a system of checks to ensure correct operation to the design specification.

communication pipe A pipe connecting the water main to the customer's external stop valve. Owned by the water undertaker.

competent person Recognised term for someone with the necessary skills, knowledge and experience to manage health and safety in the workplace.

Competent Persons Scheme Members of CPS must follow certain rules to ensure that their work complies with Building Regulations.

compliance The act of carrying out a command or requirement.

compression Back pressure of air created by water discharging down a soil pipe travelling up the stack blowing the water out of the traps.

compression fitting A mechanical fitting that requires tightening with a spanner to make a watertight joint.

compressive strength The maximum stress a material can sustain when being crushed.

condensation A process where steam turns to water.

condensing boiler A boiler that extracts all usable heat from the combustion process, cooling the flue gases to the dew point. The collected water is then evacuated from the boiler via a condensate pipe.

conduction Heat travelling through a substance with the heat being transferred from one molecule to another.

conductivity The property that enables a metal to carry heat (thermal conductivity) or electricity (electrical conductivity).

Construction (Design and Management) Regulations 2007 The principal piece of health and safety legislation specifically written for the construction industry.

contamination The introduction of a harmful substance to an area.

Control of Asbestos Regulations 2006 Legally enforceable document prohibiting the importing, supplying and use of all forms of asbestos.

Control of Lead at Work Regulations 2002 Legally enforceable document that applies to all work which exposes any person to lead in any form whereby the lead may be ingested, inhaled or absorbed into the body.

convection Heat transfer through the movement of a fluid substance, which can be water or air.

corrosion Any process involving the deterioration or degradation of metal components.

COSHH COSHH is an acronym that stands for 'Control of Substances Hazardous to Health'. Under the COSHH Regulations 2002, employers have to prevent or reduce their employees' exposure to substances that are hazardous to health.

coulomb The SI unit of electrical charge, equal to the quantity of electricity conveyed in one second by a current of 1 ampere.

creep A term that is used to describe the effects of thermal movement whereby the lead fails to return to its original position after expansion has taken place.

cross-connection When one fluid category connects with another, for example, within a mixer tap.

cuprous chloride corrosion This occurs because the chloride ions present in sodium hypochlorite solution are very aggressive when in contact with copper and copper alloys, due to the fact that the chloride can form an unstable film on the inside of the pipe. This means that even small amounts of chlorine can cause corrosion problems with copper piping.

D

Data Protection Act 1998 Gives people the right to know what information is held about them.

delivery note A document that lists the type and amount of materials that are delivered to site.

deposition The process where steam passes directly to ice.

dew point The temperature at which the moisture within a gas is released to form water droplets. When a gas reaches its dew point, the temperature has been cooled to the point where the gas can no longer hold the water and it is released in the form of water droplets.

dezincification A form of selective corrosion (often referred to as de-alloying) that happens when zinc is leached out of brass.

direct current (DC) An electrical current where the polarity or direction of the electron flow never reverses.

direct hot water storage cylinder A hot water storage vessel that does not contain a heat exchanger.

direct system of cold water A cold water system where all cold water outlets are connected to the main cold water supply.

Disability Discrimination Act 1995 Applies to companies who employ over 20 people. They are required to accommodate the needs of the disabled.

district heating A system for distributing heat generated in a centralised location for residential and commercial heating requirements.

Domestic Building Services Compliance Guide 2010 Lays down rules for minimum boiler energy efficiency requirements. Often abbreviated to DBSC Guide.

double-feed indirect hot water storage cylinder A hot water storage vessel that contains a heat exchanger in the form of a coil or an annular.

ductility A mechanical property that describes by how much solid materials can be pulled, pushed, stretched and deformed without breaking.

duty holder The person in control of a danger.

dynamic pressure The pressure of water while it is in motion.

E

economy 7 electricity A UK tariff that provides for seven hours of cheaper-rate electricity, usually between 1 am and 8 am in the summer and 12 am and 7 am in the winter (although times may vary between regions and suppliers).

effort arm In mechanics, the arm where the force is applied.

electrolyte A fluid that allows the passage of electrical current, such as water. The more impurities (such as salts and minerals) there are in the fluid, the more effective it is as an electrolyte.

elevation A drawing showing one side of a building.

Enabling Act An enabling Act allows the Secretary of State to make further laws (regulations) without the need to pass another Act of Parliament.

end feed fitting A capillary fitting for copper tubes that requires solder to be fed into it during the soldering process.

Energy Performance of Buildings (Certificates and Inspections) (England and Wales) Regulations 2007 States the requirements for clients and landlords to produce energy performance certificates when buildings are constructed, rented out or sold.

Energy Saving Trust (EST) An independent non-profit organisation set up after the 1992 Rio 'Earth Summit' that attempts to reduce energy use in the UK.

Equality Act 2010 Implemented by the Equality and Human Rights Commission (EHRC) to provide a single legal framework with clear, streamlined law that will be more effective at tackling disadvantage and discrimination.

equipotential bonding A system where all metal fixtures in a domestic property such as hot and cold water pipes, central heating pipes and gas pipes, radiators, stainless steel sinks, steel and cast iron baths and steel basins are connected together through earth bonding so that they are at the same potential voltage everywhere.

erosion corrosion Corrosion that occurs in tubes and fittings because of the fast flowing effects of fluids and gases.

erroneous Incorrect.

estimate A costing for a piece of work that is not a fixed price but can go up or down if the estimate was not accurate or the work was completed ahead of schedule.

expansion vessel A vessel divided by a membrane with air one side and water the other that allows the expansion of water to take place safely.

F

fan-assisted boiler A boiler that uses a fan to evacuate the products of combustion.

fascia bracket A clip for securing a gutter to a fascia board.

fatality Death.

feed and expansion cistern Used to feed a vented central heating system and also allows expansion of water into the cistern when the system is hot.

feed cistern Only holds the water required to supply the hot water storage vessel.

ferrous metal A metal that contains iron and is susceptible to corrosion through rusting.

filling loop A method of filling sealed central heating systems direct from the water main.

fireclay A malleable clay used for heavy-duty sanitary appliances.

flame arrester A device fitted to lead welding equipment to prevent a dangerous situation known as flame blowback.

flange A projecting flat rim or collar, which is designed to strengthen or attach to another object. Flanges can also be found on large industrial pipe installations.

flashings A term given to a small weathering, usually at an abutment.

flow rate The amount of fluid or gas that flows through a pipe or tube over a given time.

fluid category A method of water classification from 1 to 5 according to its potential level of contamination, with 5 being the most dangerous.

flushing valve A method of flushing a urinal and WCs fitted in industrial premises using water direct from the mains supply without the need for a cistern.

flux A paste used to clean oxides from the surface of the copper and to help with the flow of solder into the fitting.

footing a ladder Standing with one foot on the bottom rung, the other firmly on the ground.

force The influence on an object which, acting alone, will cause the motion of the object to change. It is measured in newtons (kgm/s^2).

forced draught Any flue that uses a fan to help evacuate the products of combustion.

fossil fuels Formed by anaerobic decomposition of buried dead carbon-based plants, these fuels are known as hydrocarbons and release a high carbon dioxide content when burnt.

Freedom of Information Act 2000 Gives people the right to ask any public body for all the information they have on any subject.

fully pumped heating systems A heating system where both hot water circulation and central heating are pumped by a central heating circulator.

G

galvanic corrosion Corrosion that occurs when two dissimilar metals are in contact with each other in the presence of an electrolyte, usually water.

Gantt chart Otherwise known as a programme of work, it is used on site to illustrate dates and lengths of time to complete particular jobs. It includes start and finish dates, labour and materials required and overall progress.

Gas Safety (Installation and Use) Regulations 1998 These cover the safe installation, maintenance and use of gas and gas appliances in private dwellings and business premises, aimed at preventing carbon monoxide (CO) poisoning, fires and explosions.

gradient curve A method of determining the fall of a 32mm waste pipe.

granular soils Gravel, sand or silt (coarse-grained soil) with little or no clay content. Although some moist granular soils exhibit apparent cohesion (grains sticking together forming a solid), they have no cohesive strength. Granular soil cannot be moulded when moist and crumbles easily when dry.

gravity feed boilers A solid fuel boiler where the fuel is automatically fed to the fire bed via gravity.

greywater recycling A method of collecting water used for bathing from baths, showers and wash basins and using it for other purposes such as WC flushing.

guard rails Erected to stop a person falling from a scaffold.

gutter profile The shape of a gutter when viewed from the side.

H

hardness The property of a material that enables it to resist bending, scratching, abrasion or cutting.

hazard Anything with the potential to cause harm (eg chemicals, working at height, a fault on electrical equipment).

hazardous substance Something that can cause ill health to people.

hazardous waste Waste that is harmful to human health, or to the environment, either immediately or over an extended period of time.

Health and Safety at Work Act 1974 The principal piece of legislation covering occupational health and safety in the UK.

health and safety file A document held by the client by which health and safety information is recorded and kept for future use.

Health and Safety Inspectors Persons employed by either the Health and Safety Executive or the local authority to enforce health and safety legislation.

heat exchanger A device or vessel that allows heat to be transferred from one water system to another without the two water systems coming into contact with each other. The transfer of heat takes place via conduction.

heat pumps An electrical device with reversible heating and cooling capability. It extracts heat from one medium at a low temperature (the source of heat) and transfers it to another at a high temperature (called the heat sink), cooling the first and warming the second.

hertz (Hz) The SI unit of frequency, measuring the number of cycles per second in alternating current.

Home Energy Conservation Act (HECA) 1995 Places obligations on local authorities to draw up plans to increase domestic energy efficiency in their area by 30% over 10–15 years.

hopper A container.

hopper head A large bucket type fitting for collecting rainwater from two or more rainwater pipes.

hot work Work that involves actual or potential sources of ignition and carried out in an area where there is a risk of fire or explosion (eg welding, flame cutting, grinding).

HTC fuse This is a special type of fuse used on printed circuit boards.

humidity The amount of moisture in the air.

hydroelectric power Electricity generated by turbines driven by the gravity movement of large amounts of water.

hydro-pneumatic A pressure intensifier that enables generation of great force.

I

ice Water in its solid state when subjected to temperatures below its freezing point.

immersion heater A hot water heater that uses an electrical heating element to heat the water. Controlled by a thermostat.

impeller A rotor used to increase the pressure and flow of a fluid.

independent boiler A freestanding boiler, usually solid fuel.

independent scaffold A scaffold that does not require the building to support it because it has two rows of vertical standards.

indirect system of cold water supply A cold water system where only the kitchen sink is connected to the mains cold water supply. All other cold water outlets are fed from a protected cistern.

induced siphonage An appliance causing the loss of trap seal of another appliance connected to the same waste pipe.

instantaneous hot water systems A system of hot water supply that heats cold water directly from the cold water main via a heat exchanger. There is no storage capacity.

integral solder ring fitting A capillary fitting for copper tubes with a ring of lead-free solder in the joint.

J

job specification A description of the installation that is being quoted for, complete with the types of materials and appliances that the installation must contain.

joule Unit of heat. 4.186 joules of heat energy (equals one calorie) is required to raise the temperature of 1g of water from 0°C to 1°C.

jumper plate A circular plate that holds a tap washer in place. It can be fixed or loose depending on the type of tap in which it is fitted.

K

kelvin (K) A unit of temperature where the lowest point, 0 Kelvin, corresponds to the point at which all molecular motion would stop. 0 Kelvin is −273° Celsius or absolute zero.

kerosene fuel oil (grade C2 28-second viscosity oil to BS 2869) A medium hydrocarbon liquid fuel. It is a residual by-product of crude oil, produced during petroleum refining. It has a high carbon content and is clear or very pale yellow in colour.

L

ladders Used to gain access to scaffolds or light work at high levels. There are three main classes: 1, 2 and 3.

latent heat A change of state as a result of temperature rise.

lead welding A type of fusion welding to join two sheets of lead.

***Legionella* bacteria (*Legionella pneumophila*)** Bacteria that breed in stagnant water. They can give rise to a lung infection called Legionnaire's disease, which is a type of pneumonia.

legislation A law or group of laws that have come into force. Health and safety legislation for the plumbing industry includes the Health and Safety at Work Act and the Electricity at Work Regulations.

level When pipework is perfectly horizontal.

lever A rigid object that can be used with a pivot point or fulcrum to multiply the mechanical force that can be applied to another, heavier object.

liability A debt or other legal obligation to compensate for harm.

liquid petroleum gas (LPG) The generic name for the family of carbon-based flammable gases that are found in coal and oil deposits deep below the surface of the Earth. They include propane, butane, methane and ethane.

local authority Ensures that all works carried out conform to the requirements of the relevant planning and building legislation.

local company A company that will send you a bill for electricity usage.

localised hot water systems Systems of hot water supply that are installed at the place where they are needed.

locking out A process by which a thermostat protects the boiler from overheating by shutting it down when a temperature of around 85°C is reached. High limit thermostats are manually resettable by pushing a small button on the boiler itself.

low-pressure, open-vented central heating systems A central heating system that is fed via a feed and expansion cistern and contains an open-vent pipe.

low-surface-temperature radiator (LST) A radiator designed to give full heat output whilst being cooler to the touch.

low-water-content boiler A boiler that contains only a small amount of water for quick water heating.

lubricant A substance, often a liquid or a grease, introduced between two moving surfaces to reduce friction.

M

malleability The property of a material, usually a metal, to be deformed by compressive strength without fracturing.

manifold A manifold, in systems for moving fluids or gases, is a junction of pipes or channels, typically bringing one into many or many into one.

manual handling The movement of items by lifting, lowering, carrying, pushing or pulling by human effort alone.

MCB (miniature circuit breaker) A type of fast-reacting, resettable fuse.

meter A display that enables a local company to take readings for your bill.

microbiological contamination Contamination by microscopic organisms, such as bacteria, viruses or fungus.

microbore system A central heating system using very small pipework, usually 8mm and 10mm, to feed the heat emitters.

molecule The smallest particle of a specific element or compound that retains the chemical properties of that element or compound.

momentum Trap seal loss caused by a large amount of water is suddenly discharged down the trap of an appliance.

multi-point hot water heater A water heater that serves more than one hot water outlet.

N

national grid The network of high-voltage cables that carries electricity around the country – pylons carry the cables well out of the way of the public.

natural gas A light hydrocarbon fuel found naturally wherever oil or coal has formed. Predominantly contains five gases – methane, ethane, butane, propane and nitrogen.

near miss Any incident that could, but does not, result in an accident.

neutral water Water that is neither hard nor soft that has a pH value of 7.

newton A unit of measurement of force (kgm/s^2).

nogging A term often used on site to describe a piece of wood that supports or braces timber joists or timber-studded walls. They are particularly common in timber floors as a way of keeping the joists rigid and at specific centres, but they can also be used as supports for appliances such as wash hand basins and radiators that are being fixed to plasterboard.

non-ferrous metal Metals that do not contain iron.

non-rising spindle Mainly found in taps, a non-rising spindle is connected to a hexagonal barrel holding the washer. It does not rise when the tap is opened.

nuclear A power station that uses atomic energy to produce steam to drive the turbines.

O

ohm The unit of electrical resistance.

one-pipe central heating system A simple ring circuit of pipework to and from the boiler and as such, there are no separate flow and return pipes.

open flue A flue that is open to the room where the appliance is fitted and relies on heat from the combustion process to create an updraught to evacuate the products of combustion. Often called natural draught.

open-vented central heating systems Systems fed from an F&E cistern in the roof space that contains a vent pipe, which is open to the atmosphere.

open-vented direct hot water storage systems A hot water storage system containing a direct cylinder.

open-vented hot water systems Systems fed from a cistern in the roof space that contains a vent pipe which is open to the atmosphere.

open-vented indirect hot water storage systems A hot water storage system containing an indirect type cylinder.

outriggers Tubes or special units that connect to the bottom of tower scaffolds at the corners, giving a greater overall base measurement and, therefore, an increase in height.

overflow pipe A method of warning of float-operated valve malfunction.

overheads On a building site, costs that include those of the site office and site/administration staff salaries.

oxygen (O_2) A very powerful oxidising agent used in gas form with acetylene when welding.

P

parallel threads A screw thread of uniform diameter used on fittings such as sockets.

parasitic circulation When hot water is pumped through one pipe below the radiators, then back to the tank for re-heating.

partially separate system A system of below-ground drainage where the foul water and some of the rainwater discharges into the foul water drain and all other rainwater discharges in a rainwater drain or soakaway.

pathogenic Causing disease.

peat A poor quality fossil fuel that has a high carbon

content but much less than coal with large amounts of ash produced during combustion.

permanently hard water Water that contains magnesium and calcium chlorides and sulphates in the solution. Cannot be softened by boiling. Alkaline, with a pH value above 7.

permitted This means being allowed to do something.

permit to work A document that gives authorisation for named persons to carry out specific work within a nominated time frame.

personal protective equipment (PPE) All equipment, including clothing for weather protection, worn or held by a person at work, which protects that person from risks to health and safety.

photovoltaic A method of generating electricity from the power of the Sun. Also known as solar arrays.

pitting corrosion This is the localised corrosion of a metal surface. It is confined to a point or small area and takes the form of cavities and pits. Pitting is one of the most damaging forms of corrosion in plumbing, especially in central heating radiators, as it is not easily detected or prevented. Copper tube, although not a ferrous metal, is relatively soft and can suffer from pitting corrosion if flux residue is allowed to remain on the tube after soldering.

planned preventative maintenance Planned maintenance, usually to a schedule, so that systems and equipment can be serviced and checked at regular intervals to ensure optimum performance.

Planning Officer Responsible for processing planning applications, listed building consent applications and conservation area consent applications.

plumb When pipework is perfectly vertical.

polybutylene A type of thermoplastic used to manufacture pipes for cold water, hot water and central heating systems.

polyethylene A type of thermoplastic used to manufacture mains cold water pipes.

polypropylene A type of thermoplastic used to manufacture cold water cisterns, WC siphons and push-fit waste and overflow pipe.

polyurethane foam A sprayed form of insulation applied to hot water storage cylinders.

portable appliance testing (PAT) A method of testing portable electrical appliances and tools to ensure that they are safe to use.

potable Potable is pronounced 'poe-table'. It comes from the French word *potable*, meaning drinkable.

power shower A cistern-fed shower mixing valve that uses a boosting pump to increase flow rate and pressure.

predetermined Decided in advance.

press-fit fittings Fitting for copper tubes that require a special electrical press tool, which crimps the fitting onto the tube to make a secure joint.

pressure Defined as force per unit area. It is measured in pascals (newtons per square metre – N/m^2).

pressure jet burner An oil burner found on oil burning central heating boilers that atomises the fuel prior to combustion.

pressure relief valve A safety valve that safeguards against over-pressurisation by allowing excess water pressure to safely discharge to drain.

primary open safety vent A pipe on a central heating system that is open to the atmosphere to provide a safety outlet should the system overheat.

primary ventilated stack A system of sanitary pipework that relies on all the appliances being closely grouped around the stack and therefore does not need an extra ventilating stack.

private water supply Drinking water source which is not provided by a licensed water undertaker.

propane (C_3H_8) A flammable gas that is heavier than air. One of the five principal gases in natural gas.

proprietary trench support A specially designed support to prevent trench collapse.

PTFE tape PTFE stands for polytetrafluoroethylene. It is a tape used to make leak-free joints in copper and low carbon steel installations.

pure metal Derived directly from the ore and containing very little in the way of impurities.

push-fit fittings Simple push-on fittings for copper tubes or polybutylene pipe.

putlog scaffold A scaffold that is not self-supporting and has only one row of vertical standards.

Q

quantity surveyor A financial consultant or accountant who advises as to how a building can be constructed within a client's budget.

quotations A fixed price for a job, which cannot vary.

R

radiation Heat transfer as thermal radiation from infrared light, visible or not, which transfers heat from one body to another without heating the space in between.

rafter bracket A bracket fixed to the roof members of a dwelling for securing a gutter when no fascia board is available.

rainwater cycle A natural process where water is continually exchanged between the atmosphere, surface water, ground water, soil water and plants. The scientific name is the hydrological cycle.

rainwater harvesting A method of collecting rainwater and using it for other purposes such as WC flushing.

refrigerant Fluorinated chemicals that are used in both liquid and gas states to create both heating and cooling effects.

regulations Rules, procedures and administrative codes set by authorities or governmental agencies to achieve an objective. They are legally enforceable and must be followed to avoid prosecution.

relative density The ratio of the density of a substance to the density of a standard substance under specific conditions.

Reporting of Injuries, Diseases and Dangerous Occurrences Regulations 2013 (RIDDOR) Places a legal duty on your employer, the self-employed and people in control of work premises to report some work-related accidents, diseases and dangerous occurrences.

repose The angle to the horizontal at which the material in the cut face is stable and does not fall away. Different materials have different angles of repose, for example, 90° for solid rock and 30° for sand.

residual current device (RCD) A fast-reacting type of fuse that detects fluctuations in current flow.

resistance arm In mechanics, the arm where the load is concentrated.

resistance to earth Opposition of a conductor to a current flow.

retro-fitting Adding installations to systems that did not have these when manufactured.

reverse osmosis A water filtration process whereby a membrane filters unwanted chemicals, particles and contaminants out of the water.

reversed central heating return system A central heating system where the return travels away from the boiler in the same direction as the flow before looping around to be connected to the return at the boiler.

rising spindle Mainly found in taps, a rising spindle is connected to the washer and jumper plate. It rises as the tap is opened.

risk The chance (large or small) of harm actually being done when things go wrong (eg risk of electric shock from faulty equipment).

risk assessment A detailed examination of any factor that could cause injury.

room-sealed appliance An appliance where the combustion process and flue gas evacuation is sealed from the space where the boiler is fitted. These can be natural draught or forced draught.

RPZ valve A reduced pressure zone valve is a backflow protection device used to protect a category 1 fluid from fluid category 4 contamination.

S

saddle The top piece of an abutment flashing.

sand cast lead sheet Lead sheet produced by traditional casting on a bed of sand.

sealed, pressurised central heating systems A central heating system fed directly from the cold water main and incorporating an expansion vessel.

secondary circulation A method of hot water circulation to prevent dead legs of cold water in hot water systems.

SEDBUK The Seasonal Efficiency of Domestic Boilers in the United Kingdom. A list of boiler efficiency ratings.

self-siphonage Water from a sanitary appliance, usually a wash basin discharging a plug of water, which creates a partial vacuum in the waste pipe between the plug of water and the water in the trap. This then pulls the water from the trap.

semi-gravity heating system A system of central heating where the hot water circulation is via gravity and the heating is pumped.

semi-permeable Allowing passage of certain small particles, but acting as a barrier to others.

sensible heat A temperature rise without a change of state.

separate system A system of below-ground drainage where rainwater and foul water discharge into separate drainage system.

service pipe A pipe that connects the external stop valve to the dwelling.

shear strength The stress state caused by opposing forces acting along parallel lines of action through the material. The action of ripping or tearing.

short circuit A short circuit is an overcurrent which is the result of a fault between two conductors which have a different potential under normal operating conditions. This situation could occur because of damage from a nail impacting conductor or from poor system design or modification.

sick building syndrome (SBS) A combination of ailments associated with an individual's place of work or residence.

single-feed self-venting indirect hot water storage

cylinder A hot water storage vessel that contains a heat exchanger that uses air entrapment to separate the primary water from the secondary water.

single point hot water heater A hot water heater that serves only one outlet. Also known as a point-of-use water heater.

siphonic WC pan A WC pan that uses a vacuum to clear the contents of the pan.

soakaway drain A specifically designed and located pit, sited away from the dwelling, which allows the water to soak away naturally to the water table.

soaker A small piece of code 3 lead used as part of an abutment weathering on a plain tiled or slated roof.

soft water Water with a high content of carbon dioxide (CO_2). Acidic, with a pH value below 7.

soil stack The lower, wet part of a sanitary pipework system, which takes the effluent away from the building.

solar collector Used with solar hot water heating, the solar collector collects the Sun's warmth and transfers it, through a heat ex-changer, to the hot water storage vessel.

solar thermal Technology that utilises the heat from the Sun to generate domestic hot water supply.

solenoid valve A solenoid valve operates with the aid of an electromagnet. When electricity is supplied to the electromagnet of the valve, the valve becomes magnetised and snaps open, allowing water to flow. Once the electricity has been switched off, the valve is no longer magnetised and a spring snaps the valve shut.

specific heat capacity The amount of heat required to change a unit mass of that substance by one degree in temperature. Measured in kJ/kg/°C.

spigot Another name for the plain end of a pipe. If the fitting we buy has a plain pipe end, we call this a spigot end.

S-plan central heating system A fully pumped heating system that uses two two-port zone valves.

stagnation Where water has stopped flowing.

Statute A major written law passed by Parliament.

steam Water that has undergone a change of state in the presence of heat.

storage cistern Designed to hold a supply of cold water to feed appliances fitted to the system.

stratification This describes how the temperature of the water varies with its depth. The nearer the water is to the top of the cistern, the warmer it will be. The deeper the water, the colder it will be. This tends to occur in layers, whereby there is a marked temperature difference from one layer to the next. The result is that water quality can vary, with the warmer water near the top being more susceptible to biological growth such as *Legionella pneumophila* (the bacterium causing Legionnaires' disease).

structural engineer Calculates the loads (wind, rain, the weight of the structure itself) and the effects of the loads on the structure.

stub stack system A system of sanitary pipework where an air admittance valve replaces the vent pipe.

sublimation A process in which ice passes directly to steam.

substation A building fitted with electrical apparatus that converts voltages from low to high or high to low.

surface water drain Used to collect rainwater and discharge it away from a dwelling directly to a water course, river or stream.

surveyor The person responsible for ensuring that the Building Regulations are followed in the planning and construction phases of a new building and extensions and conversions to existing properties.

swarf Fine chips of stone, metal, or other material produced by a machining operation.

system boiler A central heating boiler that contains an expansion vessel and pressure relief valve in a single unit.

T

tampering Interfering with something that you are unauthorised to touch.

tapered threads A standard thread cut onto the ends of pipes and blackheart malleable, male fittings to ensure a watertight, gastight or steamtight joint. The tube tightens the further it is screwed into the fitting.

temper The temper of a metal refers to how hard or soft it is.

temporary continuity bonding Provides a continuous earth to prevent an electric shock in the event of any electrical fault while removing or replacing metal pipework.

temporary hard water Water that contains minerals such as calcium carbonate (limestone). Can be softened by boiling. Alkaline, with a pH value above 7.

tensile strength A measure of how well or badly a material reacts to being pulled or stretched until it breaks.

terminal The entry of the cable to a fixed position where it is known to terminate/fix.

thermal A power station that uses coal to produce steam to drive the turbines.

thermal envelope The part of a building that is enclosed within walls, floor and roof, and that is thermally insulated in accordance with the requirements of the Building Regulations.

thermistor A resistor that varies with temperature.

thermo-mechanical cylinder control valves A non-electrical method of controlling secondary hot water temperature. Works in a similar way to a thermostatic radiator valve.

thermocouple A connection between two different metals that produces an electrical voltage when subjected to heat.

thermometer A device for measuring temperature.

thermoplastic A type of plastic made from polymer resins that becomes liquid form when heated and hard when cooled.

thermosetting plastic Rigid plastics, resistant to higher temperatures than thermoplastics.

toe boards A board placed around a platform or a sloping roof to prevent personnel or materials from falling.

torque The property of force that is exhibited when an object rotates around its axis.

tower scaffold A small, temporary structure for holding workers and materials during the construction or repair of a building. They can be static or mobile.

toxic Poisonous.

trade foreman The leader of tradesmen on site. For instance, a plumbing foreman is the plumber who is running the plumbing installation on site. The plumbing supervisor would have many sites to visit, and each one would have a plumbing foreman.

tripping Turning off, or breaking of a circuit, as a result of a fault occurring.

turbidity Turbidity refers to how clear or cloudy water is as a result of the amount of total suspended solids it contains – the greater the amount of total suspended solids (TSS) in the water, the cloudier it will appear. Cloudy water can therefore be said to be turbid.

two-pipe central heating system A system having two pipes, a flow and a return, which are connected to the boiler.

underfloor heating A method of using concealed underfloor pipework to warm a dwelling.

units of power These are read as kWh (kilowatts per hour).

unplasticised polyvinyl chloride (PVCu) A type of thermoplastic used for waste pipes, soil pipes,

underground drainage, gutters and rainwater pipes. Can be solvent welded.

unvented hot water storage systems Systems fed directly from the cold water main that are not open to the atmosphere and contain an expansion vessel or expansion bubble.

vaporising burner An oil burner found on some oil-fired appliances that warms the fuel to vaporise it prior to combustion.

velocity The measurement of the rate at which an object changes its position.

vent stack The upper part of a sanitary pipework system that introduces air into the system to help prevent loss of trap seal.

ventilated discharge branch system A sanitary pipework system used on larger installations where there is a risk of trap seal loss because the waste pipe lengths are excessive.

venturi boost mixing valves A shower valve using the principle of a venturi tube for mixing hot and cold water to produce a showering temperature.

venturi tube A pipe that is suddenly reduced in size creating a reduction in pressure but an increase in velocity, in accordance with Bernoulli's principle.

vitreous china Clay material with an enamelled surface used to manufacture bathroom appliances.

volt The unit of electrical potential.

W

waste carrier's licence A licence required by the local authority for anyone transporting waste materials.

Waste Management Duty of Care Code of Practice Legislation that aims to ensure that producers of waste take responsibility for making sure that their waste is managed without harm to human health or to the environment.

water A compound constructed from two hydrogen atoms and one oxygen atom. The most abundant compound on earth. It can be fresh water or saline (salt) water.

water course A river, stream or other flowing natural water source.

Water Supply (Water Fittings) Regulations 1999 These relate to the supply of safe, clean, wholesome drinking water to properties and dwellings, specifically targeting the prevention of contamination, waste, undue consumption, misuse and erroneous metering.

water undertaker A water company in the UK. A supplier of treated, wholesome water.

watt SI unit for power. It is equivalent to one joule per second (1 J/s), or in electrical units, one volt ampere (1 V·A).

wavering out Trap seal loss caused by wind blowing across the top of a vent stack.

WC cistern Used to flush a WC.

whiteheart fittings A type of fitting for low carbon steel pipe with a parallel female thread.

wind turbine A method of generating electricity from a turbine connected to a large propeller driven by the wind.

working drawings All plans, elevations and details needed by a contractor and trades to complete a building.

Y-plan central heating system A fully pumped heating system that uses one three-port mid-position valve.

zero carbon fuel A fuel where the net carbon dioxide emissions from all the fuel used is zero.

INDEX